AF334594

Peroxynitrite Detection in Biological Media
Challenges and Advances

RSC Detection Science Series

Editor-in-Chief:
Professor Michael Thompson, *University of Toronto, Canada*

Series Editors:
Dr Sub Reddy, *University of Surrey, Guildford, UK*
Professor Damien Arrigan, *Curtin University, Australia*

Titles in the Series:
1: Sensor Technology in Neuroscience
2: Detection Challenges in Clinical Diagnostics
3: Advanced Synthetic Materials in Detection Science
4: Principles and Practice of Analytical Techniques in Geosciences
5: Microfluidics in Detection Science: Lab-on-a-chip Technologies
6: Electrochemical Strategies in Detection Science
7: Peroxynitrite Detection in Biological Media: Challenges and Advances

How to obtain future titles on publication:
A standing order plan is available for this series. A standing order will bring
delivery of each new volume immediately on publication.

For further information please contact:
Book Sales Department, Royal Society of Chemistry, Thomas Graham
House, Science Park, Milton Road, Cambridge, CB4 0WF, UK
Telephone: +44 (0)1223 420066, Fax: +44 (0)1223 420247
Email: booksales@rsc.org
Visit our website at www.rsc.org/books

Peroxynitrite Detection in Biological Media
Challenges and Advances

Edited by

Serban F. Peteu
Michigan State University, East Lansing, Michigan, USA
Email: serbanfpeteu@gmail.com; peteu@msu.edu

Sabine Szunerits
University of Lille, Lille, France
Email: Sabine.Szunerits@iri.univ-lille1.fr

and

Mekki Bayachou
Cleveland State University, Cleveland, Ohio, USA
Email: m.bayachou@csuohio.edu

RSC Detection Science Series No. 7

Print ISBN: 978-1-78262-085-3
PDF eISBN: 978-1-78262-235-2
ISSN: 2052-3068

A catalogue record for this book is available from the British Library

Published by The Royal Society of Chemistry,
Thomas Graham House, Science Park, Milton Road,
Cambridge CB4 0WF, UK

Registered Charity Number 207890

For further information see our web site at www.rsc.org

Printed in the United Kingdom by CPI Group (UK) Ltd, Croydon, CR0 4YY, UK

Preface

Professor Joseph S. Beckman recently unveiled some lesser known facts about his seminal work on peroxynitrite: "For over a year, our work had been rejected after lengthy review times by *Science*, *Nature* and the *Journal of Biological Chemistry*. In contrast, the editors of *Archives* allowed us to publish a series of five papers that established peroxynitrite as a biological oxidant, showed how tyrosine nitration can be used to quantify peroxynitrite production and demonstrated that peroxynitrite was quite toxic to bacteria." (*Arch. Biochem. Biophys.*, 2009, **484**, 114). It is perhaps interesting to mention that this ground-breaking paper has now garnered over 1200 citations. This is indeed remarkable for a paper initially judged to be "not of sufficient interest for the readers."

As an indication of the excitement produced by this discovery, the five most cited peroxynitrite articles garnered over 14 000 citations, as illustrated in Figure 1.

Although the peroxynitrite lifetime is a fraction of a second at physiological pH, it can damage a wide array of molecular components in cells, including DNA and proteins, due to its intrinsic high nitrating properties and surrogate oxidative effects through other radical-generating molecules such as carbon dioxide. Furthermore, abnormal levels of peroxynitrite are clinically correlated with pathogenic effects, including neurodegenerative, cardiovascular or chronic inflammatory diseases, and diabetes complications.

The establishment of peroxynitrite as a culprit in many devastating diseases brings an urgent need and great interest to develop specific detection and quantification tools. Thus the main objectives of this tome are to critically discuss, arguably for the first time in one book, the challenges and latest advancements in peroxynitrite quantification in biological media.

In Chapter 1, Willem H. Koppenol (from the Institute of Inorganic Chemistry, Swiss Federal Institute of Technology, Zürich, Switzerland) introduces

RSC Detection Science Series No. 7
Peroxynitrite Detection in Biological Media: Challenges and Advances
Edited by Serban F. Peteu, Sabine Szunerits, and Mekki Bayachou
© The Royal Society of Chemistry 2016
Published by the Royal Society of Chemistry, www.rsc.org

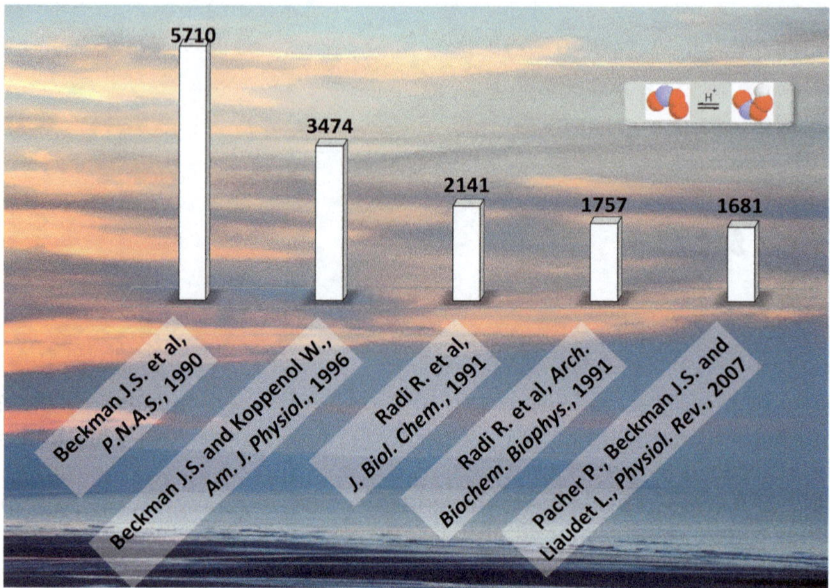

Figure 1   The first five of the most cited peroxynitrite papers garnered over 14 000 citations.

the readers to peroxynitrite basics. When it became clear in the early 1990s that peroxynitrite is biologically significant, Willem Koppenol with Joseph Beckman and others (W. H. Koppenol, J. J. Moreno, W. A. Pryor, H. Ischiropoulos and J. S. Beckman, *Chem. Res. Toxicol.*, 1992, **5**, 834) reviewed its thermodynamic and kinetic properties. This first part of the book outlines what is still relevant and discusses some of the important aspects that have evolved since then.

Chapter 2 by Sabine Borgmann (from the Fakultät für Chemie und Biochemie, Ruhr-Universität, Bochum, Germany) highlights the complex chemical nature of peroxynitrite for biologists and immunologists, explains the biological impact and complexity of peroxynitrite in biological matrices for chemists and offers guided principles for optimizing quantitative assays.

Chapter 3 is by Mekki Bayachou and colleagues (from the Department of Chemistry, Cleveland State University, and Department of Pathology, Lerner Research Institute, The Cleveland Clinic, Cleveland, USA). This chapter reviews the methods of peroxynitrite synthesis or generation *in situ*. The advantages and drawbacks of the various methods are discussed in the context of the use of prepared authentic peroxynitrite samples in the development and validation of sensors and probes.

Chapter 4 is by Sabine Szunerits and Rabah Boukherroub (from the Institut d'Electronique, Microélectronique et de Nanotechnologie, University Lille 1, Villeneuve d'Ascq, France) with Serban Peteu (from the Chemical Engineering and Materials Science Department, Michigan State University, USA). It explores the state of the art in the development and use of peroxynitrite sensitive matrices for electrochemical sensing. The electrochemical methods will be further discussed in Chapters 5 to 9.

Chapter 5 by Sophie Griveau and Fethi Bedioui (from the Unité de Technologies Chimiques et Biologiques pour la Santé, Université Paris Descartes, Paris, France) outlines the challenges and perspectives of electrochemical detection schemes for peroxynitrite in biological solutions.

This is followed by the contribution of Christian Amatore, Manon Guille-Collignon and Fréderic Lemaitre (from the Ecole Normale Supérieure, Paris, France). Their Chapter 6 deals with real-time monitoring of peroxynitrite by stimulation of macrophages with ultramicroelectrodes. This unique configuration involving platinized carbon fiber ultramicroelectrodes is of particularly great interest since it allows one to quantify in real time the very harmful and unstable peroxynitrite anion within the oxidative burst at the single cell level.

Many of these species interfere with fluorescence and electrochemical detection methods for peroxynitrite. Therefore, to better understand the role of peroxynitrite *in situ*, a separation method to isolate peroxynitrite from potential interferences is necessary. Chapter 7 by Susan Lunte, Joseph Siegel, Richard de Campos, Dulan Gunasekara (from the Ralph N. Adams Institute for Bioanalytical Chemistry, University of Kansas, USA), and José da Silva (from the Chemistry Institute, State University of Campinas, São Paulo, Brazil) reviews the electrophoretic methods developed to separate and subsequently detect peroxynitrite as well as its metabolites and degradation products.

Chapter 8 by R. Mark Worden, Ying Liu and Serban Peteu (from the department of Chemical Engineering and Materials Science, Michigan State University, East Lansing, Michigan, USA) investigates peroxynitrite–biomembrane interactions using biomimetic interfaces. The mechanisms by which peroxynitrite damages membrane lipids are reviewed and the biomimetic interface platforms and characterization methods are critically discussed. Furthermore, a novel application is proposed to measure the effect of peroxynitrite exposure on biomembrane electrochemical resistance.

In Chapter 9, Serban Peteu and co-workers from the Institut d'Electronique, Microélectronique et de Nanotechnologie, Lille, France (Sabine Szunerits and Rabah Boukherroub) and from Bucharest, Romania (Alina Vasilescu, International Centre for Biodynamics; Valentina Dinca, Mihaela Filipescu, Laurentiu Rosen, Maria Dinescu, the National Institute of Laser, Plasma and Radiation Physics; and Ioana S. Hosu, the National Institute of Chemistry) highlight recent approaches to enhance the selectivity and sensitivity of peroxynitrite detection.

The next two chapters deal with new optical methods for peroxynitrite detection, as these have grown exponentially in the last 5 years. Chapter 10 by Zhijie Chen, Tan Truong and Hui-wang Ai (from the Department of Chemistry and the Cell, Molecular and Developmental Biology Graduate Program, University of California Riverside, USA) outlines recent achievements in the use of fluorescent probes for the detection of peroxynitrite. Indeed, among the various methods for peroxynitrite detection in biological media, fluorescent probes remain one of the most efficient approaches, owing to their sensitivity and specificity for peroxynitrite in live cells.

Chapter 11 by Peng Li and Keli Han (from the State Key Laboratory of Molecular Reaction Dynamics, Dalian Institute of Chemical Physics, China) reports on the use of reversible near-infrared fluorescent probes for peroxynitrite monitoring. The design, synthesis, characterization and imaging applications of cyanine-based, selenium–tellurium modulated fluorescent probes is also described.

What Does the Future Hold for Peroxynitrite?

One thing seems certain: the interest to selectively detect peroxynitrite in biological media will continue, as it comes at least in part from its involvement in the pathophysiology of inflammation, cardiovascular disease, neurodegeneration and diabetes, to name a few. The authors themselves already offer their own outlook in their chapters. Several trends have been revealed in the detection of peroxynitrite and associated nitro-oxidative species. Microsensor arrays and other biomimetic platforms are urgently needed to simultaneously detect multiple analytes in real time, at the cell or tissue level. By coupling several detection techniques, such as electrochemical and fluorescence, one can shed light and extract valuable information on peroxynitrite dynamics with enhanced sensitivity and selectivity.

For *in vivo* applications, fluorescent imaging frequently lacks good spatial resolution due to limited penetration depth in animal tissue. The exploration of photoacoustic (PA) probes, ensuring highly specific molecular imaging with unprecedented performance, from millimeters to centimeters of tissue with resolutions in the 20–200 μm range is a promising development in this direction. The principle of PA probes is based on the generation of acoustic waves following the absorption of ultrashort light pulses allowing imaging beyond the optical diffusion limit by integrating optical excitation with ultrasonic detection. Semiconducting polymer nanoparticles (SPN) were introduced as a new class of near-infrared photoacoustic contrast agents for *in vivo* PA molecular imaging of peroxynitrite (K. Pu, A. J. Shuhendler, J. V. Jokerst, J. Mei, S. S. Gambhir, Z. Bao and J. Rao, *Nature Nanotechnol.*, 2014, **9**, 233).

Other very recent efforts have targeted real-time imaging of oxidative and nitrosative stress in whole animals, using polymeric nanosensors, integrating fluorescence resonance energy transfer (FRET) and chemiluminescence resonance energy transfer (CRET). Peroxynitrite and other reactive nitro-oxidative species can be detected in a dose-dependent manner in the liver of mice within minutes of drug administration, preceding histological changes, protein nitration or DNA strand break induction (A. J. Shuhendler, K. Pu, L. Cui, J. P. Uetrecht and J. Rao, *Nature Biotechnol.*, 2014, **12**, 373).

In conclusion, there is no shortage of new and innovative detection methods and it will be exciting to follow their future progress in monitoring peroxynitrite dynamics *in vivo*. One goal is to discriminate between the physiological and pathological levels of peroxynitrite in different organs to facilitate the design of better therapies to counteract the nitro-oxidative effects of this species.

As book editors, we are thankful to the Royal Society of Chemistry who embraced our project and put their prestige behind it. We were also lucky to involve so many eminent authors to share their knowledge and know-how. With them all, we hope that this content will be of interest for researchers, students and anyone else with an interest in the quickly evolving peroxynitrite story. We trust that you will enjoy your reading.

Serban F. Peteu
Sabine Szunerits
Mekki Bayachou

With deep appreciation to our families for all
of their support

Contents

RSC Detection Science Series No. 7
Peroxynitrite Detection in Biological Media: Challenges and Advances
Edited by Serban F. Peteu, Sabine Szunerits, and Mekki Bayachou
© The Royal Society of Chemistry 2016
Published by the Royal Society of Chemistry, www.rsc.org

Peroxynitrite: The Basics

WILLEM H. KOPPENOL[*a]

[a]Institute of Inorganic Chemistry, Department of Chemistry and Applied Biosciences, Swiss Federal Institute of Technology, Zürich 8093, Switzerland
*E-mail: koppenol@inorg.chem.ethz.ch

1.1 History

1.1.1 Before 1990

In this section, I will focus on the early discovery of peroxynitrite, its ability to nitrate aromatic compounds, its sensitivity to carbon dioxide and early determination with permanganate.

More than a 100 years ago, Baeyer and Villiger[1] proposed that a "Nitrosopersäure" (ROONO) was formed as an intermediate in reactions of nitrite, ethyl nitrite, and amyl nitrite (R^1ONO) with hydrogen peroxide and ethyl hydroperoxide (R^2OOH). They came to the conclusion that an adduct between R^1ONO and R^2OOH was formed that yielded R^1OH and R^2ONO_2. Had a direct oxidation of the nitrite taken place, R^1ONO_2 and R^2OH would have been the products. In the case of the reaction between nitrous acid and hydrogen peroxide, nitrate (NO_3^-) is ultimately formed and the accompanying formula shows an adduct between peroxynitrous acid and water.[1] Although they did not provide direct evidence for the structure of peroxynitrous acid, we may credit them with the discovery of this reactive species. In 1907, Raschig[2] prepared a solution of bromide in hydrogen peroxide and one of bromide in nitrous acid. Both solutions stayed clear, but upon mixing, dibromine was formed as

RSC Detection Science Series No. 7
Peroxynitrite Detection in Biological Media: Challenges and Advances
Edited by Serban F. Peteu, Sabine Szunerits, and Mekki Bayachou

Published by the Royal Society of Chemistry, www.rsc.org

deduced from the reddish-brown color and the smell. He assumed that two hydrogen peroxide molecules reacted with one nitrous acid and proposed the formula HNO_4 and the name "Übersalpetersäure"; apparently he was not aware of the publication of Baeyer and Villiger.[1] In 1922, Trifonow[3,4] explored the properties of the short-lived product of the reaction between nitrous acid and hydrogen peroxide for analytical purposes. He concluded that "naszente Persalpetersäure" (nitric acid *in statu nascendi*) is capable of oxidizing aniline and nitrating aromatic compounds. He proposed using these reactions for the detection of nitrite and aromatic compounds due to the intensively colored products. In 1929, Gleu and Roell[5] reported that the reaction of azide with ozone results in a deep orange–red solution that smells of hypochlorite and sometimes nitrogen dioxide. Although stable in alkaline solution, the color disappears rapidly upon neutralization by addition of hydrogen carbonate, or by lowering of the pH. In spite of considerable efforts, they were unable to isolate the new unstable compound, but their experiments allowed them to exclude hydrogen peroxide as the oxidant, and they concluded that they were dealing with peroxynitrous acid. Gleu and Hubold described in their paper of 1935 a simple synthesis of peroxynitrite from hydrogen peroxide and nitrite at low pH: one mixes nitrite and hydrogen peroxide, adds acid followed by base within 2 s. If done correctly, the deep yellow color of peroxynitrite is observed.[6] Use of a quenched-flow reactor improves the yield.[7] Kortüm and Finckh[8] showed in 1941 that the absorption maximum of peroxynitrite anion ($ONOO^-$) in the UV ultraviolet range is close to that of NO_3^-, but the intensity is about 15 times higher and the band is much broader. The initiation of the polymerization of methyl acrylate and the hydroxylation and nitration of aromatic compounds by peroxynitrous acid, reported in 1952 by Halfpenny and Robinson,[9] was rationalized in terms of homolysis of the O–O bond in ONOOH. However, in 1954, Anbar and Taube[10] studied the reactions of peroxynitrite labeled with two O^{18}, and found doubly labeled NO_3^- as a product of intramolecular rearrangement of peroxynitrite and, in the presence of an excess of unlabeled nitrite, singly labeled nitrite and nitrate as products of O^{18} transfer from peroxynitrite to nitrite, a result that is not easily explained by homolysis. The first kinetics study of peroxynitrous acid involving its isomerization to NO_3^- appeared in 1962[11] and the second in 1969.[12] The latter gives a pK_a of 6.6 and a rate of isomerization of 0.10 s^{-1}, obtained at a temperature of 1 °C and an ionic strength of 0.5 M. It also mentioned that peroxynitrite vanishes quickly in the presence of carbonate or borate. Formation of peroxynitrite from nitrate in solution by ultraviolet light was demonstrated in 1964.[13] The same study also showed that permanganate oxidized peroxynitrite.[13] Hughes and Nicklin reported in 1968 the extinction coefficient of peroxynitrite at 302 nm of 1670 ± 50 M^{-1} cm^{-1},[14] which is within the error of that obtained by Bohle and coworkers in 1994 with pure tetramethylammonium peroxynitrite, 1705 ± 10 M^{-1} cm^{-1}.[15] Blough and Zafiriou made a very important observation in 1985 of a yellow color after mixing a mostly anaerobic alkaline solution of superoxide with that of nitrogen monoxide. They concluded that superoxide and nitrogen monoxide react to form peroxynitrite.[16]

As nitrogen monoxide was identified in 1987 as an "endothelium-derived relaxing factor",[17,18] and SOD extends the life of nitrogen monoxide,[19] the finding of Blough and Zafiriou could be relevant to physiology!

We see that a few properties and reactions that have been discovered can be used to quantitate peroxynitrite: its yellow color, its ability to nitrate aromatic compounds and the reduction of purple permanganate to green manganate. An excellent review on the "older" chemistry of peroxynitrite by Edwards and Plumb appeared in 1993.[20] It also discusses the role of peroxynitrite in atmospheric chemistry.

From 1901 until 1990, *ca.* 40 papers on peroxynitrite appeared. Since then, the number of publications has sharply increased to well over 12 000 as a recent search (November 2014) on the Web of Science showed.

1.1.2　1990 and Later

Up to 1990, oxidative injury to tissues was explained with a modification of the Haber–Weiss cycle that included catalysis by iron.[21] In essence, the presence of iron(II) and hydrogen peroxide was postulated; injury was caused by the Fenton reaction,[22] which, at neutral pH, yields the hydroxyl radical[23] or a higher oxidation state of iron,[24] as reviewed by Koppenol and Bounds.[25] However, Beckman *et al.*[26] pointed out that "generation of strong oxidants by the iron-catalyzed Haber–Weiss reaction is not an entirely satisfactory explanation for superoxide dismutase (SOD)-inhibitable injury *in vivo*." Since SOD protects, superoxide must be damaging itself—which it is not[27]—or react to form a more reactive species. Instead of a reaction between some undefined iron(III) complex and superoxide, Beckman *et al.*,[26] referring to the work of Blough and Zafiriou,[16] proposed that superoxide reacts very quickly with nitrogen monoxide to form peroxynitrite. As the reaction in question is a radical–radical reaction, this assumption is quite reasonable. Thus, formation of peroxynitrous acid is kinetically far more feasible than the "iron-catalyzed Haber–Weiss reaction".[26] Beckman *et al.* also described some scavenger studies to find out whether peroxynitrous acid underwent homolysis to produce nitrogen dioxide and hydroxyl radicals. They concluded that "a potent oxidant similar to HO$^{\bullet}$ in reactivity" was formed.[26] Furthermore, they pointed out that peroxynitrous acid, being long-lived, can diffuse over several cell diameters. For that reason it is a more selective and toxic oxidant than the hydroxyl radical. It is of importance that it also nitrates aromatic compounds. The question of whether nitrogen monoxide and superoxide production under pathological conditions could be high enough to generate peroxynitrite was posed,[26] and we now know that this is so.

First, what are the basics of peroxynitrite chemistry? When it became clear in the early 1990s that peroxynitrite is biologically relevant, we prepared an overview of its thermodynamic and kinetic properties.[28] Much of what we published is still relevant, but some important details have changed. Nevertheless, the paper is still being cited and has now garnered over 1000 citations. Peroxynitrous acid isomerizes with a rate constant of 1.1 s^{-1}. Earlier and slightly higher rate constants have been reviewed.[29] The pK_a of peroxynitrous acid is 6.5 to 6.8, depending on the ionic strength and temperature.[30] The peroxynitrite anion is fairly stable, but at pH values equal to the pK_a and higher, there is decomposition to nitrite and dioxygen, a reaction that proceeds *via* an adduct between peroxynitrite and peroxynitrous acid.[29,31,32] The spectrum of the anion has a broad maximum at 302 nm ($\varepsilon = 1705 \pm 10$ M^{-1} cm^{-1}).[15] Peroxynitrous acid is a powerful one- and two-electron oxidizing agent; the calculated one- and two-electrode potentials are, respectively: $E^{\circ\prime}(ONOOH, H^+/NO_2^{\bullet}, H_2O) = +1.6 \pm 0.1$ V and $E^{\circ\prime}(ONOOH, H^+/NO_2^-, H_2O) = +1.3 \pm 0.1$ V, both at pH 7.[30] These electrode potentials, as all others, are given relative to the normal hydrogen electrode. Experimentally, a one-electron electrode potential over *ca.* $+1$ V[33] was measured by cyclic voltammetry in a cell in which alkaline peroxynitrite was repeatedly mixed with acid to generate the short-lived peroxynitrous acid ($t_{1/2} = 0.63$ s at 25 °C). The lack of pH dependence indicates that a transient $HOONO^{\bullet-}$ was formed. It is clear that peroxynitrous acid is a potent one-electron oxidant. The notion that 30% of peroxynitrous acid undergoes homolysis to nitrogen dioxide and the hydroxyl radical is widespread in the literature.[34] We reviewed all evidence, pro and contra, extensively with the conclusion that, if any, homolysis is limited to at most 5%.[35] It may be appropriate to illustrate this point with an example. Above, decomposition to nitrite and dioxygen was mentioned. One can, in principle, explain these products by two homolyses:[36–38]

$$ONOOH \rightarrow NO_2^{\bullet} + HO^{\bullet} \tag{1.1}$$

$$ONOO^- \rightarrow NO^{\bullet} + O_2^{\bullet-} \tag{1.2}$$

followed by:

$$NO^{\bullet} + NO_2^{\bullet} \rightarrow N_2O_3\ (+\ H_2O) \rightarrow 2NO_2^- + 2H^+ \tag{1.3}$$

$$HO^{\bullet} + H^+ + O_2^{\bullet-} \rightarrow H_2O + O_2 \tag{1.4}$$

If this is so, then the *rate* of nitrite and dioxygen formation is limited by the slowest process, which is the reaction in eqn (1.2), that is, it will not depend on the concentration of peroxynitrite.[38] The rate constant of the reaction in eqn (1.2), as determined by the reduction of permanganate by superoxide, is 0.020 s^{-1}.[39] Furthermore, the *relative yield* of nitrite and dioxygen should also not depend on the peroxynitrite concentration. However, experimentally, we found that both the rate and the relative yield increase with the peroxynitrite concentration.[31,32,40] Furthermore, we showed that peroxynitrate is an intermediate:[32]

$$ONOOH + ONOO^- \rightarrow O_2NOO^- + NO_2^- + H^+ \tag{1.5}$$

which itself decomposes:

$$O_2NOO^- \rightarrow NO_2^- + O_2 \tag{1.6}$$

Thus, we concluded that peroxynitrous acid behaves like other peracids.[35]

Biochemically far more important is the reaction with carbon dioxide. In 1993, Radi's group rediscovered that peroxynitrite is not stable in the presence of hydrogen carbonate.[41] Two years later, Lymar and Hurst[42] found that the peroxynitrite anion reacts rapidly with dissolved carbon dioxide, which leads to $ONOOCO_2^-$, nitrosoperoxycarbonate or 1-carboxylato-2-nitrosodioxidane. This compound itself, or its homolysis products, nitrogen dioxide and trioxidocarbonate($\cdot^-$)[43,44] are strongly oxidizing: $E^\circ(NO_2^\cdot/NO_2^-) = +1.04$ V and $E^\circ(CO_3^{\cdot-}/CO_3^{2-}) = +1.57$ V.[45] Although the extent of homolysis is generally assumed to be *ca.* 30%, we estimated it, using nitrogen monoxide as a scavenger of nitrogen dioxide and trioxidocarbonate($\cdot^-$), at *ca.* 4%.[46] Thus, peroxynitrite, itself an oxidant with an electrode potential greater than +1 V, may cause the formation of equally strong, or even stronger, oxidants. Reduction of peroxynitrite on a microelectrode with modified surfaces has been used to detect it in biological samples.[47,48] In 2001, Amatore and coworkers showed that it is possible to oxidize peroxynitrite at a potential of +0.5 V,[49] quite close to the potential of +0.44 V estimated in 1992.[28] This method has been used to detect and quantify peroxynitrite in single fibroblasts[49] and macrophages.[50] Polymerized hemin on a carbon electrode has also been used to oxidize, and thereby, detect peroxynitrite.[51] Chemiluminescence has been used to detect peroxynitrite as well. This technique is sensitive, but in the case of dichlorodihydrofluorescein and dihydrorhodamine, the reaction is zero-order in the indicator molecule,[52] which detracts from its usefulness. Detection based on fluorescence is often not specific for a particular oxidant, and great care should be taken in the interpretation of results.[53,54] A direct reaction has been established for weakly fluorescent boronates that become strongly fluorescent after reaction with peroxynitrite.[55,56] As these compounds also react with hydrogen peroxide, there is again the issue of specificity. It is not important that the reaction of boronate with peroxynitrite is much faster than with hydrogen peroxide; it is the product of the rate constant and concentration that counts. In the following chapters, much more will be said about these techniques.

Tetramethylammonium peroxynitrite crystallizes in the *cis*-conformation[57] and Raman spectroscopy studies indicate that this is also the conformation in solution.[58]

1.2 Peroxynitrite *In Vivo*

Can peroxynitrite be formed *in vivo*? Under most conditions, superoxide is disposed of by Cu,Zn-SOD, present in the cytosol at a concentration of *ca.* 10 μM. Is that enough to prevent the formation of peroxynitrite? Let us assume that the concentration of nitrogen monoxide is *ca.* 10 nM. If we compare the products of the rate constant of superoxide with Cu,Zn-SOD and with nitrogen monoxide ($k_{SOD} = 2 \times 10^9$ M^{-1} s^{-1},[59,60] $k_{NO^\cdot} = 1.6 \times 10^{10}$ M^{-1} s^{-1},[61] respectively), and the concentrations just mentioned, we get 2×10^4 s^{-1} and 1.6×10^2 s^{-1}, respectively. Thus only 1% of superoxide ends up as peroxynitrite. However, near activated macrophages the concentration of nitrogen monoxide may be in the micromolar region, say 10 μM, and in that case *ca.* 90% of all superoxide is converted to peroxynitrite. This calculation is simplistic, but it illustrates why the SOD concentration has to be 5–10 μM. There is another reason why SOD is essential. Although superoxide reacts quite quickly with its protonated form, this reaction is second order in superoxide: the initial phase of the reaction is very fast but the end stage proceeds very slowly, and a concentration of zero superoxide is "never" reached. It is the reaction with SOD, which is first order in both superoxide and in proteins, that finishes off superoxide.

We have now established that formation of peroxynitrite *in vivo* is kinetically feasible. Evidence for its formation in activated macrophages was detected by its ability to nitrate tyrosine 108 of Cu,Zn-SOD.[62,63] At that time it was also shown that peroxynitrous acid oxidized thiols[64] and lipids.[65] Furthermore, peroxynitrous acid not only nitrates tyrosine, it also hydroxylates it.[66] In atherosclerotic tissue, nitrotyrosine was detected by immunohistochemistry,[67] which is an important observation as it shows that peroxynitrite plays a role in that disease. Antibodies against nitrotyrosine are now commercially available. Furthermore, nitrotyrosine can be detected by high-performance liquid chromatography.[66] Interestingly, the nitration of tyrosine is a reaction that is zero order in tyrosine, and the yield is low,[66] and only somewhat higher in the presence of carbon dioxide. By contrast, tryptophan is nitrated in a bimolecular reaction, but with a low rate constant.[68,69] To the best of my knowledge, no immunological detection method for nitrated tryptophan has been developed.

As shown in a review I co-authored with Beckman,[70] the bad properties of nitrogen monoxide are nearly all those of peroxynitrite. In the title of that review, we refer to nitrogen monoxide, superoxide and peroxynitrite as the good, the bad and the ugly. Even before endothelium-derived relaxing factor was identified as nitrogen monoxide, it was known that superoxide removed it.[19] However, it is not just the removal of a signaling molecule that is important, the reaction product is a powerful oxidizing and nitrating agent. While formation of peroxynitrous acid is thus harmful under normal conditions, it is used by activated macrophages to remove microorganisms and other inflammatory insults. For the same reason, activated neutrophils make another inorganic compound, bleach or oxidochlorate(1⁻).[71] Thus, anywhere inflammation occurs, one may expect to find oxidized, nitrated and chlorinated biomolecules. Unsurprisingly, there is a large number of diseases where formation of peroxynitrite has been established.[72]

1.3   Challenges to the Detection of Peroxynitrite

For the development of a sensor it is necessary to test it with pure peroxynitrite. As mentioned above, peroxynitrite can be synthesized by mixing hydrogen peroxide with nitrous acid, followed by rapid quenching with base.[6,7] Oxygenation of hydroxylamine,[73] ozonization of an azide solution[5,74] and treating solid potassium superoxide with nitrogen monoxide[75] have also been used. None of these syntheses results in the preparation of a pure product: common contaminants are the decay products nitrite and nitrate, and, dependent on the synthesis, remaining reactants, nitrite, hydroxylamine, azide and hydrogen peroxide. *In vivo* one may encounter hydrogen peroxide and nitrite. Only biomimetic synthesis in ammonia reported by Bohle and coworkers[15,76] results in a pure preparation of tetramethylammonium peroxynitrite. Contaminants are important because they may interfere with the detection of peroxynitrite; their presence requires control experiments.

Detection of the "footprint" of peroxynitrite, nitrated tyrosine, is well-established. Although useful for diagnostic purposes, this method cannot be used for time-resolved detection in cells or tissues. The remaining techniques are fluorescence and electrochemistry. Sensitivity and selectivity are important issues. The steady-state concentration will be very low, presumably in the nanomolar region. As one would like to determine the target of peroxynitrite in cells or tissues, detection should not disturb that concentration, that is, not more than 10% of the peroxynitrite should react with the reporter molecule or with the electrode. The issue of selectivity has been brought up for methods based on fluorescence,[53] but it applies equally to electrochemical detections. The importance of reliable methods to detect peroxynitrite in tissues and biological samples has recently been stressed by Chen *et al.*[77]

1.4 Nomenclature

For reasons unknown, many working in the field of radical research use the abbreviations ROS and RNS, which stand for "reactive oxygen species" and "reactive nitrogen species", respectively. Generally, superoxide, the hydroxyl radical, singlet dioxygen and hydrogen peroxide are "ROS", and nitrogen monoxide, nitrogen dioxide and peroxynitrite belong to "RNS". However, as already discussed, superoxide and nitrogen monoxide are not reactive, and neither is hydrogen peroxide. These acronyms are thus very misleading. In 2013, a commentary in the *Biophysical Journal* contained the sentence: "Superoxide can be quite damaging because it has an extremely high affinity for electrons, ripping them away from nearby proteins, lipids, and nucleic acids *via* oxidation."[78] In the 30 year old compilation of rate constants for reactions of superoxide by Bielski and coworkers[27] one finds that superoxide does not react with amino acids. It behaves as a mild reductant, while its hydronated form may act as an oxidant. By addressing these species collectively as ROS and RNS one ignores the efforts of many chemists that have characterized the different thermodynamic and kinetic properties of these molecules. The use of these acronyms has also been lamented in a recent editorial in *Free Radical Biology and Medicine*.[54] Furthermore, these species have their own names that should be used: for $O_2^{\bullet-}$ one can still use the venerable name superoxide; the systematic name is dioxide($^\bullet1^-$). For other species named in this chapter the names are, with the allowed one in italics: O_2, dioxygen; $HO_2^\bullet$, hydrogen dioxide or hydridodioxygen($^\bullet$); H_2O_2, *hydrogen peroxide*, dihydridodioxygen or dioxidane; $HO^\bullet$, *hydroxyl radical*, hydridooxygen($^\bullet$) or oxidanyl; $NO^\bullet$, nitrogen monoxide or oxidonitrogen($^\bullet$); $NO_2^\bullet$, nitrogen dioxide or dioxidonitrogen($^\bullet$); $ONOO^-$, *peroxynitrite* or (dioxido)oxidonitrate(1^-); ONOOH, *peroxynitrous acid* or (hydridodioxido)oxidonitrogen; $ONOOH^{\bullet-}$, (hydridodioxido)oxidonitrate($^{•}$1$^{−}$); $O_2NOO^{−}$ *peroxynitrate* or (dioxido)dioxidonitrate(1$^{−}$); $CO_3^{•−}$, trioxidocarbonate(1$^{•−}$); $ONOOCO_2^{−}$, nitrosoperoxycarbonate or 1-carboxylato-2-nitrosodioxidane.[79,80] The reader will have noticed that the systematic names are based on the selection of a central atom and that the attached atoms are named as negatively charged groups (even if they are not). Furthermore, one can often use names based on -ane nomenclature: methane for CH_4, oxidane for H_2O, azane for NH_3, *etc.* The advantage of the new nomenclature recommendations is that the name tells you what the chemical composition is. Names such as nitric oxide or nitrous oxide do not do that; use of that nomenclature leads to the name "pernitric oxide" for $NO_2^{•}$, which nobody uses, and there would be no name for $NO_3^{•}$.

Acknowledgements

Research on peroxynitrite has been supported by the Swiss Federal Institute of Technology and the Swiss National Research Foundation.

References

1. A. Baeyer and V. Villiger, *Ber. Dtsch. Chem. Ges.*, 1901, **34**, 755–762.
2. F. Raschig, *Ber. Dtsch. Chem. Ges.*, 1907, **40**, 4580–4588.
3. I. Trifonow, *Z. Anorg. Allg. Chem.*, 1922, **124**, 123–135.
4. I. Trifonow, *Z. Anorg. Allg. Chem.*, 1922, **124**, 136–139.
5. K. Gleu and E. Roell, *Z. Anorg. Allg. Chem.*, 1929, **179**, 233–266.
6. K. Gleu and R. Hubold, *Z. Anorg. Allg. Chem.*, 1935, **223**, 305–317.
7. J. W. Reed, H. H. Ho and W. L. Jolly, *J. Am. Chem. Soc.*, 1974, **96**, 1248–1249.
8. G. Kortüm and B. Finckh, *Z. Phys. Chem.*, 1941, **48B**, 32–49.
9. E. Halfpenny and P. L. Robinson, *J. Chem. Soc.*, 1952, 928–938.
10. M. Anbar and H. Taube, *J. Am. Chem. Soc.*, 1954, **76**, 6243–6247.
11. J. D. Ray, *J. Inorg. Nucl. Chem.*, 1962, **24**, 1159–1162.
12. W. G. Keith and R. E. Powell, *J. Chem. Soc. A*, 1969, 90.
13. H. M. Papée and G. L. Petriconi, *Nature*, 1964, **204**, 142–144.
14. M. N. Hughes and H. G. Nicklin, *J. Chem. Soc. A*, 1968, 450–452.
15. D. S. Bohle, B. Hansert, S. C. Paulson and B. D. Smith, *J. Am. Chem. Soc.*, 1994, **116**, 7423–7424.
16. N. V. Blough and O. C. Zafiriou, *Inorg. Chem.*, 1985, **24**, 3502–3504.
17. L. J. Ignarro, J. M. Buga, K. S. Wood, R. S. Byrns and G. Chaudhuri, *Proc. Natl. Acad. Sci. U. S. A.*, 1987, **84**, 9265–9269.
18. R. M. J. Palmer, A. G. Ferrige and S. Moncada, *Nature*, 1987, **327**, 524–526.
19. R. J. Gryglewski, R. M. J. Palmer and S. Moncada, *Nature*, 1986, **320**, 454–456.
20. J. O. Edwards and R. C. Plumb, *Prog. Inorg. Chem.*, 1993, **41**, 599–635.
21. W. H. Koppenol, *Redox Rep.*, 2001, **6**, 229–234.
22. W. H. Koppenol, *Free Radical Biol. Med.*, 1993, **15**, 645–651.
23. Z. Maskos, J. D. Rush and W. H. Koppenol, *Free Radical Biol. Med.*, 1990, **8**, 153–162.

24. H. Bataineh, O. Pestovsky and A. Bakac, *Chem. Sci.*, 2012, **3**, 1594–1599.
25. W. H. Koppenol and P. L. Bounds, Redox-active metals: Iron and copper, in *Principles of Free Radical Biomedicine*, ed. K. Pantopoulos and H. M. Schipper, Nova Science Publishers, Inc., Hauppage, NY, 2011, pp. 91–111.
26. J. S. Beckman, T. W. Beckman, J. Chen, P. A. Marshall and B. A. Freeman, *Proc. Natl. Acad. Sci. U. S. A.*, 1990, **87**, 1620–1624.
27. B. H. J. Bielski, D. E. Cabelli, R. L. Arudi and A. B. Ross, *J. Phys. Chem. Ref. Data*, 1985, **14**, 1041–1100.
28. W. H. Koppenol, J. J. Moreno, W. A. Pryor, H. Ischiropoulos and J. S. Beckman, *Chem. Res. Toxicol.*, 1992, **5**, 834–842.
29. C. Molina, R. Kissner and W. H. Koppenol, *Dalton Trans.*, 2013, **42**, 9898–9905.
30. W. H. Koppenol and R. Kissner, *Chem. Res. Toxicol.*, 1998, **11**, 87–90.
31. R. Kissner and W. H. Koppenol, *J. Am. Chem. Soc.*, 2002, **124**, 234–239.
32. D. Gupta, B. Harish, R. Kissner and W. H. Koppenol, *Dalton Trans.*, 2009, 5730–5736.
33. C. Kurz, X. Zeng, S. Hannemann, R. Kissner and W. H. Koppenol, *J. Phys. Chem. A*, 2005, **109**, 965–969.
34. S. Goldstein and G. Merényi, *Methods Enzymol.*, 2008, **436**, 49–61.
35. W. H. Koppenol, P. L. Bounds, T. Nauser, R. Kissner and H. Ruegger, *Dalton Trans.*, 2012, **41**, 13779–13787.
36. J. W. Coddington, J. K. Hurst and S. V. Lymar, *J. Am. Chem. Soc.*, 1999, **121**, 2438–2443.
37. S. V. Lymar, R. F. Khairutdinov and J. K. Hurst, *Inorg. Chem.*, 2003, **42**, 5259–5266.
38. M. Kirsch, H.-G. Korth, A. Wensing, R. Sustmann and H. de Groot, *Arch. Biochem. Biophys.*, 2003, **418**, 133–150.
39. M. Sturzbecher, R. Kissner, T. Nauser and W. H. Koppenol, *Inorg. Chem.*, 2007, **46**, 10655–10658.
40. R. Kissner, T. Nauser, P. Bugnon, P. G. Lye and W. H. Koppenol, *Chem. Res. Toxicol.*, 1997, **10**, 1285–1292.
41. A. Denicola, B. A. Freeman, M. Trujillo and R. Radi, *Arch. Biochem. Biophys.*, 1996, **333**, 49–58.
42. S. V. Lymar and J. K. Hurst, *J. Am. Chem. Soc.*, 1995, **117**, 8867–8868.
43. R. Meli, T. Nauser and W. H. Koppenol, *Helv. Chim. Acta*, 1999, **82**, 722–725.
44. M. G. Bonini, R. Radi, G. Ferrer-Sueta, A. M. D. Ferreira and O. Augusto, *J. Biol. Chem.*, 1999, **274**, 10802–10806.
45. D. A. Armstrong, R. E. Huie, S. V. Lymar, W. H. Koppenol, G. Merényi, P. Neta, D. M. Stanbury, S. Steenken and P. Wardman, *BioInorg. React. Mech.*, 2013, **9**, 59–61.
46. R. Meli, T. Nauser, P. Latal and W. H. Koppenol, *J. Biol. Inorg. Chem.*, 2002, **7**, 31–36.
47. J. Xue, X. Y. Ying, J. S. Chen, Y. H. Xian, L. T. Jin and J. Jin, *Anal. Chem.*, 2000, **72**, 5313–5321.
48. W. C. A. Koh, J. I. Son, E. S. Choe and Y. B. Shim, *Anal. Chem.*, 2010, **82**, 10075–10082.

49. C. Amatore, S. Arbault, D. Bruce, P. de Oliveira, M. Erard and M. Vuillaume, *Chem.–Eur. J.*, 2001, **7**, 4171–4179.
50. C. Amatore, S. Arbault, C. Bouton, K. Coffi, J.-C. Drapier, H. Ghandour and Y. Tong, *ChemBioChem*, 2006, **7**, 653–661.
51. S. F. Peteu, T. Bose and M. Bayachou, *Anal. Chim. Acta*, 2013, **780**, 81–88.
52. J. Glebska and W. H. Koppenol, *Free Radical Biol. Med.*, 2003, **35**, 676–682.
53. B. Kalyanaraman, V. Darley-Usmar, K. J. A. Davies, P. A. Dennery, H. J. Forman, M. B. Grisham, G. E. Mann, K. Moore, L. J. Roberts II and H. Ischiropoulos, *Free Radical Biol. Med.*, 2012, **52**, 1–6.
54. H. J. Forman, O. Augusto, R. Brigelius-Flohe, P. A. Dennery, B. Kalyanaraman, H. Ischiropoulos, G. E. Mann, R. Radi, L. J. Roberts II, J. Vina and K. J. A. Davies, *Free Radical Biol. Med.*, 2015, **78**, 233–235.
55. A. Sikora, J. Zielonka, M. Lopez, J. Joseph and B. Kalyanaraman, *Free Radical Biol. Med.*, 2009, **47**, 1401–1407.
56. J. Zielonka, A. Sikora, M. Hardy, J. Joseph, B. P. Dranka and B. Kalyanaraman, *Chem. Res. Toxicol.*, 2012, **25**, 1793–1799.
57. M. Wörle, P. Latal, R. Kissner, R. Nesper and W. H. Koppenol, *Chem. Res. Toxicol.*, 1999, **12**, 305–307.
58. J.-H. M. Tsai, J. G. Harrison, J. C. Martin, T. P. Hamilton, M. van der Woerd, M. J. Jablonsky and J. S. Beckman, *J. Am. Chem. Soc.*, 1994, **116**, 4115–4116.
59. D. Klug, J. Rabani and I. Fridovich, *J. Biol. Chem.*, 1972, **247**, 4839–4842.
60. E. M. Fielden, P. B. Roberts, R. C. Bray, D. J. Lowe, G. N. Mautner, G. Rotilio and L. Calabrese, *Biochem. J.*, 1974, **139**, 49–60.
61. T. Nauser and W. H. Koppenol, *J. Phys. Chem. A*, 2002, **106**, 4084–4086.
62. H. Ischiropoulos, L. Zhu and J. S. Beckman, *Arch. Biochem. Biophys.*, 1992, **298**, 446–451.
63. C. D. Smith, M. Carson, M. van der Woerd, J. Chen, H. Ischiropoulos and J. S. Beckman, *Arch. Biochem. Biophys.*, 1992, **299**, 350–355.
64. R. Radi, J. S. Beckman, K. M. Bush and B. A. Freeman, *J. Biol. Chem.*, 1991, **266**, 4244–4250.
65. R. Radi, J. S. Beckman, K. M. Bush and B. A. Freeman, *Arch. Biochem. Biophys.*, 1991, **288**, 481–487.
66. M. S. Ramezanian, S. Padmaja and W. H. Koppenol, *Chem. Res. Toxicol.*, 1996, **9**, 232–240.
67. J. S. Beckman, Y. Z. Ye, P. G. Anderson, J. Chen, M. A. Accavitti, M. M. Tarpey and C. R. White, *Biol. Chem. Hoppe-Seyler*, 1994, **375**, 81–88.
68. S. Padmaja, M. S. Ramezanian, P. L. Bounds and W. H. Koppenol, *Redox Rep.*, 1996, **2**, 173–177.
69. B. Alvarez, H. Rubbo, M. Kirk, S. Barnes, B. A. Freeman and R. Radi, *Chem. Res. Toxicol.*, 1996, **9**, 390–396.
70. J. S. Beckman and W. H. Koppenol, *Am. J. Physiol. Cell Physiol.*, 1996, **271**, C1424–C1437.
71. J. K. Hurst and S. V. Lymar, *Acc. Chem. Res.*, 1999, **32**, 520–528.
72. P. Pacher, J. S. Beckman and L. Liaudet, *Physiol. Rev.*, 2007, **87**, 315–424.
73. M. N. Hughes and H. G. Nicklin, *J. Chem. Soc. A*, 1971, 164–168.

74. W. A. Pryor, R. Cueto, X. Jin, W. H. Koppenol, M. Ngu-Schwemlein, G. L. Squadrito, P. L. Uppu and R. M. Uppu, *Free Radical Biol. Med.*, 1995, **18**, 75–83.
75. W. H. Koppenol, R. Kissner and J. S. Beckman, *Methods Enzymol.*, 1996, **269**, 296–302.
76. D. S. Bohle, P. A. Glassbrenner and B. Hansert, *Methods Enzymol.*, 1996, **269**, 302–311.
77. X. Chen, H. Chen, R. Deng and J. Shen, *Biomed. J.*, 2014, **37**, 120–126.
78. J. J. Saucerman, *Biophys. J.*, 2013, **105**, 1287–1288.
79. W. H. Koppenol, *Pure Appl. Chem.*, 2000, **72**, 437–446.
80. N. G. Connelly, T. Damhus, R. M. Hartshorn and A. T. Hutton, *Nomenclature of Inorganic Chemistry. IUPAC Recommendations 2005*, Royal Society of Chemistry, Cambridge, 2005.

Quantifying Peroxynitrite: Bridging the Gap Between Chemistry, Biology and Immunology

SABINE BORGMANN*[a]

[a]Research Department Interfacial Systems Chemistry (RD IFSC),
Ruhr-Universität Bochum, Fakultät für Chemie und Biochemie,
NC 02/168, Universitätsstr. 150, 44801 Bochum, Germany
*E-mail: sabine.borgmann@rub.de

2.1 General Remarks on the Concept of "Redox Signaling"

Small chemical molecules can serve as transmitters and modulators of cellular signaling in biological and immunological processes. They can act within a biological cell or can be produced within a cell or in the vicinity of the cell membrane and diffuse to surrounding cells. This extracellular chemical signaling in the microenvironment of biological cells is vital for cell–cell interactions and for a functional organism. If this signaling process is out of balance, it can result in pathological conditions for the local cells or even the entire organism.

Reactive nitrogen and oxygen species (RNS and ROS, respectively) are very fascinating species among these chemical signaling molecules.[1,2] Due to their

RSC Detection Science Series No. 7
Peroxynitrite Detection in Biological Media: Challenges and Advances
Edited by Serban F. Peteu, Sabine Szunerits, and Mekki Bayachou

Published by the Royal Society of Chemistry, www.rsc.org

high reactivity and low biological half-life it is not trivial to distinguish which of the numerous biological effects is affected by which molecule(s). Basically, signaling specificity is achieved by the different chemical reactivities of the involved RNS and/or ROS.[3] A systematic investigation of the complex "redox signaling" within and in the microenvironment of biological cells is not trivial. Even the quantification of these RNS and ROS represents a true analytical challenge. Table 2.1 compiles a list of typical RNS and ROS molecules.

The effects of these chemical signaling molecules depend on the nature, concentration, and spatial and temporal distribution of the soluble chemical signals, as well as the identity, location, local microenvironment and state of the responding cells. As a general concept of "redox signaling" one can consider that there is most likely a "redox balance" between the exposure to "oxidative stress" and the actions of "antioxidant protection mechanisms".[4] Figure 2.1(a) visualizes the general concept of the relationship between the oxidative burden generated by RNS/ROS and the different types of "redox signaling". Certain basal concentration levels of RNS and ROS are vital for the functionality of physiological processes ("physiological redox signaling")—especially at "low" concentration levels. Moderate perturbations of these oxidant levels represent normal "signaling events". With increasing oxidant

Table 2.1 Typical RNS and ROS molecules grouped in radical and non-radical species.

RNS	ROS
Radicals	
$NO^{\bullet}$ nitric oxide	$O_2^{\bullet -}$ superoxide
$NO_2^{\bullet}$ nitrogen dioxide	$^{\bullet}OH$ hydroxyl
	$RO_2^{\bullet}$ peroxyl
	$RO^{\bullet}$ alkoxyl
	$HO_2^{\bullet}$ hydroperoxyl
	$CO_3^{\bullet -}$ carbonate
Non-radicals	
$ONOO^-$ peroxynitrite	O_3 ozone
$ONOOH$ peroxynitrous acid	$^1O_2$ singlet oxygen
$ROONO$ alkyl peroxynitrite	H_2O_2 hydrogen peroxide
NO^- nitroxyl anion	$HOCl^-$ hypochlorous acid
NO^+ nitrosyl cation	$ONOO^-$ peroxynitrite
NO_2^+ nitronium anion	
HNO_2 nitrous acid	
N_2O_3 dinitrogen trioxide	
N_2O_4 dinitrogen tetroxide	
NO_2Cl nitryl chloride	

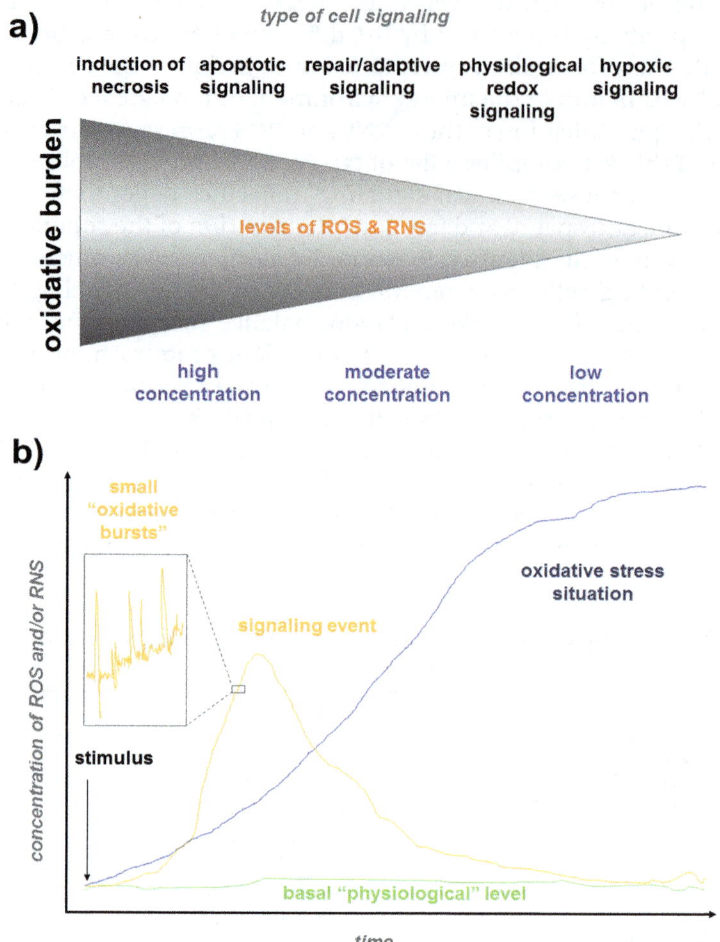

Figure 2.1 Simplified schematic overview of the different types of "redox signaling"[1,4] that may occur *via* RNS and ROS in biological systems. (a) The idea is that the mediated signaling type is dependent on the concentration of RNS and/or ROS and the duration of their presence. At low concentrations, physiological signaling that is vital for the functionality of the biological system takes place. At moderate concentration levels, the biological cells employ adaptive and repair signaling mechanisms. At high concentration levels, toxicological effects occur either *via* apoptotic or necrotic signaling. (b) The time course of these different "redox signaling" types may differ significantly. Compared with low concentrations of RNS/ROS at a baseline "physiologic" level, short-term oxidative bursts may occur and lead to vital "redox signaling" events, whereas toxic levels of RNS/ROS can accumulate over time in pathological conditions. It seems that it is dependent on the production rate of these species if the effects induced by RNS/ROS are beneficial or pathological.

burden, severe "oxidative stress" occurs and impairs cell function and viability. This may finally result in apoptotic and necrotic signaling.

The actual production rates and concentration profiles (fluxes) of RNS and/or ROS are dependent on a variety of factors such as cell type and function; metabolic, immunologic and cell cycle stage; cellular interactions and cell-to-cell variations; as well as the local microenvironment. The underlying hypothesis is that the fate of the induced signaling effects seems to be dose and time dependent. Figure 2.1b gives a schematic overview of how these different signals ("oxidative bursts") may look on a temporal sequence. The production rate of RNS and ROS may be small and low in fluctuations at a basal "physiological" level. During a physiological "redox signaling" event after a certain stimulus, there may be a short-term increase of one or several RNS/ROS levels. Since the time resolution of many methods employed for quantification is often not sufficient enough, it is not clear if these rises in RNS/ROS are accompanied by tiny spikes of "oxidative bursts" similar to neurotransmitter signaling, as suggested by data from Amatore *et al.*[5] When higher concentration levels of RNS/ROS are produced over longer time scales, apoptotic or necrotic signaling mechanisms are induced. However, it is important to keep in mind that the time courses of the fluxes of RNS/ROS have often not yet been reliably determined. Furthermore, the redox signaling mechanisms seem to be very complex, *e.g.* the balance between oxidative stress/signaling and antioxidant responses. How the actual redox regulation of biological systems occurs is still under investigation and controversially discussed.

2.2 The Basic Chemical Properties of Peroxynitrite Compared with Other RNS/ROS as Signaling Molecules

The focus of this chapter is peroxynitrite ($ONOO^-$), which is represented in the *cis* configuration in Figure 2.2. Chapter 1 of this book refers in detail to the chemistry and properties of peroxynitrite. In addition, the chemistry and biochemistry of peroxynitrite has been thoroughly reviewed elsewhere.[6-12] Peroxynitrite plays an important role in redox biology, in health and disease,[13,14] and in the immune system.[15] In this chapter certain features of the molecule peroxynitrite are highlighted and put into perspective. The idea is to focus on the biochemical and immunological aspects of $ONOO^-$ that are of particular relevance for a chemist to have a basic idea and understanding of the underlying biology. This defines the complexity of the actual task and conditions to quantify $ONOO^-$. For a biologist or immunologist, certain chemical and analytical principles of peroxynitrite detection are emphasized in order to facilitate a better understanding of the basic challenges to realize reliable and reproducible quantification of $ONOO^-$.

Nitric oxide ($NO^{\bullet}$) and superoxide anion ($O_2^{\bullet -}$) react at nearly diffusion-controlled rates to peroxynitrite.[16] The fast reaction rate[17] implies that peroxynitrite is formed *in vitro* and *in vivo* despite the potential presence of RNS

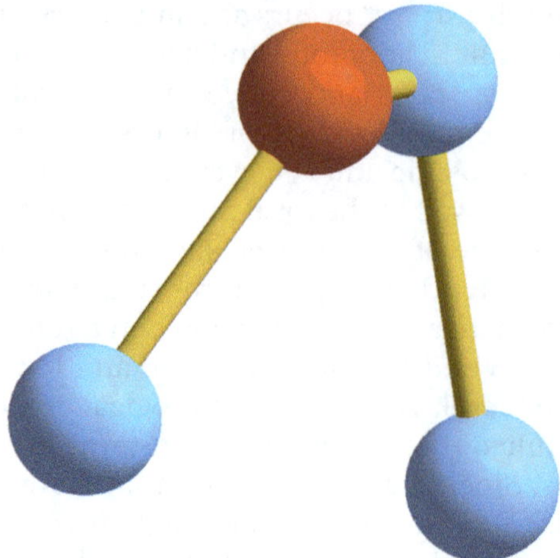

Figure 2.2   Schematic representation of peroxynitrite, ONOO⁻, in the *cis* conformation, which is the more stable than the *trans* rotamer.

and ROS scavengers, antioxidant molecules and enzymes. This reaction rate ensures that enzymatic removal of $O_2^{•-}$ by superoxide dismutase (SOD) can be outcompeted provided that a sufficient source of $NO^•$ is available in the vicinity of the $O_2^{•-}$ source. The history of the discovery of peroxynitrite has been reviewed.[18,19] In 1990, it was published for the first time that ONOO⁻ could act as a biological oxidant.[16,20] Besides its oxidative activity, ONOO⁻ is also a powerful nitrating agent.

One important feature of RNS and ROS is their relatively short half-life in aqueous solutions, which is a result of the high reactivity of these species. Note that peroxynitrite exhibits much higher chemical reactivity compared with $NO^•$ and $O_2^{•-}$.[16,20,21] Figure 2.3 summarizes the dimensions of typical reported chemical half-lives of different RNS and ROS.

ONOO⁻ can be produced inside and outside of biological cells. ONOO⁻ as well as peroxynitrous acid (ONOOH) can diffuse or be transported *via* anion channels through membranes.[22,23] Taking into account that the half-life of peroxynitrite in biological systems was determined to be in the millisecond region,[24,25] peroxynitrite is able to diffuse several micrometers (in the region of one to a couple of cell diameters). This means it can intracellularly diffuse through cell compartments and extracellularly towards and inside neighboring cells.[26] Employing a theoretical model, it was shown that the presence of carbon dioxide only partially alters the reactivity and diffusion of intravascularly generated, extracellular peroxynitrite.[25] Another theoretical model was developed to predict intracellular concentrations and diffusion rates of nitric oxide, superoxide, and peroxynitrite inside cells in a stirred suspension cell culture exposed to nitric oxide and/or peroxynitrite.[27] Peroxynitrite is able to diffuse up to a few micrometers, similar to nitric oxide,[28] in biological systems. Whereas $O_2^{•-}$ diffusion lies within about a cell diameter. The most reactive molecule, the hydroxyl radical $(OH^•)$, diffuses only a few ångströms.[29] Hydrogen peroxide (H_2O_2)[30] on the

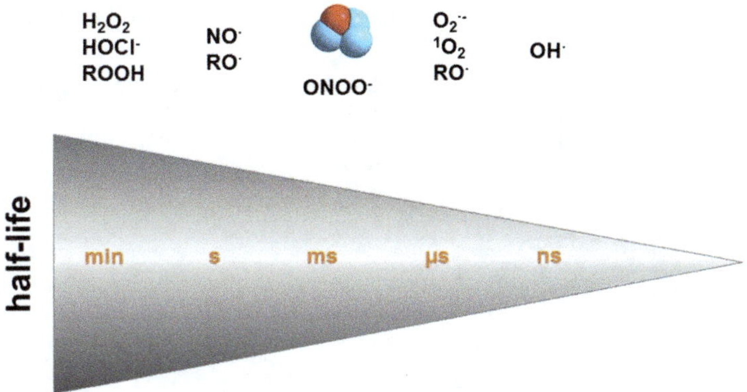

Figure 2.3 Simplified representation of the chemical half-lives of typical ROS and RNS in aqueous and biological systems.

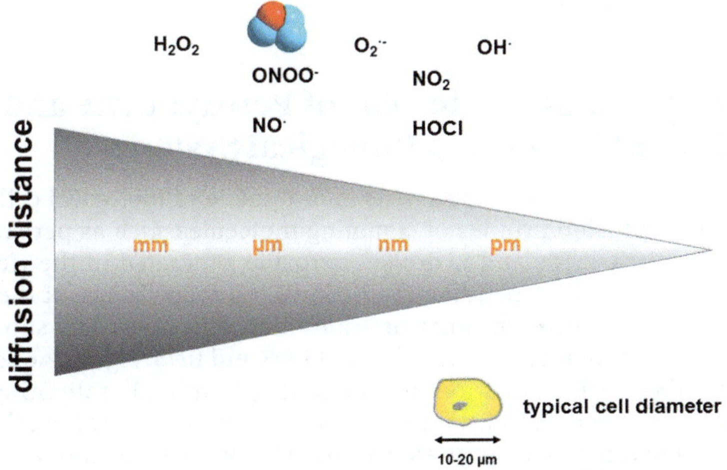

Figure 2.4 Simplified representation of the dimensions of diffusion distances of ROS and RNS in aqueous and biological systems.

other hand is a molecule that is more stable than $O_2^{\bullet-}$. It can diffuse over longer distances[26] given that it is not consumed, *e.g.* by catalase (CAT) and myeloperoxidase, before it diffuses through cell membranes. The dimensions of typical diffusion distances of some RNS and ROS are summarized in Figure 2.4.

Furthermore, it is important to consider the ability of peroxynitrite and other RNS/ROS to permeate cell membranes. Biomembranes can compartmentalize the action radii of these molecules. Note that ONOOH has a higher permeability through cell membranes than $ONOO^-$.[31] In Figure 2.5 a comparison of $ONOO^-$ and ONOOH with several other RNS and ROS with respect to their permeability through pure lipid membranes is given.[31]

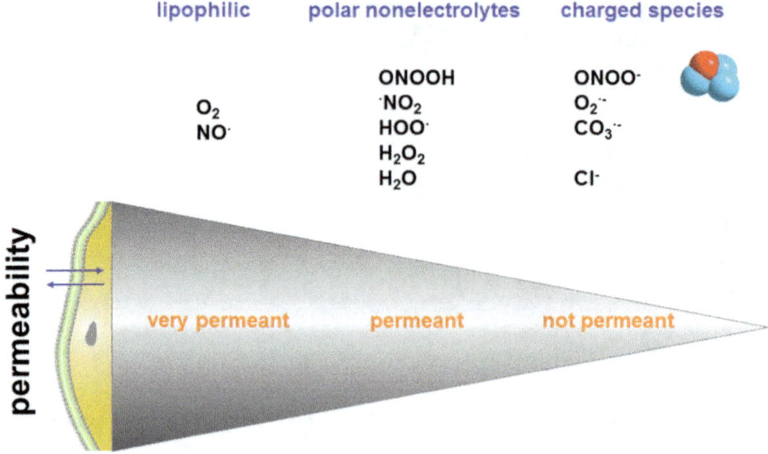

Figure 2.5   Overview of the membrane permeability of ROS and RNS. Data derived from ref. 31.

2.3   The Complex Interplay of Peroxynitrite and Other RNS/ROS in Biological Systems

Once RNS and ROS are produced within or outside of biological cells, there are complex diffusion profiles of signaling molecules such as peroxynitrite, superoxide, nitric oxide and hydrogen peroxide generated in the vicinity of these biological cells. Depending on the local microenvironment these molecules may react with each other or additional molecules. The subsequent reaction products may result in a variety of ROS and RNS. Figure 2.6 schematically describes such a situation in a simplified fashion, highlighting which species can be formed and what their chemical interplay may look like.

H_2O_2 is generated by dismutation from $O_2^{\cdot-}$ by SOD and can act as a signaling molecule.[26,32] The actions of H_2O_2 are biologically contained by CAT, glutathione peroxidase, and peroxiredoxins through degradation to water. In general, H_2O_2 is relatively inert to reaction with biomolecules compared with other ROS. However, it reacts with seleno-, thiol or heme peroxidases, or other transition metal centers.[26] The damaging effect of H_2O_2 is facilitated *e.g. via* transformation into hydroxyl radicals ($^{\cdot}OH$) by Fenton chemistry, which are the strongest oxidants in biological systems.[26] $^{\cdot}OH$ can also be generated by radiation, reactions at transition metal ions, the decomposition of peroxynitrite, and decomposition of ozone.[33]

$O_2^{\cdot-}$ is generated in biological cells *e.g.* by the multi-enzyme complex NADPH oxidase (NOX),[34] the mitochondrial respiratory chain, xanthine oxidase (XO), cytochrome P-450 enzymes, and autooxidation of biomolecules.[35,36] It can act as an oxidant and reductant.[33]

$NO^{\cdot}$ is synthesized from the amino acid L-arginine[37] by nitric oxide synthase (NOS)[38] and serves as a physiological messenger and signaling molecule.[39–41] Note that inflammatory cells also generate nitric oxide as an important part of

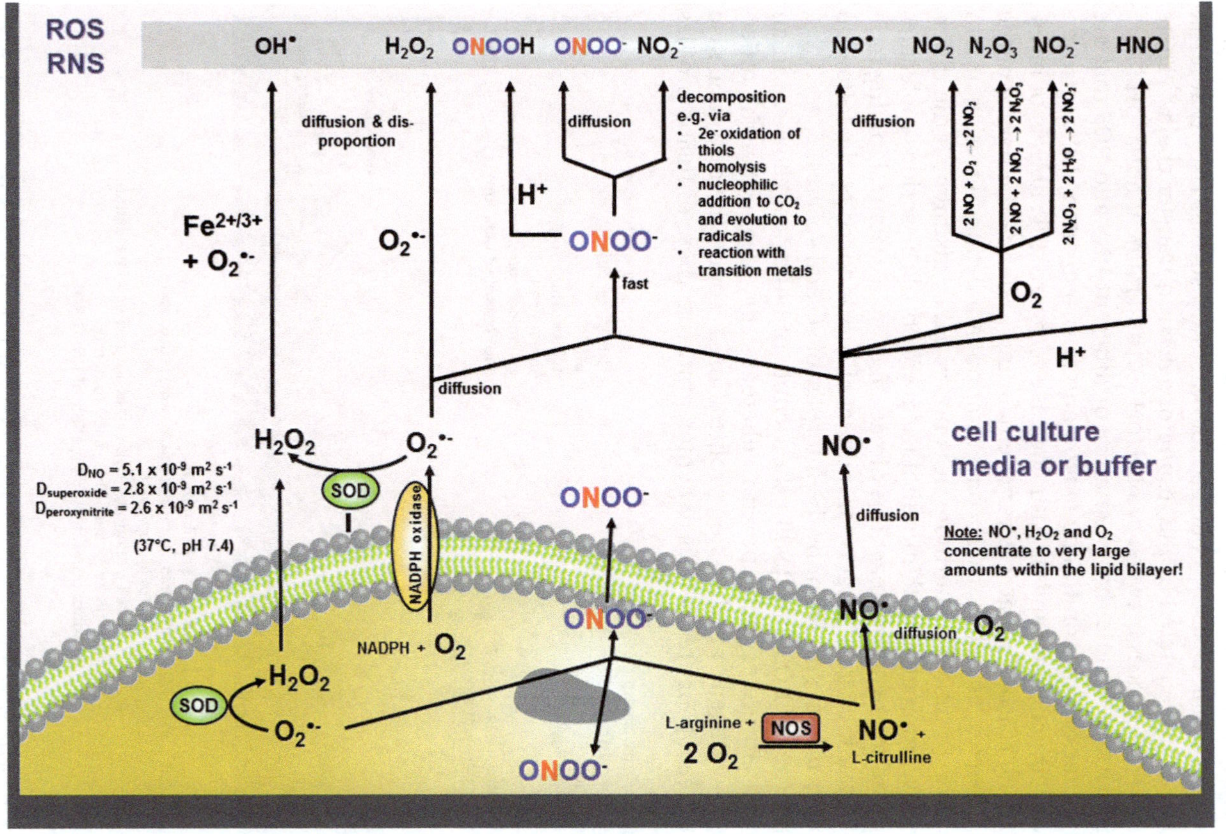

Figure 2.6 A variety of ROS and RNS play a role in intra- and inter-cellular communication, and may also chemically react and interact with each other. A selection of typical biochemical and chemical reactions of NO•, $O_2^{•-}$, H_2O_2 and $ONOO^-$ is displayed (adapted from ref. 72). The diffusion coefficients were taken from ref. 148.

the host defense mechanism.[42,43] Nitric oxide is a poor oxidant but reacts with superoxide to peroxynitrite (ONOO⁻). The chemistry of NO• and ONOO⁻ is complex.[6,44–49] Peroxynitrite is a potent oxidant and nitrating agent. It can react with a great variety of biomolecules.[8,10,12,24] Mitochondria are an important source not only for NO• and ONOO⁻,[50] but also for other ROS such as H_2O_2 that result from superoxide generated by the respiratory electron transfer chain.[52] There are local pH differences between cytosol (pH 7.0) and mitochondria (pH 8.0).[53]

It is important to note that the nature of different RNS and ROS that are formed under certain conditions *in vitro* and *in vivo* is often still controversial, and their beneficial and/or pathological effects are still subjects of investigation. One reason is that it is very demanding to define the nature of the involved species, as well as their concentration at a defined location and in a certain time frame. Furthermore, it is not trivial to distinguish the different chemical, biological and immunological effects of the various molecules involved. Are the resulting reactions direct effects of a certain signaling molecule on their biological targets, or indirect effects induced by the follow-up metabolites of that species or by the interplay of several species?

The mixture of ROS and RNS in the vicinity of biological cells, *e.g.* in an *in vitro* cell culture experiment with adherent cells, means that the different species will generate diffusion profiles/fluxes around the cells (and also within the cells). Figure 2.7 tries to demonstrate that these diffusion profiles

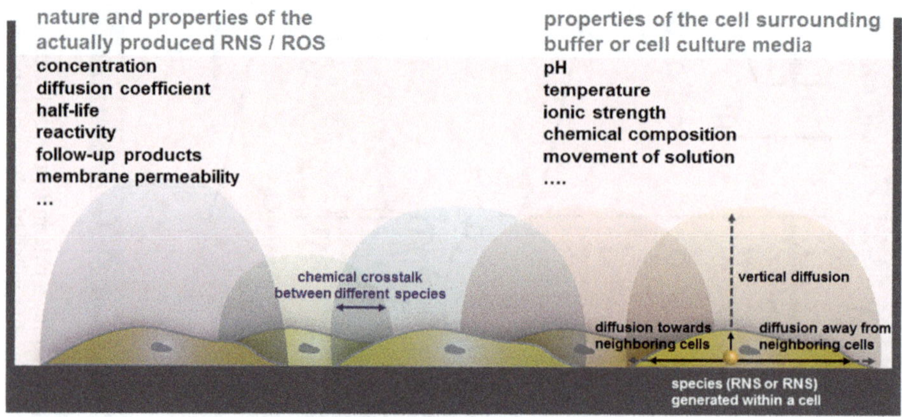

Figure 2.7   Complex diffusion profiles of different ROS and RNS in the vicinity of a cell monolayer (modified from ref. 72). The state that the biological cells are in defines the nature, amount and timeline of the produced species. The local microenvironment (surrounding cells and cell culture medium) around the cells impacts the fluxes and diffusion profiles of ROS and RNS. A variety of parameters influence the chemical crosstalk between the species involved as well as cell–cell signaling.

depend on the state of the biological cells, the reactivity of the molecules involved, their diffusion profiles and the local microenvironment.

$ONOO^-$ is reported to be more toxic to cells than H_2O_2, $O_2^{\bullet-}$, or $NO^{\bullet}$.[54,55] For example, it can generate oxidative and nitrative damage to DNA, lipids and proteins.[10,15] Together with other ROS and RNS, peroxynitrite is involved in the chemical biology of inflammation.[4,56] As one example of a host defense mechanism, neutrophils and macrophages produce RNS and ROS as oxidants in order to damage and kill invading microorganisms. Note that the first evidence for peroxynitrite formation in biological cells was detected in freshly isolated rat alveolar macrophages when they produced nitric oxide and superoxide.[57] Hydrogen peroxide and nitrite are the major decomposition products of superoxide and peroxynitrite in phagosomes.[58]

Figure 2.8 highlights, as one example of the immune response, how macrophages produce $NO^{\bullet}$ and $O_2^{\bullet-}$, which quantitatively form $ONOO^-$.[4,57,59,60] The inducible NOS (iNOS) in macrophages is activated by lipopolysaccharide (LPS) of gram-negative bacteria or by proinflammatory cytokines produced by infected host cells, for example. Activated macrophages release a variety

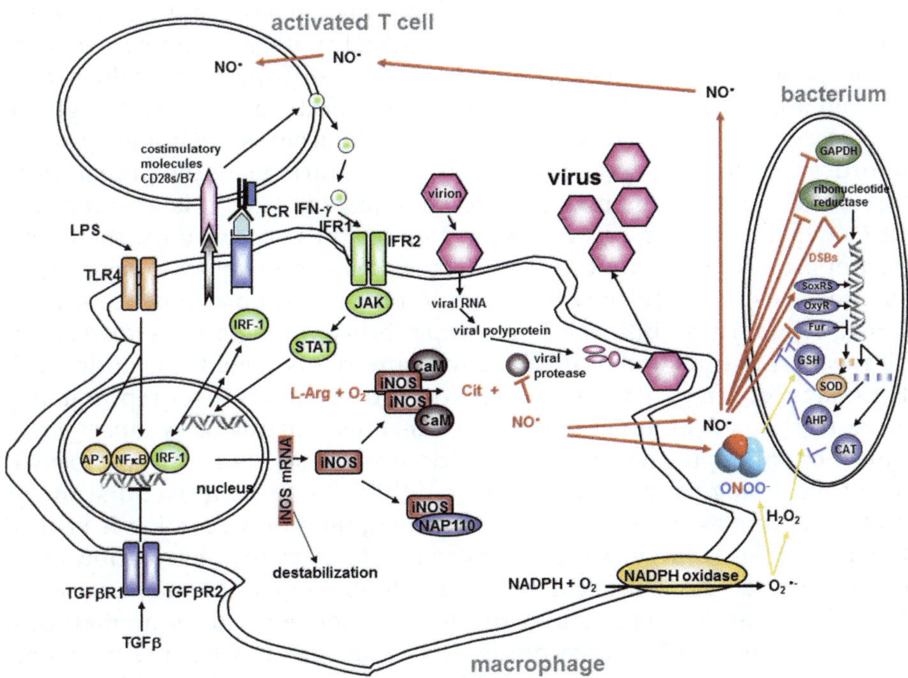

Figure 2.8   Schematic overview of the production of ROS and RNS by activated macrophages with a focus on $NO^{\bullet}$ signaling during an immune response (data taken from ref. 59 and 60, refer to these publications for more detailed description of the involved signaling pathways). Note that the species formed, such as H_2O_2, $NO^{\bullet}$ and $ONOO^-$, induce toxic damage to the intruding bacterium.

of molecules—among them nitric oxide and superoxide—to inhibit the replication of pathogens such as bacteria, fungi and viruses. The resulting cocktail of different RNS and ROS, including H_2O_2 and $ONOO^-$, is able to penetrate bacterial and fungal cell walls and is a highly effective toxic agent for pathogens.[61] Peroxynitrite can induce cell death *via* apoptosis and necrosis.[62] The actions and biochemical targets of peroxynitrite in disease conditions are reviewed in detail by Pacher *et al.*[13]

2.4   Implications of Selecting a Model System for *In Vitro*, *In Vivo* and *Ex Vivo* Experiments

Some basic considerations have to be made when selecting a suitable model system for *in vitro* or *in vivo* experiments. Figure 2.9 summarizes the different workflow steps of designing an experimental study to quantify $ONOO^-$ in biological systems. Once the biological question of interest is raised, in our case the quantification of the amount of extracellular $ONOO^-$ produced from biological cells, it is important to decide on the cell culture or sampling conditions and the methodology for $ONOO^-$ detection. Once the initial experimental strategy is defined, it needs to be reviewed for statistical soundness, taking into account that an appropriate number of biological and methodological replicates and controls are made. Now a proof-of-principle study can be performed that allows testing and optimization of the selected experimental parameters. It might be necessary to optimize the sampling or sample processing/storage process in order to avoid changes in the redox status during sample preparation. Cell culture conditions may have to be adjusted. After it is ensured that $ONOO^-$ quantification can be realized under these conditions, the detection approach can be optimized further. Besides improving key criteria such as sensitivity, selectivity, dynamic range, response time, reproducibility, long-term stability and biocompatibility among other efforts, overall measuring time and cost of the measurements need to be addressed. For instance, bringing the tip of a peroxynitrite microsensor reproducibly close to the cell monolayer into the so-called "artificial synapse" position[110] by employing a controlled positioning device, as already performed for other biological relevant analytes,[72,110,127] could be a useful strategy. After a routine assay protocol is established, method validation is critical before starting a larger set of measurements. A critical evaluation must be achieved at the analytical as well as the biological/biochemical level. One bias-escaping approach is the use of independent assay methods. Note that bias is difficult to address in the design of experiments, in part because it is often unintentional. Bias can only be minimized by careful control and planning of the experiments. Finally, biostatistics and chemometrics help to properly interpret the data obtained to address the biological question.

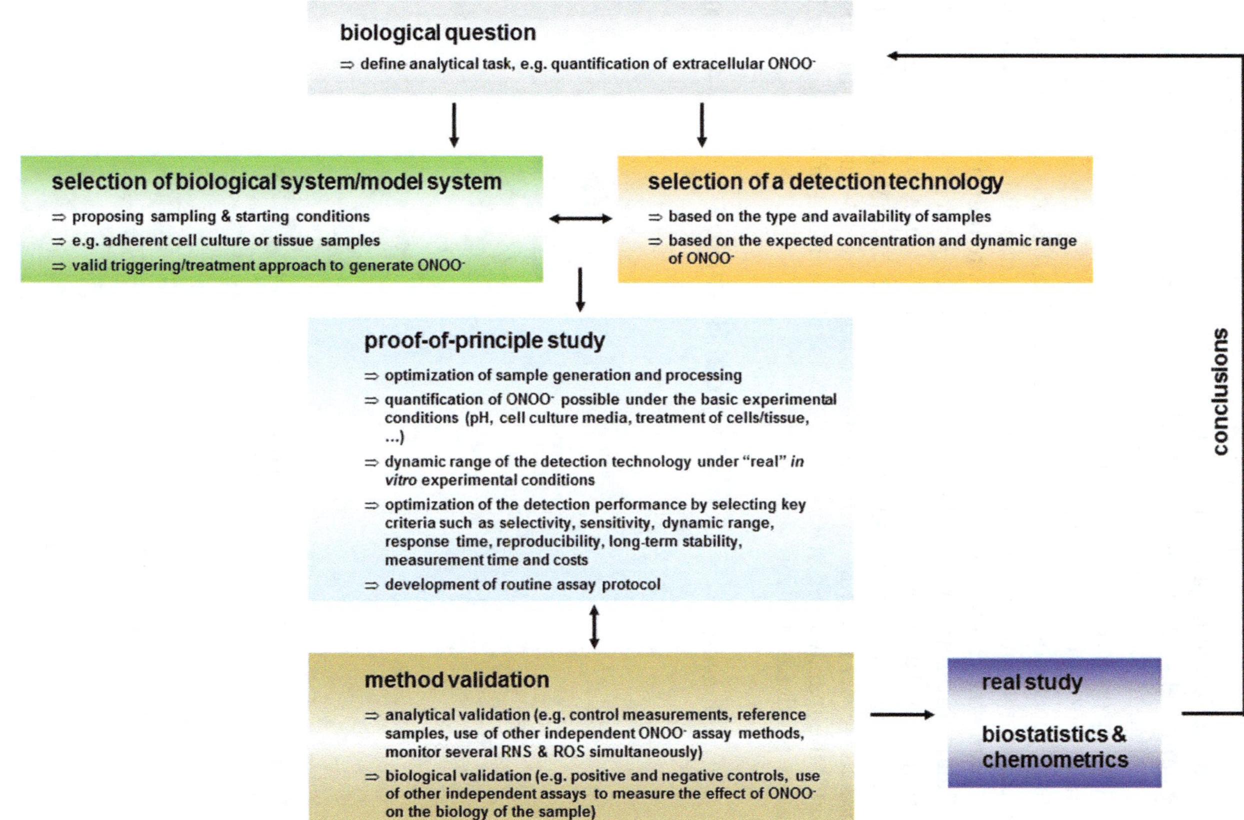

Figure 2.9 Workflow of developing a study design to quantify ONOO⁻ in biological systems.

2.4.1   Use of Biological Sample Materials

Biological sample material (Figure 2.10) can be collected from humans, animals, insects, plants, cells and the environment. Human material can be sampled during diagnostic procedures and therapeutic treatments, autopsies, and donation of organs and/or tissues. Tissues, organs, hair, skin, nail clippings and cells are "solid" material. Body fluids include blood, serum, plasma, saliva, mucosa, sweat, urine, semen, cerebrospinal fluid, milk and feces. The performance and quality of sample processing and storage is of great importance to obtain representative biological material with respect to the high reactivity of RNS and ROS. Standardized procedures should be employed. Many components of biological sample materials are relevant for $ONOO^-$ chemistry. Here, some abundant molecules are mentioned, including intracellular concentrations of about 5–10 mM thiols (with glutathione as the main component) and 5–10 µM SOD are reported.[29,63] Due to significant CO_2 (1 mM in cells,[13] 1.3 mM in blood plasma and 12–30 mM in blood[64]) and bicarbonate levels (12 mM in intracellular fluid and 25–30 mM in blood plasma),[64] the reactivity of $ONOO^-$ with CO_2 is relevant and probably a significant route for detoxification *in vivo*.[9,65,66]

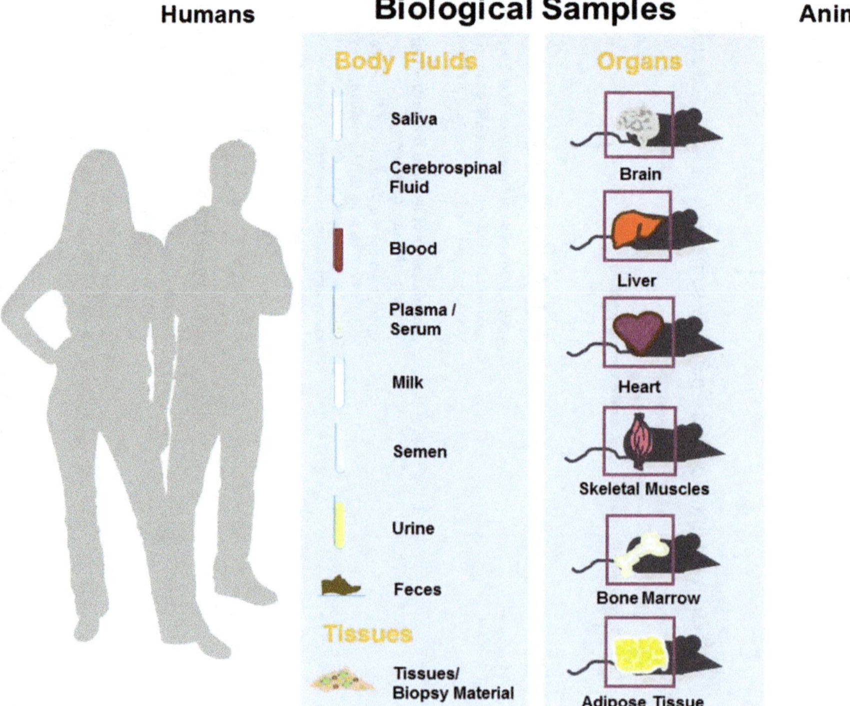

Figure 2.10   Selection of typical biological samples employed for biological and immunological studies.

Note that it is also possible to mimic and create simulated or artificial biological fluids.[67] This might be helpful for testing and validating analytical approaches before starting complex studies. With respect to the scope of the envisaged experiment, it might also be useful, if available, to use standard reference blood or plasma (healthy *vs.* a defined "diseased" state).

2.4.2 Performing Cell Culture Studies

For *in vitro* experiments cell cultures are often used. For generating primary cell cultures, cells are sampled from a tissue and proliferated under appropriate cell culture conditions. These conditions are dependent on the cell type of interest. The incubator conditions should be such that the pH, osmotic pressure, gas saturation and temperature are regulated to a selected level. A cell culture medium, which typically consists of essential nutrients such as amino acids, carbohydrates, vitamins and minerals, growth factors, hormones, buffer, and gases (O_2, CO_2), is required. The composition of some typical cell culture media is given in Table 2.2.

For primary cultures, cells are isolated from a tissue sample and further proliferated and subcultured. These cell lines have a limited life span. Immortal cells, however, are continuous cell lines because they are able to divide indefinitely. Most cell types must grow on a solid or semi-solid substrate. These adherent cells can grow as mono- or multi-layers, and may consist of one or multiple cell types (Figure 2.11). Non-adherent cells float freely in the cell culture medium in suspension, and may require sufficient mixing to avoid clogging at the bottom of the cell culture flask.

Cell culture experiments are performed because they provide a simplified experimental setting where at least several variables can be properly controlled, and the samples can be manipulated relatively easily.[68] In addition, the amount of other sample material is often limited and can only be used for selected targeted studies. Furthermore, researchers aim to reduce the use of animals for experiments. Thus, cell culture studies are well-established tools in biochemical and medical research.

On the other hand, one has to keep in mind that *in vitro* cell culture experiments intrinsically do not reflect *in vivo* situations holistically. Thus, data interpretation has to be done with appropriate care. For instance, one has to consider that the cell culture process itself generates oxidative stress[69] (*e.g.* by light and shear stress during the changing of medium and passaging of cells; or temperature changes between moving the cells from the incubator to ambient air). Immortal cell lines, which are cancer cell lines, are often used. Thus, one can raise the general question whether the behavior of these cancer cells indeed represents *in vivo* physiological behavior. Another aspect is that cell cultures often use 95% air:5% CO_2 conditions, which could be considered hyperoxic compared with many *in vivo* conditions. During the handling of the cells, they are exposed to light, which does not occur *in vivo* in most cases. Cell culture media are solutions that mimic

Table 2.2   Formulations of selected cell culture media and physiological buffers.[156]

Minimum essential medium eagle (MEM; pH 7.0–7.4)	
Compound	Concentration (g L^{-1})
Inorganic salts	
Calcium chloride	0.2
Magnesium sulfate (anhydrous)	0.09767
Potassium chloride	0.4
Potassium phosphate monobasic (anhydrous)	—
Sodium bicarbonate	—
Sodium chloride	6.8
Sodium phosphate dibasic (anhydrous)	—
Sodium phosphate monobasic (anhydrous)	0.122
Amino acids	
L-Alanine	—
L-Alanyl- L-glutamine	—
L-Arginine · HCl	0.126
L-Asparagine · H_2O	—
L-Aspartic acid	—
L-Cystine · 2HCl	0.0313
L-Glutamic acid	—
L-Glutamine	0.292
Glycine	—
L-Histidine · HCl · H_2O	0.042
L-Isoleucine	0.052
L-Leucine	0.052
L-Lysine · HCl	0.0725
L-Methionine	0.015
L-Phenylalanine	0.032
L-Proline	—
L-Serine	—
L-Threonine	0.048
L-Tryptophan	0.01
L-Tyrosine · 2Na · $2H_2O$	0.0519
L-Valine	0.046
Vitamins	
Choline chloride	0.001
Folic acid	0.001
myo-Inositol	0.002
Niacinamide	0.001
D-Pantothenic acid (hemicalcium)	0.001
Pyridoxal · HCl	0.001
Riboflavin	0.0001
Thiamine · HCl	0.001
Other	
Glucose	1
Phenol red · Na	0.011
Add	
L-Glutamine	—
Sodium bicarbonate	2.2

Dulbecco's modified eagle's medium (DMEM; pH 7.0–7.4)	
Compound	Concentration ($g\ L^{-1}$)
Inorganic salts	
Calcium chloride	0.2
Ferric nitrate · $9H_2O$	0.0001
Magnesium sulfate (anhydrous)	0.09767
Potassium chloride	0.4
Sodium bicarbonate	3.7
Sodium chloride	6.4
Sodium phosphate monobasic (anhydrous)	0.109
Amino acids	
L-Arginine · HCl	0.084
L-Cystine · 2HCl	—
Glycine	0.03
L-Histidine · HCl · H_2O	0.042
L-Isoleucine	0.105
L-Leucine	0.105
L-Lysine · HCl	1.46
L-Methionine	—
L-Phenylalanine	0.066
L-Serine	0.042
L-Threonine	0.095
L-Tryptophan	0.016
L-Tyrosine · 2Na · $2H_2O$	0.10379
L-Valine	0.094
Vitamins	
Choline chloride	0.004
Folic acid	0.004
myo-Inositol	0.0072
Niacinamide	0.004
D-Pantothenic acid (hemicalcium)	0.004
Pyridoxal · HCl	—
Pyridoxine · HCl	0.004
Riboflavin	0.0004
Thiamine · HCl	0.004
Other	
D-Glucose	4.5
Phenol red · Na	0.0159
Pyruvic acid · Na	0.11
Add	
L-Glutamine	0.584

RPMI-1640 medium (pH 7.0–7.4)	
Compound	Concentration ($g\ L^{-1}$)
Inorganic salts	
Calcium nitrate · $4H_2O$	0.1
Magnesium sulfate (anhydrous)	0.04884
Potassium chloride	0.4
Sodium bicarbonate	2
Sodium chloride	6
Sodium phosphate dibasic (anhydrous)	0.8

(*continued*)

Table 2.2   (*continued*)

RPMI-1640 medium (pH 7.0–7.4)	
Compound	Concentration (g L^{-1})
Amino acids	
L-Alanyl-L-Glutamine	—
L-Arginine	0.2
L-Asparagine (anhydrous)	0.05
L-Aspartic acid	0.02
L-Cystine · 2HCl	0.0652
L-Glutamic acid	0.02
L-Glutamine	—
Glycine	0.01
L-Histidine	0.015
Hydroxy-L-proline	0.02
L-Isoleucine	0.05
L-Leucine	0.05
L-Lysine · HCl	0.04
L-Methionine	0.015
L-Phenylalanine	0.015
L-Proline	0.02
L-Serine	0.03
L-Threonine	0.02
L-Tryptophan	0.005
L-Tyrosine · 2Na · 2H$_2$O	0.02883
L-Valine	0.02
Vitamins	
D-Biotin	0.0002
Choline chloride	0.003
Folic acid	0.001
myo-Inositol	0.035
Niacinamide	0.001
p-Aminobenzoic acid	0.001
D-Pantothenic acid (hemicalcium)	0.00025
Pyridoxine · HCl	0.001
Riboflavin	0.0002
Thiamine · HCl	0.001
Vitamin B$_{12}$	0.000005
Other	
D-Glucose	2
Glutathione (reduced)	0.001
Phenol red · Na	0.0053
Add	
L-Glutamine	0.3
Sodium bicarbonate	—

Optional additional components to cell culture media	
Compound	Concentration
Fetal calf serum (FSC)	10%
Penicillin	100 U ml^{-1}
Streptomycin	100 U ml^{-1}

Hanks' balanced salt solution (HBSS; pH 6.7–7.8)	
Compound	Concentration (g L^{-1})
Inorganic salts	
$CaCl_2 \cdot 2H_2O$	0.185
$MgSO_4$ (anhydrous)	0.09767
KCl	0.4
KH_2PO_4	0.06
NaCl	8
Na_2HPO_4 (anhydrous)	0.04788
Other	
D-Glucose	1
Phenol red · Na	0.11
Add	
$NaHCO_3$	0.35

Krebs–Ringer bicarbonate buffer (pH with sodium bicarbonate 7.0–7.6)	
Compound	Concentration (g L^{-1})
Inorganic salts	
Magnesium chloride · $6H_2O$	0.1
Potassium chloride	0.34
Sodium chloride	7
Sodium phosphate dibasic (anhydrous)	0.1
Sodium phosphate monobasic (anhydrous)	0.18
Other	
D-Glucose	1.8
Add	
Sodium bicarbonate	1.26

Phosphate buffered saline (PBS; pH 7.4)	
Compound	Concentration (g L^{-1})
Inorganic salts	
NaCl	8
KCl	0.2
Na_2HPO_4	1.44
KH_2PO_4	0.24

physiological conditions only to a certain extent. Components that are part of physiological extracellular antioxidant mechanisms may not all be necessarily present. Furthermore, their composition and concentration might even be dynamic.

2.4.3 Use of Scavengers, Pharmaceuticals or Enzymes to Alter Peroxynitrite Signaling Pathways

"Antioxidant" enzymes such as CATs, peroxidases and SODs are part of natural enzymatic detoxification mechanisms in biological systems together with the actions of small antioxidant molecules such as glutathione, vitamin C,

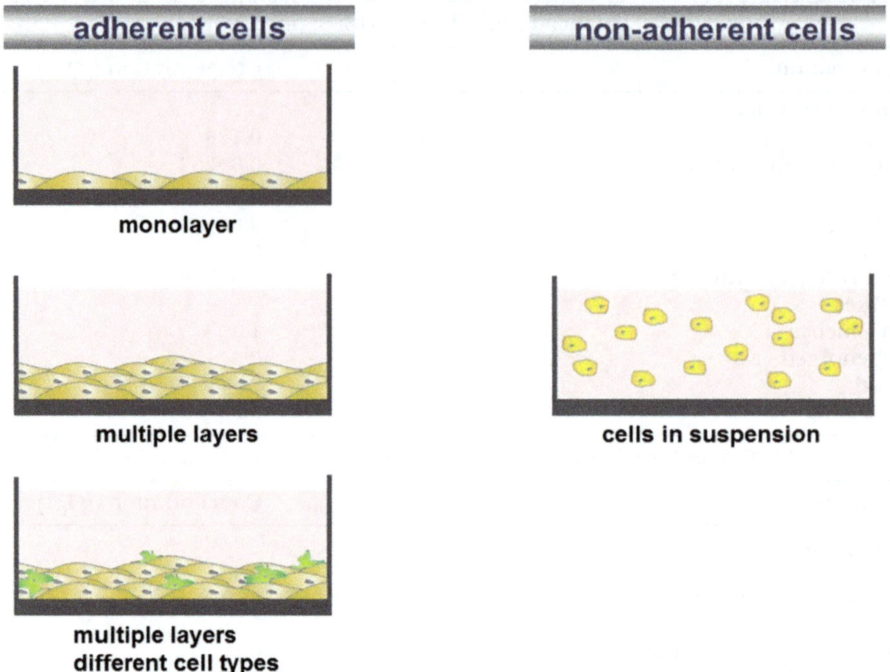

Figure 2.11 Schematic representations of adherent and non-adherent growing cells in cell culture.

vitamin E, β-carotenes and ubiquinone (Table 2.3).[4] Some of these molecules or enzymes are also valuable candidates to identify the effects of the different RNS/ROS, however, note that not all of them are selective towards a certain molecule, and some of them also interact with each other.

Antioxidants and pharmaceuticals can serve as biological controls for peroxynitrite formation in validation experiments or as drugs to modulate the actions of peroxynitrite.[13,70] One approach is to interfere with the peroxynitrite signaling pathways by reducing the amount of precursors for $ONOO^-$, namely nitric oxide and superoxide. As NOS inhibitors for all three isoforms of NOS, N-substituted analogs of L-arginine, which inhibit the biosynthesis of $NO^•$ from L-arginine by NOS can be used.[38] More specific NOS inhibitors are also available.[38,71] SOD can be added to remove $O_2^{•-}$ by catalyzing its dismutation to H_2O_2. Allopurinol and oxypurinol inhibit the generation of superoxide and H_2O_2 from xanthine oxidase. CATs and peroxidases detoxify H_2O_2 to water.

The other approach is to use scavengers and decomposition catalysts for $ONOO^-$. Due to the high reactivity of $ONOO^-$ it is not trivial to obtain a selective, robust and fast reacting scavenger for this molecule that is also not toxic for the system under study and may be tuned to cross cell

Table 2.3 Molecules and enzymes that play a role in biological antioxidant defense systems.

Antioxidants or scavengers for RNS and ROS	
Molecule/enzyme	Abbreviation
Dopamine	
Glutathione	GSH/GSSG
Nicotinamide adenine dinucleotide (phosphate)	NAD(P)H
Coenzyme Q_{10} (ubiquinone/ubiquinol)	CoQ_{10}
Uric acid	UA
Vitamin A (retinol) and carotenes	
Vitamin C (ascorbic acid)	AA
Vitamin E (α-tocopherol)	

Enzymes that play a role in reducing oxidative stress and detoxifying mechanisms			
Molecule/enzyme	Abbreviation	EC number	Reaction
Superoxide dismutase	SOD	1.15.1.1	$2H^+ + 2O_2^{\cdot -} \rightarrow H_2O_2 + O_2$
Catalase	CAT	1.11.1.6	$2H_2O_2 \rightarrow O_2 + 2H_2O$
Glutathione peroxidase	GPx	1.11.1.9	$ROOH + 2GSH \rightarrow ROH + GS\text{-}SG$; $ONOO^- + 2GSH \rightarrow NO_2^- + GS\text{-}SG$
Peroxiredoxin	Prx	1.11.1.15	$Prx(red) + H_2O_2 \rightarrow Prx(ox) + 2H_2O$;
Thioredoxin	Trx		$Prx(ox) + Trx(red) \rightarrow Prx(red) + Trx(ox)$ $ONOOH + PrxS^- \rightarrow NO_2^- + PrxSOH$

Enzymes that generate RNS or ROS			
Molecule/enzyme	Abbreviation	EC number	Reaction
Xanthine oxidase	XO	1.17.3.2	Hypoxanthine $+ H_2O + O_2 \rightarrow$ xanthine $+ H_2O_2$; Xanthine $+ H_2O + O_2 \rightarrow$ uric acid $+ H_2O_2$
Nitric oxide synthase	NOS	1.14.13.39	$3NADPH + 2arginine^+ + 3O_2 + H^+ \rightarrow 3NADP^+ + 3citrulline + 2NO^{\cdot} + 2OH^-$
NADPH oxidase	NOX	1.6.3.1	$NADPH + 2O_2 \rightarrow NADP^+ + H^+ + 2O_2^{\cdot -}$
Lipoxygenase	LOX	1.13.11.12	Fatty acid $+ O_2 \rightarrow$ fatty acid hydroperoxide

membranes. A detailed list of potential candidates as $ONOO^-$ scavengers is beyond the scope of this chapter, but detailed summaries of the application of several drugs, antioxidants and scavengers that have been hypothesized to alter the $ONOO^-$ signaling pathways can be found in two extensive reviews.[13,70]

2.5   General Challenges of Detecting RNS and ROS: The Complexity of the Task to Quantify Peroxynitrite in Biological Samples

Today it is still a general challenge to monitor highly reactive species such as RNS and ROS in biological systems.[72–87] As outlined in detail above, the complex chemistry, short half-lives and high reactivity of the molecules, as well as the complexity of the biological samples themselves are limiting their reliable quantification in real time.

An "ideal" method to quantify ROS and RNS in biological systems would exhibit the following characteristics:

- Direct, label-free detection of the analyte of interest
- Discrimination between different ROS/RNS metabolites
- Measurements *in vitro* and *in vivo* possible
- High selectivity
- High sensitivity
- Low detection limit (down to picomolar or femtomolar range)
- Linear measuring range from picomolars to micromolars
- Fast response time
- Real-time measurements
- Self-calibrating
- Miniaturization possible
- Suitable long-term stability
- Sufficient biological compatibility and environmentally friendly
- Easy to operate
- Low costs
- Compatible for high-throughput experiments and/or clinical studies

2.5.1   Overview of the Different Quantification Approaches

For the analyte of interest, peroxynitrite, there are several detection principles available:[88] both direct and indirect approaches. Each one of the analytical methods has its specific merits and drawbacks, and technological limitations. Some of the principal advantages and disadvantages for RNS and ROS detection, in general, are outlined in Figure 2.12. Please note that this list is not comprehensive.

Direct quantitative electron spin resonance (ESR)—also named electron paramagnetic resonance (EPR)—and spin-trapping studies are powerful tools to monitor one-electron oxidation and free-radical intermediates caused by $ONOO^-$ in complex biological samples.[58,85,89,90]

Optical techniques for detecting peroxynitrite are highlighted in Chapter 5 (fluorescent probes) and Chapter 6 (near-infrared and luminescent probes), and discussed in the literature.[91–96] They are powerful techniques,

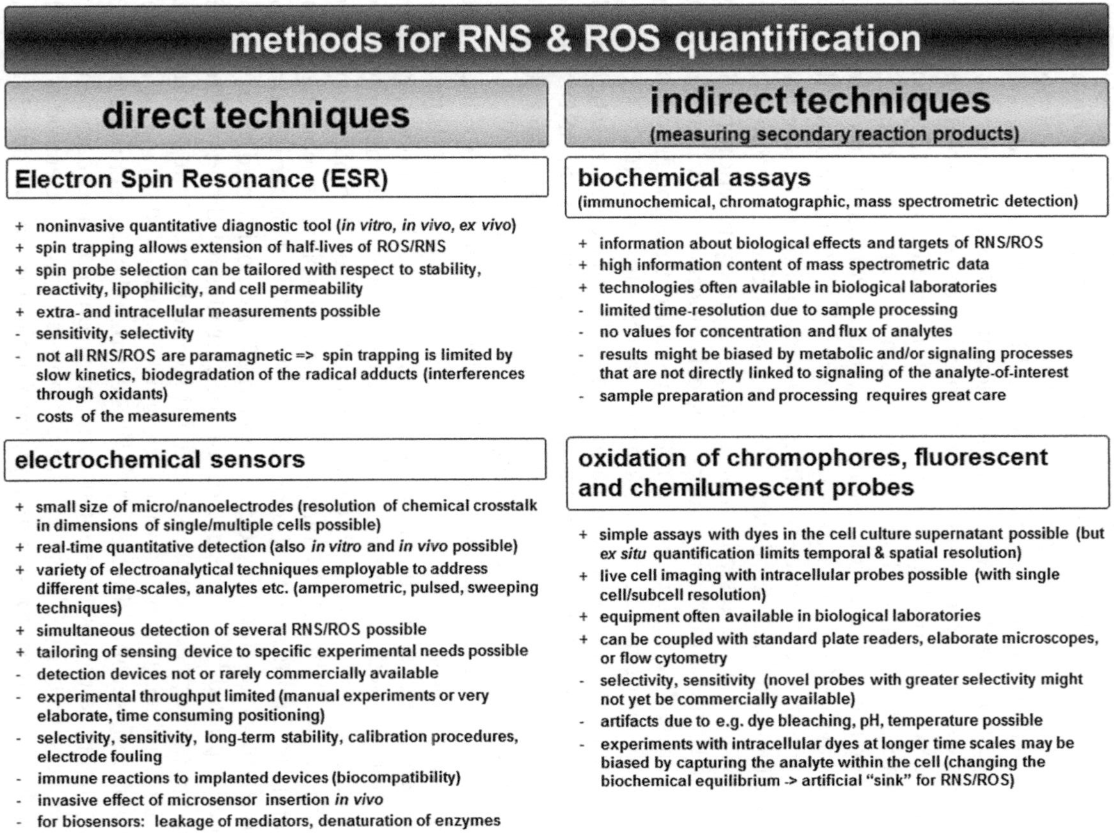

Figure 2.12   Comparison of methods for the direct and indirect quantification of peroxynitrite.

but despite great progress, still suffer from selectivity issues, which limit the unambiguous attribution of obtained results to ONOO⁻. Furthermore, if an optical probe is inserted into biological cells it has to be evaluated if physiological behavior is altered or toxic effects occur. This holds true especially for experiments that monitor RNS and ROS over longer time periods or that require relatively high intracellular amounts of dyes. It needs to be checked whether the probes may change the chemical equilibrium of the biological system. The dyes could become an artificial "sink" for RNS/ROS. Thus, the dynamics of redox signaling could be artificially altered. Light-induced stress or interferences by local redox and pH status may also occur.

Biochemical assays monitor the biochemical reactions upstream or downstream of ONOO⁻ generation. Generally, the time resolution of these methods is limited and does not reach real time. Tyrosine nitration was established as a biomarker for *in vivo* peroxynitrite formation, and can be monitored by immunoassays, HPLC or mass spectrometry.[57,97–99] Furthermore, the proteins involved in the diverse signaling cascades can be monitored by state-of-the-art biochemical techniques. The change in concentration of precursors of ONOO⁻ can be tested. In addition, secondary reaction products of ONOO⁻ can be measured in the bulk of the cell supernatant by indirect (often optical) assays. The results of the different indirect approaches will give a broad spectrum of biochemical information relevant to ONOO⁻ signaling, but suffer from the fact that bias cannot be excluded due to metabolic and/or signal transduction processes that are not directly related to ONOO⁻ or are induced by artifacts.[100] In addition, there is a great variety of assays available to evaluate general antioxidant activity in biological systems.[101,102]

Peroxynitrite anion (ONOO⁻) can be electrochemically oxidized and detected upon release from biological cells.[103–108] The electrochemical detection of peroxynitrite features in Chapters 4, 7 and 8 of this book, whereas electrophoretic methods are highlighted in Chapter 9. Note that only a limited number of studies cover the electrochemical detection of peroxynitrite in the literature.[103,104,109–115] There is a trend towards simultaneous detection of several RNS and ROS in the literature.[5,72,106,107,116–120] The use of ultramicroelectrode arrays for simultaneous detection of nitric oxide and peroxynitrite has been reported.[121,122] The use of microelectrodes for direct electrochemical oxidation of peroxynitrite is straightforward because the formal redox potential E° is +0.20 V *vs.* normal hydrogen electrode (NHE) and it decomposes with sub-second kinetics.[104,123,124] Electroanalytical techniques offer a wide dynamic range, fast response time, and combine adequate sensitivity with decent selectivity. The small size of nano- and micro-electrodes as well as chip devices is ideal for *in vitro* and, with limitations, applicable for *in vivo* experimental settings. The technological advantages and challenges of using microelectrodes and microbiosensors have been reviewed with a focus on neurophysiology.[49,110,125–127] The formal potentials (E° *vs.* NHE) of a selection of RNS and ROS are displayed in Figure 2.13.

It is important for biologists and immunologists to understand the working principle of the selected analytical technique(s) for ONOO⁻ detection in

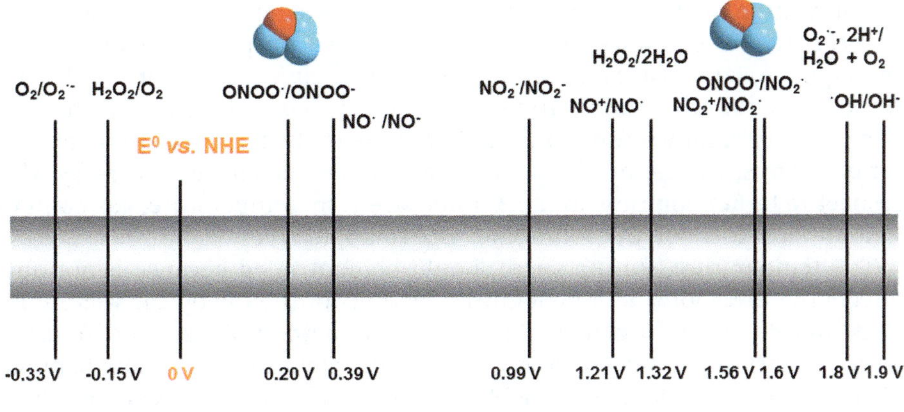

Figure 2.13 Summary of formal potentials (*E° vs.* NHE) of selected ROS and RNS. Values derived from Peteu *et al.*, Buettner and Indira Priyadarsini.[77,149,150]

detail. It might be advantageous to work together with analytical/physical chemists in order to employ the most optimal experimental assays for the biological/immunological question in mind. This ensures a maximum outcome of information and minimizes compromised or biased results. Note that many established detection approaches have already been used *in vitro* and *in vivo* whereas many of the novel approaches have only been tested so far in proof-of-concept studies. This may currently limit their applicability to clinical and pharmaceutical studies. But with more thorough future testing in *in vitro* and *in vivo* settings, the amount of information generated by these approaches will certainly increase. When it comes to *in vitro* and *in vivo* experiments, it is indispensable for the chemists to collaborate with biologists, *e.g.* an immunologist or neurophysiologist, because special knowledge is required to handle cell cultures or animals properly, and to maintain a reproducible sample quality and processing. In addition, they know the biochemistry and signal transduction of their systems best, which is of great importance to understand the actions of peroxynitrite in more depth.

2.5.2 Parameters Impacting the Performance of ONOO⁻ Quantification

Designing a suitable *in vitro* or *in vivo* experimental setting requires that the many variables that may have an impact on the performance of ONOO⁻ production and quantification are carefully taken into account. In the following section, some parameters of special importance are highlighted.

Chapter 3 of this book is dedicated to the advantages and drawbacks of the synthesis of peroxynitrite. A reliable source of ONOO⁻ is indispensable to serve as a calibration standard to validate the performance of the quantification methods, but also as a positive control during biological experiments. The employment of proper calibration procedures is a basic prerequisite. For *in vitro* and *in vivo* experiments, it is important to calibrate not only in simplified measuring buffers but it has to be ensured that the analytical technique is functional and accurate under real measurement conditions. Thus, the calibration conditions should match the real assay with respect to buffer composition, pH, ionic strength, temperature, gas content and physical force on the solution/device (if applicable).

The temperature, for instance, should be controlled because it not only affects the metabolism and chemical reactions of the biological system but also impacts the performance of the detection method. Thus, temperature is a very important parameter for it not only influences the cell metabolism but also the chemistry of RNS and ROS (*e.g.* solubility, diffusion, reaction kinetics). The solubility of gases such as O_2 and $NO^{\bullet}$ is temperature dependent. The performance of electroanalytical methods is temperature dependent as well. Detection approaches and calibration methods should be tested under physiological, constant temperature conditions (37 °C) to mimic actual *in vitro* or *in vivo* conditions.

Note that the pH, ionic strength and nature of the solvent impact the electrochemical properties of electroactive species in solution. For instance, dopamine with a two-electron, two-proton redox couple is an example of a prominent signaling molecule in neurophysiology that is electrochemically active. It shows a linear dependency between its formal potential and pH in accordance with the Nernst equation with a slope of about 59 mV per pH unit.[128] Thus, with decreasing pH the oxidation and reduction peaks of the cyclic voltammograms shift to higher potentials. The equilibrium of $ONOO^{-}$ with its corresponding ONOOH has a pK_a of 6.8.[129] This explains why under physiological conditions (pH 7.4) the anionic form of peroxynitrite is present at 80%.

In biological systems there may be a variety of molecules present that can directly react with peroxynitrite and, thus, (bio)chemically interfere with the chosen detection approach. One also has to consider that peroxynitrite can directly react with several of the molecules present in biological cells and/or cell culture media or buffers. The composition of some typical cell culture media is given in Table 2.2. It has been reported that $ONOO^{-}$ can react with glucose, fructose, urate, mannitol, thiols, glutathione, proteins and lipids, among others.[130–137] The resulting reaction products, *e.g.* *S*-nitrosothiols (RSNO), nitrites (RONO) and nitrates ($RONO_2$), may become NO donors in follow-up reactions in biological systems.

The sensitivity of detecting RNS and ROS is impacted and sometimes biased by the composition of the cell culture media or minimal buffers (refer also to Table 2.2). Note that added human serum albumin (HSA) or fetal calf serum (FCS) is able to react with peroxynitrite (*e.g.* protein nitration). These proteins can also react with other RNS and ROS. These proteins and other media components may also directly interact or react with RNS and ROS, and may generate further secondary radical molecules. These effects will alter the nature and concentration of the present species and impact their diffusion profiles. In addition, these proteins can also contribute to electrode fouling processes on the sensing device.

HEPES, as a buffer component in tissue culture medium, triggers, together with light-induced riboflavin radicals and traces of metal ions, additional superoxide generation from molecular oxygen.[138] The increased superoxide level generates subsequently higher levels of peroxynitrite upon reaction with nitric oxide (Figure 2.14). The increased levels of peroxynitrite and hydrogen peroxide may be toxic for biological cells. It has also been reported that $ONOO^-$ can directly react with HEPES or MOPS to generate a relatively stable $NO^\bullet$ donor.[139] In addition, it was shown that cytotoxicity generated by 3-morpholinosydnonimine-*N*-ethylcarbamide (SIN-1), a compound that releases $NO^\bullet$ and $O_2^{\bullet-}$, was higher in the presence of HEPES buffer compared with in the absence of HEPES.[140] Thus, it is important to keep in mind the reported effects that are dependent on the chosen buffer system and the degree of exposure to light, and if necessary, to perform suitable control experiments.

In addition, it has been reported that H_2O_2 can be produced from several cell culture media by reducing agents such as phenolic compounds (*e.g.* dopamine) and ascorbic acid.[141–143] This can lead to artifacts if not considered

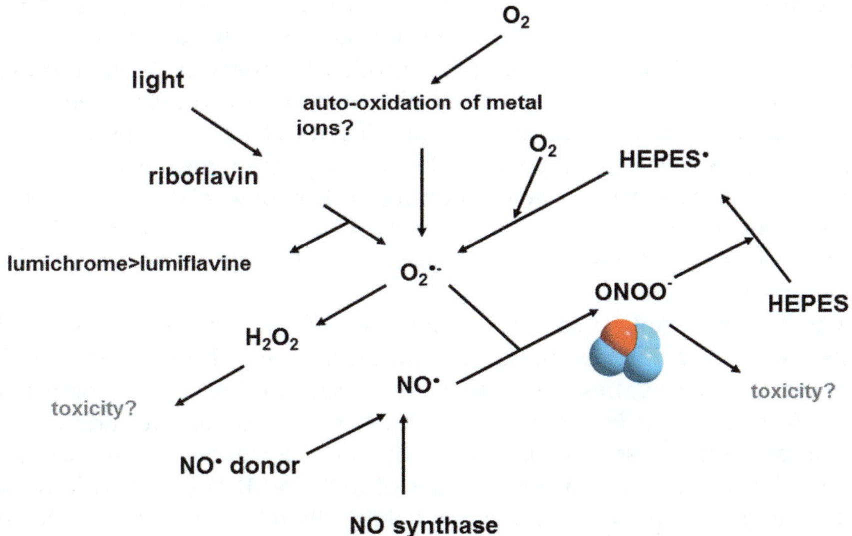

Figure 2.14 Model of possible interactions between the HEPES buffer, riboflavin and light with ROS and RNS. $O_2^{\bullet-}$ can be generated by riboflavin and/or traces of metal contaminants. Subsequently, $ONOO^-$ is generated by the reaction of $O_2^{\bullet-}$ with $NO^\bullet$. $ONOO^-$ is able to oxidize HEPES to a HEPES radical. If the HEPES radicals react with O_2, the $O_2^{\bullet-}$ levels are further increased. This results in an increase in the resulting molecules of H_2O_2 and $ONOO^-$, which may be the dominant species responsible for cytotoxicity and other effects. Adapted from Keynes and Griffiths.[138]

properly. As one control experiment, it is possible to use CAT to diminish the effect from H_2O_2.

Furthermore, some of the molecules present in biological samples, such as ascorbic acid, uric acid, dopamine, norepinephrine, epinephrine, serotonin, 3,4-dihydroxyphenyl-acetic acid, 5-hydroxyindole-3-acetic acid, acetaminophen (or paracetamol), nitrite, nitric oxide and H_2O_2, may even be electroactive and can potentially interfere with the electrochemical detection of peroxynitrite due to their redox potential. If these molecules contribute to the overall current detected, the selectivity is very much dependent on the potential window and electroanalytical technique applied. In Figure 2.15 formal potentials ($E°$ *vs.* NHE) of some biologically and chemically relevant substances that may potentially interfere with ONOO⁻ detection are summarized. Several of the naturally occurring antioxidants can be electrochemically detected.[144,145] Note that several molecules also impact each other chemically or modulate interconnected signaling pathways relevant to their own formation or to the formation of RNS/ROS.

One approach to reduce the impact of electrochemically relevant interference is the use of permselective membranes as additional layers on chemical and biosensors. This enhances the selectivity by reducing the diffusion of interfering substances to the transducer surface. This can be achieved by charge repulsion (surface charge of the membrane), size exclusion (pore size of the membrane), preferring certain non-covalent interactions (hydrophilicity or hydrophobicity of the membrane) and significantly extending the travel time of the interferent towards the electrode (thickness of the membrane). Another approach is the use of chemical sensor architectures to electrocatalytically modify the electron transfer kinetics of ONOO⁻. The applied potential and type of electroanalytical technique (*e.g.* amperometry, differential pulse techniques, fast scan cyclic voltammetry) are powerful tools to tune the sensors' sensitivity and selectivity, as well as the response and overall assay time. Critical guidelines for validation of the selectivity of *in vivo* chemical microsensors have been introduced.[146]

It is, however, not so easy to define a set of parameters for testing the impact of certain interferences. First of all, it is not always clear at which concentration levels these molecules are present in different biological systems. What are the proper concentration values for a set of interferences that should be used to assess the selectivity of an ONOO⁻ assay? Figure 2.16 tries to compile typical concentration values of ROS and RNS reported in the literature in comparison to some other potentially relevant molecules for the biology of redox signaling. To date, the large dynamic range (from femtomolars to micromolars) of reported RNS and ROS concentration levels is one of the greatest challenges for designing suitable methods for quantification besides attaining suitable selectivity.

For the signaling molecule NO˙, it has been questioned whether the reported physiological concentration values are indeed realistic or overestimated.[147] The current methodologies were critically reviewed, and the difficulties were elaborated to define where physiological concentration levels end and pathology levels start. If it is actually the case that physiological

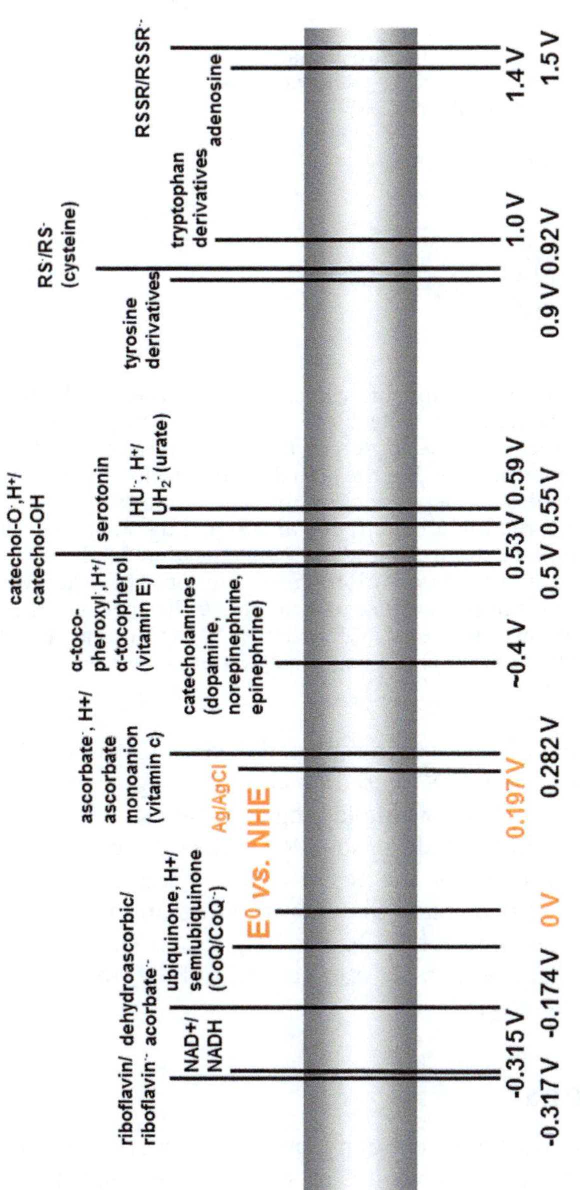

Figure 2.15 Summary of formal potentials (*E*° *vs.* NHE) of selected biological molecules that may interfere with electrochemical quantification of peroxynitrite. Values derived from Robinson *et al.*[126] and Buettner.[149]

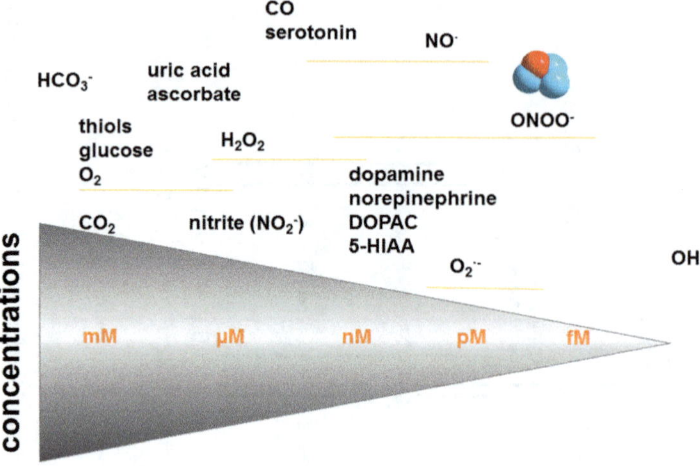

Figure 2.16   Simplified compilation of typical concentration levels of ROS and RNS and some other molecules in biological samples such as blood, plasma, serum, urine, different tissue samples, extracellular fluids and cytoplasm. Note that the actual concentration levels of the different molecules are dependent on the type of sample material, the cell types, and the metabolic, redox and signaling state, *etc.* Thus, only dimensions of the magnitude of concentration levels are given as a general reference. In addition, great care must be taken when comparing concentration values of different studies because the different experimental conditions and analytical techniques/assays employed may lead to (slightly) varying results. Values derived from multiple studies.[2,10,12,51,52,65,87,106,126,130,141,147,151–155]

NO· levels are significantly lower than reported in several studies, this would have direct implications on ONOO⁻ research and the design of suitable detection approaches. To avoid further long-lasting controversial discussions, it is now time for scientists in the field of RNS/ROS research to develop a set of standardized experiments to test the performance of the versatile analytical approaches and biochemical assays that quantify RNS and ROS in biological systems. Such reference scenarios could also help to address the issue of lab-to-lab variations of results, which are often a result of the overall complexity of detecting redox signaling.

2.6   Conclusions

To further support substantially the conceptual framework of redox cellular signaling, more quantitative knowledge about RNS/ROS, especially ONOO⁻, is essential. Thus, the lack of evidence with respect to controversial standpoints and hypotheses is mainly due to the lack of reliable, selective and sensible methods to directly detect ONOO⁻ and other RNS/ROS. The urgent need to actually quantify ONOO⁻ and related species in biological system still remains. However, with the progress being made today and hopefully in the future, especially with respect to simultaneously detecting multiple analytes in real time, this field of research is looking forward to exciting discoveries and a better understanding of redox signaling.

The remaining questions are still the following: how do the $ONOO^-$ fluxes from different sources look *in vitro* and *in vivo* for different biological systems? Is it possible to distinguish between physiological and pathological levels? How does $ONOO^-$ exert signaling specificity with respect to the resulting biological responses? Is it possible to unambiguously differentiate between the influence of $ONOO^-$, its precursors nitric oxide and superoxide, or the decomposition products of $ONOO^-$?

Electroanalytical approaches are, despite their intrinsic limitations, very promising tools for *in vitro* and *in vivo* assessment of $ONOO^-$ fluxes. Due to the limited number of studies reported so far, this research field has great promise for future developments once more effort is put into performing substantial proof-of-principle studies in biological systems. Much can be learned from other works related to neurophysiology, implanted (bio)sensors, and electroanalytical quantification of other RNS and ROS. In particular, the capability to perform real-time measurements is very attractive and will lead to a higher degree of information about the production and decomposition rates of peroxynitrite.

However, answering the biological and immunological questions related to peroxynitrite-based redox signaling also requires the use and improvement of the entire arsenal of $ONOO^-$ quantification methods in order to benefit from the complementary sets of information that can be obtained. A significant step forward would be to define a set of standardized experimental settings as reference experiments that can serve as platforms of performance tests for all available quantification techniques relevant to biological systems. Control experiments for the biology of the system but also a standardized assessment of the sensitivity and selectivity towards biologically relevant interferences need to be defined. This would make meaningful comparisons of the performance of the different assay systems much easier and rational optimization strategies much more straight forward.

Acknowledgements

The author is grateful to L. Koch and B. Mallik for help with some of the graphics. The many valuable discussions with current and former colleagues are gratefully acknowledged.

References

1. W. Dröge, *Physiol. Rev.*, 2002, **82**, 47.
2. B. D'Autreaux and M. B. Toledano, *Nat. Rev. Mol. Cell Biol.*, 2007, **8**, 813.
3. C. Nathan, *J. Clin. Invest.*, 2003, **111**, 769.
4. H. J. Forman and M. Torres, *Mol. Aspects Med.*, 2001, **22**, 189.

5. C. Amatore, S. Arbault and A. C. W. Koh, *Anal. Chem.*, 2010, **82**, 1411.
6. W. H. Koppenol, J. J. Moreno, W. A. Pryor, H. Ischiropoulos and J. S. Beckman, *Chem. Res. Toxicol.*, 1992, **5**, 834.
7. J. S. Beckman, J. Chen, H. Ischiropoulos and J. P. Crow, *Methods Enzymol.*, 1994, **233**, 229.
8. W. A. Pryor and G. L. Squadrito, *Am. J. Physiol.*, 1995, **268**, L699.
9. G. Ferrer-Sueta and R. Radi, *ACS Chem. Biol.*, 2009, **4**, 161.
10. C. Ducroq, B. Blanchard, B. Pignatelli and H. Ohshima, *Cell. Mol. Life Sci.*, 1999, **55**, 1068.
11. C. Szabó, H. Ischiropoulos and R. Radi, *Nat. Rev. Drug Discovery*, 2007, **6**, 662.
12. M. P. Murphy, M. A. Packer, J. L. Scarlett and S. W. Martin, *Gen. Pharmacol.*, 1998, **31**, 179.
13. P. Pacher, J. S. Beckman and L. Liaudet, *Physiol. Rev.*, 2007, **87**, 315.
14. B. Kalyanaraman, *Redox Biol.*, 2013, **1**, 244.
15. R. P. Patel, J. McAndrew, H. Sellak, C. R. White, H. Jo, B. A. Freeman and V. M. Darley-Usmar, *Biochim. Biophys. Acta*, 1999, **1411**, 385.
16. J. S. Beckman, T. W. Beckman, J. Chen, P. A. Marshall and B. A. Freeman, *Proc. Natl. Acad. Sci. U. S. A.*, 1990, **87**, 1620.
17. R. Kissner, J. S. Beckman and W. H. Koppenol, *Methods Enzymol.*, 1999, **301**, 342.
18. W. H. Koppenol, *Redox Rep.*, 2001, **6**, 339.
19. J. S. Beckman, *Arch. Biochem. Biophys.*, 2009, **484**, 114.
20. J. S. Beckman, *Nature*, 1990, **345**, 27.
21. J. S. Beckman and W. H. Koppenol, *Am. J. Physiol. Cell Physiol.*, 1996, **271**, C1424.
22. S. S. Marla, J. Lee and J. T. Groves, *Proc. Natl. Acad. Sci. U. S. A.*, 1997, **94**, 14243.
23. A. Denicola, J. M. Souza and R. Radi, *Proc. Natl. Acad. Sci. U. S. A.*, 1998, **95**, 3566.
24. R. Radi, *Chem. Res. Toxicol.*, 1998, **11**, 720.
25. N. Romero, A. Denicola, J. M. Souza and R. Radi, *Arch. Biochem. Biophys.*, 1999, **368**, 23.
26. C. C. Winterbourn, *Nat. Chem. Biol.*, 2008, **4**, 278.
27. N. Nalwaya and W. M. Deen, *Chem. Res. Toxicol.*, 2003, **16**, 920.
28. J. R. Lancaster, *Proc. Natl. Acad. Sci. U. S. A.*, 1994, **91**, 8137.
29. J. S. Beckmann, *Ann. N. Y. Acad. Sci.*, 1994, **738**, 69.
30. C. C. Winterbourn, *Methods Enzymol.*, 2013, **528**, 3.
31. N. Möller, J. R. Lancaster and A. Denicola, The Interaction of Reactive Oxygen and Nitrogen Species with Membranes, in *Free Radical Effects on Membranes, Current Topics of Membranes*, ed. S. Matalon, Academic Press, London, 2008, vol. 61, pp. 23–43.
32. S. G. Rhee, *Free Radical Biol. Med.*, 2003, **35**, S7.
33. O. Augusto and S. Miayamoto, Oxygen Radicals and Related Species, in *Principles of Free Radical Biomedicine*, ed. K. Pantopoulos and H. Schippe, Nova Science Pub. Inc., Hauppauge, 2011, ch. 2, vol. 1, pp. 19–42.

34. D. I. Brown and K. K. Griendling, *Free Radical Biol. Med.*, 2009, **47**, 1239.
35. K. M. Holmström and T. Finkel, *Nat. Rev. Mol. Cell Biol.*, 2014, **15**, 411.
36. J.-M. Li and A. M. Shah, *Am. J. Physiol.*, 2004, **287**, R1014.
37. G. Y. Wu and S. M. Morris, *Biochem. J.*, 1998, **336**, 1.
38. W. K. Alderton, C. E. Cooper and R. G. Knowles, *Biochem. J.*, 2001, **357**, 593.
39. C. J. Lowenstein, J. L. Dinerman and S. H. Snyder, *Ann. Intern. Med.*, 1994, **120**, 227.
40. E. Anggard, *Lancet*, 1994, **343**, 1199.
41. L. J. Ignarro, *Angew. Chem., Int. Ed.*, 1999, **38**, 1882.
42. C. Bogdan, *Nat. Immunol.*, 2001, **2**, 907.
43. M. B. Grisham, D. Jourd'Heuil and D. A. Wink, *Am. J. Physiol.*, 1999, **276**, G315.
44. L. A. Ridnour, D. D. Thomas, D. Mancardi, M. G. Espey, K. M. Miranda, N. Paolocci, M. Feelisch, J. Fukuto and D. A. Wink, *Biol. Chem.*, 2004, **385**, 1.
45. D. Mancardi, L. A. Ridnour, D. D. Thomas, T. Katori, C. G. Tocchetti, M. G. Espey, K. M. Miranda, N. Paolocci and D. A. Wink, *Curr. Mol. Med.*, 2004, **4**, 723.
46. S. Pfeiffer, B. Mayer and B. Hemmens, *Angew. Chem., Int. Ed.*, 1999, **38**, 1715.
47. G. Ferrer-Sueta and R. Radi, *ACS Chem. Biol.*, 2009, **4**, 161.
48. S. Carballal, S. Bartesaghi and R. Radi, *Biochim. Biophys. Acta*, 2014, **1840**, 768.
49. B. Bucur, *Curr. Neuropharmacol.*, 2012, **10**, 197.
50. M. A. Packer, J. L. Scarlett, S. W. Martin and M. P. Murphy, *Biochem. Soc. Trans.*, 1997, **25**, 831.
51. L. Valdez, S. Alvarez, S. Lores Arnáiz, F. Schöfer and A. Boveris, *Free Radical Biol. Med.*, 2000, **29**, 349.
52. A. Boveris and E. Cadenas, *IUBMB Life*, 2000, **50**, 245.
53. B. Alberts, D. Bray, J. Lewis, M. Raff, K. Roberts and J. D. Watson, *Molecular Biology of the Cell*, Garland Publishing Inc., New York, 1994, vol. 3, pp. 665–666.
54. L. Zuh, C. Gunn and J. S. Beckmann, *Arch. Biochem. Biophys.*, 1992, **298**, 452.
55. L. Brunelli, J. P. Crow and J. S. Beckmann, *Arch. Biochem. Biophys.*, 1995, **316**, 327.
56. P. C. Dedon and S. R. Tannenbaum, *Arch. Biochem. Biophys.*, 2004, **423**, 12.
57. H. Ischiropoulos, L. Zhu and J. S. Beckman, *Arch. Biochem. Biophys.*, 1992, **298**, 446.
58. S. Dikalov, M. Skatchkov and E. Bassenge, *Biochem. Biophys. Res. Commun.*, 1997, **230**, 54.
59. C. J. Lowenstein and E. Padalko, *J. Cell Sci.*, 2004, **117**, 2865.
60. E. Karpuzoglu and S. A. Ahmed, *Nitric Oxide*, 2006, **15**, 177.
61. F. C. Fang, *J. Clin. Invest.*, 1997, **99**, 2818.

62. L. Virag, E. Szabo, P. Gergely and C. Szabo, *Toxicol. Lett.*, 2003, **140–141**, 113.
63. M. P. Murphy, M. A. Packer, J. L. Scarlett and S. W. Martin, *Gen. Pharmacol.*, 1998, **31**, 179.
64. C. Ducrocq, B. Blanchard, B. Pignatelli and H. Ohshima, *Cell. Mol. Life Sci.*, 1999, **55**, 1068.
65. A. Denicola, B. A. Freeman, M. Trujillo and R. Radi, *Arch. Biochem. Biophys.*, 1996, **333**, 49.
66. R. Meli, T. Nauser and W. H. Koppenol, *Helv. Chim. Acta*, 1999, **82**, 722.
67. M. R. C. Marques, R. Loebenberg and M. Almukainzi, *Dissolution Technol.*, 2011, **18**, 15.
68. J. R. Masters and G. N. Stacey, *Nat. Protoc.*, 2007, **2**, 2276.
69. B. Halliwell, *FEBS Lett.*, 2003, **540**, 3.
70. A. Olmos, R. M. Giner and S. Máñez, *Med. Res. Rev.*, 2007, **27**, 1.
71. P. Mukherjee, M. A. Cinelli, S. Kang and R. B. Silverman, *Chem. Soc. Rev.*, 2014, **43**, 6814.
72. S. Borgmann, *Anal. Bioanal. Chem.*, 2009, **394**, 95.
73. T. Rassaf, M. Feelisch and M. Kelm, *Free Radical Biol. Med.*, 2004, **36**, 413.
74. I. P. Kaur and T. Geetha, *Mini-Rev. Med. Chem.*, 2006, **6**, 305.
75. Z. H. Taha, *Talanta*, 2003, **61**, 3.
76. S. Archer, *FASEB J.*, 1993, **7**, 349.
77. S. F. Peteu, R. Boukherroub and S. Szunertis, *Biosens. Bioelectron.*, 2014, **58**, 359.
78. F. Bedioui and N. Villeneuve, *Electroanalysis*, 2003, **15**, 5.
79. M. M. Tarpey, D. A. Wink and M. B. Grisham, *Am. J. Physiol.*, 2004, **286**, R431.
80. M. M. Tarpey and I. Fridovich, *Circ. Res.*, 2001, **89**, 224.
81. M. Yuasa and K. Oyaizu, *Curr. Org. Chem.*, 2005, **9**, 1685.
82. L. M. Fan and J.-M. Li, *J. Pharmacol. Toxicol. Methods*, 2014, **70**, 40.
83. R. Rodriguez-Rodriguez and U. Simonsen, *Curr. Anal. Chem.*, 2012, **8**, 485.
84. C. Calas-Blanchard, G. Catanante and T. Noguer, *Electroanalysis*, 2014, **26**, 1277.
85. B. Halliwell and M. Whiteman, *Br. J. Pharmacol.*, 2004, **142**, 231.
86. S. G. Rhee, T. S. Chang, W. Jeong and D. Kang, *Mol. Cells*, 2010, **29**, 539.
87. B. J. Privett, J. H. Shin and M. H. Schoenfisch, *Chem. Soc. Rev.*, 2010, **39**, 1925.
88. R. Radi, G. Peluffo, M. N. Alvarez, M. Naviliat and A. Cayota, *Free Radical Biol. Med.*, 2001, **30**, 463.
89. D. Pietraforte and M. Minetti, *Biochem. J.*, 1997, **325**, 675.
90. J. Vasquez-Vivar, M. A. Santos, V. B. C. Junqueira and O. Augusto, *Biochem. J.*, 1996, **314**, 869.
91. X. Chen, H. Chen, R. Deng and J. Shen, *Biomed. J.*, 2014, **37**, 120.
92. J. P. Crow, *Nitric Oxide*, 1997, **1**, 145.
93. P. Wardman P, *Free Radical Biol. Med.*, 2007, **43**, 995.

94. J. Zielonka, A. Sikora, M. Hardy, J. Joseph, B. P. Dranka and B. Kalyanaraman, *Chem. Res. Toxicol.*, 2012, **25**, 1793.

95. C. Lu, G. Song and J.-M. Lin, *Trends Anal. Chem.*, 2006, **25**, 985.

96. G. Bartosz, *Clin. Chim. Acta*, 2006, **368**, 53.

97. H. Ischiropoulos, L. Zhu, J. Chen, M. Tsai, J. C. Martin, C. D. Smith and J. S. Beckman, *Arch. Biochem. Biophys.*, 1992, **198**, 431.

98. J. S. Beckman, H. Ischiropoulos, L. Zhu, M. van der Woerd, C. Smith, J. Chen, J. Harrison, J. C. Martin and M. Tsai, *Arch. Biochem. Biophys.*, 1992, **298**, 438.

99. C. D. Smith, M. Carson, M. van der Woerd, J. Chen, H. Ischiropoulos and J. S. Beckmann, *Arch. Biochem. Biophys.*, 1992, **299**, 350.

100. B. Halliwell, K. Zhao and M. Whiteman, *Free Radical Res.*, 1999, **31**, 651.

101. M. N. Alam, N. J. Bristi and M. Rafiquzzaman, *Saudi Pharm. J.*, 2013, **21**, 143.

102. C. López-Alarcón and A. Denicola, *Anal. Chim. Acta*, 2013, **763**, 1.

103. C. Amatore, S. Arbault, C. Bouton, K. Coffi, J. C. Drapier, H. Ghandour and Y. H. Tong, *ChemBioChem*, 2006, **7**, 653.

104. C. Amatore, S. Arbault, D. Bruce, P. de Oliveira, M. Erard and M. Vuillaume, *Chem.–Eur. J.*, 2001, **7**, 4171.

105. R. Kubant, C. Malinski, A. Burewicz and T. Malinski, *Electroanalysis*, 2006, **18**, 410.

106. F. Bedioui, D. Quinton, S. Griveau and T. Nyokong, *Phys. Chem. Chem. Phys.*, 2010, **12**, 9976.

107. S. Griveau and F. Bedioui, *Anal. Bioanal. Chem.*, 2013, **405**, 3475.

108. S. Arbault, H. Ghandour, Y. H. Tong, K. Coffi, C. Bouton, J. C. Drapier and C. Amatore, *Nitric Oxide*, 2006, **14**, A21.

109. D. Quinton, S. Griveau and F. Bedioui, *Electrochem. Commun.*, 2010, **12**, 1446.

110. C. Amatore, S. Arbault, M. Guille and F. Lemaitre, *Chem. Rev.*, 2008, **108**, 2585.

111. C. Amatore, S. Arbault, D. Bruce, P. De Oliveira, M. Erard and M. Vuillaume, *Faraday Discuss.*, 2000, **116**, 319.

112. J. Xue, X. Ying, J. Chen, Y. Xian, L. Jin and J. Jin, *Anal. Chem.*, 2000, **72**, 5313.

113. R. Kubant, C. Malinski, A. Burewicz and T. Malinski, *Electroanalysis*, 2006, **18**, 410.

114. D. Quinton, A. Girard, L. T. T. Kim, V. Raimbault, L. Griscom, F. Razan, S. Griveau and F. Bedioui, *Lab Chip*, 2011, **11**, 1342.

115. R. Oprea, S. F. Peteu, P. Subramanian, W. Qi, E. Pichonat, H. Happy, M. Bayachou, R. Boukherrouba and S. Szunerits, *Analyst*, 2013, **138**, 4345.

116. C. Lu, J.-M. Heldt, M. Guille-Collignon, F. Lemaître, G. Jaouen, A. Vessières and C. Amatore Christian, *ChemMedChem*, 2014, **9**, 1286.

117. Y. Li, C. Sella, F. Lemaître, M. Guille-Collignon, L. Thouin and C. Amatore, *Electroanalysis*, 2013, **25**, 895.

118. A. Nassi, L. T. O. Thi Kim, A. Girard, L. Griscom, F. Razan, S. Griveau, L. Thouin and F. Bedioui, *Microchim. Acta*, 2012, **179**, 337.

119. Y. Wang, J.-M. Noel, J. Velmurugan, W. Nogala, M. V. Mirkin, C. Lu, M. Guille-Collignon, F. Lemaitre and C. Amatore, *Proc. Natl. Acad. Sci. U. S. A.*, 2012, **109**, 11534.

120. R. Hu, M. Guille, S. Arbault, C. J. Lin and C. Amatore, *Phys. Chem. Chem. Phys.*, 2010, **12**, 10048.

121. L. T. O. Thi Kim, V. Escriou, S. Griveau, A. Girard, L. Griscom, F. Razan and F. Bedioui, *Electrochim. Acta*, 2014, **140**, 33.

122. Y. Li, C. Sella, F. Lemaître, M. Guille-Collignon, L. Thouin and C. Amatore, *Electrochim. Acta*, 2014, **144**, 111.

123. W. H. Koppenol, *Free Radical Biol. Med.*, 1998, **25**, 385.

124. W. H. Koppenol, *Methods Enzymol.*, 1996, **268**, 7.

125. G. S. Wilson and M. A. Johnson, *Chem. Rev.*, 2008, **108**, 2462.

126. D. L. Robinson, A. Hermans, A. T. Seipel and R. M. Wightman, *Chem. Rev.*, 2008, **108**, 2554.

127. A. Schulte, M. Nebel and W. Schuhmann, *Annu. Rev. Anal. Chem.*, 2010, **3**, 299.

128. I. L. Medintz, M. H. Stewart, S. A. Trammell, K. Susumu, J. B. Delehanty, J. C. Mei, J. S. Melinger, J. B. Blanco-Canosa, P. E. Dawson and H. Mattoussi, *Nat. Mat.*, 2010, **9**, 676.

129. R. Radi, J. S. Beckman, K. M. Bush and B. A. Freeman, *J. Biol. Chem.*, 1991, **266**, 4244.

130. J. P. Eiserich, R. P. Patel and V. B. O'Donnell, *Mol. Aspects Med.*, 1998, **19**, 221.

131. M. A. Moro, V. M. Darley-Usmar, D. A. Goodwin, N. G. Read, R. Zamora-Pino, M. Feelisch, M. W. Radomski and S. Moncada, *Proc. Natl. Acad. Sci. U. S. A.*, 1994, **91**, 6702.

132. M. A. Moro, V. M. Darley-Usmar, I. Lizasoain, Y. Su, R. G. Knowles, M. W. Radomski and S. Moncada, *Br. J. Pharmacol.*, 1995, **116**, 1999.

133. L. M. Villa, E. Salas, V. M. Darley-Usmar, M. W. Radomski and S. Moncada, *Proc. Natl. Acad. Sci. U. S. A.*, 1994, **91**, 12383.

134. P. Liu, C. E. Hock, R. Nagele and P. Y. Wong, *Am. J. Physiol.*, 1997, **272**, H2327.

135. M. Wu, K. A. J. Pritchard, P. M. Kaminski, R. P. Fayngersh, T. H. Hintze and M. S. Wolin, *Am. J. Physiol.*, 1994, **266**, H2108.

136. K. A. Skinner, C. R. White, R. B. Patel, S. Tan, S. Barnes, M. Kirk, V. Darley-Usmar and D. A. Parks, *J. Biol. Chem.*, 1998, **273**, 24491.

137. C. R. White, D. Moellering, R. P. Patel, M. Kirk, S. Barnes and V. M. Darley-Usmar, *Biochem. J.*, 1997, **328**, 517.

138. R. G. Keynes, C. Griffiths and J. Garthwaite, *Biochem. J.*, 2003, **369**, 399.

139. K. Schmidt, S. Pfeiffer and B. Mayer, *Free Radical Biol. Med.*, 1998, **24**, 859.

140. E. E. Lomonosova, M. Kirsch, U. Rauen and H. de Groot, *Free Radical Biol. Med.*, 1998, **24**, 522.

141. B. Halliwell, M. V. Clement, J. Ramalingam and L. H. Long, *IUBMB Life*, 2000, **50**, 251.

142. M. V. Clement, J. Jeyakumar, L. H. Long and B. Halliwell, *Antioxid. Redox Signaling*, 2001, **3**, 157.

143. N. Arakawa, S. Nemoto, E. Suzuki and M. Otsuka, *J. Nutr. Sci. Vitaminol.*, 1994, **40**, 219.
144. P. A. Kilmartin, *Antioxid. Redox Signaling*, 2001, **3**, 941.
145. R. Oliveira, F. Bento, C. Sella, L. Thouin and C. Amatore, *Anal. Chem.*, 2013, **85**, 9057.
146. P. E. M. Phillips and R. M. Wightman, *Trends Anal. Chem.*, 2003, **22**, 509.
147. C. N. Hall and J. Garthwaite, *Nitric Oxide*, 2009, **21**, 92.
148. B. Chen and W. M. Deen, *Chem. Res. Toxicol.*, 2001, **14**, 135.
149. G. R. Buettner, *Arch. Biochem. Biophys.*, 1993, **300**, 535.
150. K. Indira Priyadarsini, Redox Reactions of Antioxidants- Contributions from Radiation Chemistry of Aqueous Solutions, in *Charged Particle and Photon Interactions with Matter: Recent Advances, Applications, and Interfaces*, ed. Y. Hatano, Y. Katsumura and A. Mozumder, CRC Press, Boca Raton, 2010, vol. 22, pp. 595–622.
151. P. R. Gardner and I. Fridovich, *J. Biol. Chem.*, 1991, **266**, 1478.
152. T. Malinski and T. Ziad, *Nature*, 1992, **358**, 676.
153. G. Ellman and H. Lysko, *Anal. Biochem.*, 1979, **93**, 98.
154. P. Liu, B. Xu, J. Quilley and P. Y.-K. Wong, *Am. J. Physiol. Cell Physiol.*, 2000, **279**, C1970.
155. B. Halliwell, M. V. Clement and L. H. Long, *FEBS Lett.*, 2000, **486**, 10.
156. http://www.sigmaaldrich.com.

Methods of Peroxynitrite Synthesis in the Context of the Development and Validation of Peroxynitrite Sensors

MEKKI BAYACHOU*[a,b], GHAITH ALTAWALLBEH[a], HAITHAM KALIL[a], SARAH WOJCIECHOWSKI[a], AND TIYASH BOSE[a]

[a]Department of Chemistry, Cleveland State University, 2399 Euclid Ave, Cleveland, Ohio 44115, USA; [b]Department of Pathobiology, Lerner Research Institute, The Cleveland Clinic, Cleveland, Ohio 44195, USA
*E-mail: m.bayachou@csuohio.edu

3.1 Background

Over the past 25 years, more than 7000 peroxynitrite-related papers based on the data available in "PubMed" were published after the discovery of its multifaceted roles in pathophysiology. Peroxynitrite ($ONOO^-$) was first identified in 1904.[1] Chemists, including Halfpenny and Robinson in 1952,[2,3] Hughes and Nicklin in 1968,[4,5] Keith and Powell in 1969,[6] and Mahoney in 1970,[7] studied the chemistry of this reactive analyte, but the work of Beckman *et al.* in 1990 is considered a key milestone in bringing the attention back to this molecule through its suggested biological roles.[8] The IUPAC

RSC Detection Science Series No. 7
Peroxynitrite Detection in Biological Media: Challenges and Advances
Edited by Serban F. Peteu, Sabine Szunerits, and Mekki Bayachou

Published by the Royal Society of Chemistry, www.rsc.org

nomenclature system recommends oxoperoxonitrate for peroxynitrite and hydrogen oxoperoxonitrate for its conjugated peroxynitrous acid. On a fundamental level, the unpaired electrons of superoxide anion and nitric oxide radical recombine to form a new bond $(ON-OO^-)$. Two conformational isomers of peroxynitrite are possible as depicted in Figure 3.1. Vibrational spectroscopy shows that the *cis* isomer is more stable than the *trans* isomer. The *cis* isomer is also the reactive species responsible for nitration of targets such as proteins.[9]

Since the pK_a of peroxynitrous acid (ONOOH) is in the range of 6.5–6.8, $ONOO^-$ is protonated at physiological pH. The protonation equilibrium is given in eqn (3.1). At pH = 7.4 about 80% of peroxynitrite is in the anionic form.

$$ONOO^- + H^+ \rightleftharpoons ONOOH \tag{3.1}$$

The conjugate acid, peroxynitrous acid, is unstable and decays rapidly mostly through a rearrangement to give nitrate (eqn (3.2)):

$$ONOOH \rightarrow [ONOOH] \rightarrow NO_3^- + H^+ \tag{3.2}$$

In biological systems, peroxynitrite is formed mainly through the reaction of nitric oxide $(NO^{\bullet})$ with superoxide anion $(O_2^{\bullet -})$.[8,10–13] NO is generated in a two-step catalytic oxidation of L-arginine by nitric oxide synthase (NOS). Superoxide ion, on the other hand, is a byproduct of side reactions of many oxidases and complexes of the respiratory chain. Interestingly, the uncoupling of endothelial NOS (eNOS) also generates superoxide ion within biological sites that also generate nitric oxide. The diffusion-controlled reaction of nitric oxide and superoxide ion $(\sim 1.6 \times 10^{10} \text{ M}^{-1} \text{ s}^{-1})$[14] is extremely fast, and is therefore operational even in the presence of natural catalytic detoxification pathways by superoxide dismutases provided that nitric oxide is at the micromolar level. Proinflammatory conditions enhance the levels of superoxide ion leaks, which in turn increase the levels of peroxynitrite.

Peroxynitrite reacts with many cellular targets as well as available electrophiles such as CO_2 to form nitrosoperoxycarbonate adduct $(ONOOCO_2^-)$. The homolysis of the O–O bond in both ONOOH and nitrosoperoxycarbonate $(ONOOCO_2^-)$ generates very reactive radicals such as $^{\bullet}OH$, $^{\bullet}NO_2$, and $CO_3^{\bullet -}$. For this and other reasons, peroxynitrite was identified as a potent cytotoxic compound in the early 1990s.[8,15] Peroxynitrite is relatively short-lived (<1 s) due to its multiple decay pathways.[12,16,17] Despite its short physiological half-life, peroxynitrite can still affect cells within an area of up to 10 μm in radius.[17]

Assessing peroxynitrite's deleterious effects and examining hypotheses of its potential signaling roles cannot be achieved without first accurately

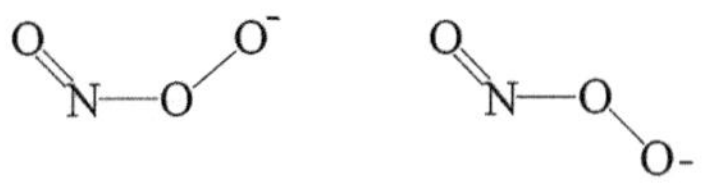

Figure 3.1   *Cis* and *trans* conformational isomers of the peroxynitrite anion.

measuring and monitoring its concentration. This task is, however, inherently difficult due to its low submicromolar concentrations under physiologic conditions coupled with its high reactivity. The development and validation of analytical methods to measure and monitor the formation and reactivity of peroxynitrite under physiological conditions rely primarily on the availability of suitable authentic sources of this analyte. In the following section, we give a brief overview of major methods of peroxynitrite synthesis or *in situ* generation that have been used over the years.

3.2   Overview of Major Methods of Peroxynitrite Synthesis

Several methods have been developed and used for the preparation of authentic peroxynitrite samples. However, most methods of preparation imply the presence of unavoidable contaminants such as hydrogen peroxide (H_2O_2), sodium azide, nitrate, or nitrite. Methods that generate relatively high concentrations of peroxynitrite with minimum or no secondary interferents are the ultimate goal. In this section we will summarize, compare, and contrast various published synthesis methods and their potential limitations.

3.2.1   Peroxynitrite from the Reaction of Nitrite with Acidified Hydrogen Peroxide

In acidic solutions, the reaction of nitrous acid with hydrogen peroxide yields the short-lived peroxynitrous acid, which is then quenched to give peroxynitrite anion. The latter is relatively stable in alkaline solutions.[18] This method uses a quenched-flow reactor to give relatively high yields of sodium peroxynitrite. The method is based upon mixing two streams of reactant solutions. The resulting reaction flow is quenched downstream with a flow of high concentration sodium hydroxide (NaOH; Scheme 3.1).

Typically, an aqueous solution of 0.6 M sodium nitrite is rapidly mixed with 0.7 M H_2O_2 containing 0.6 M hydrochloric acid (HCl). The reaction flow is immediately quenched with the same volume of 1.5 M NaOH. The reaction is performed on ice.[2,5,19,20]

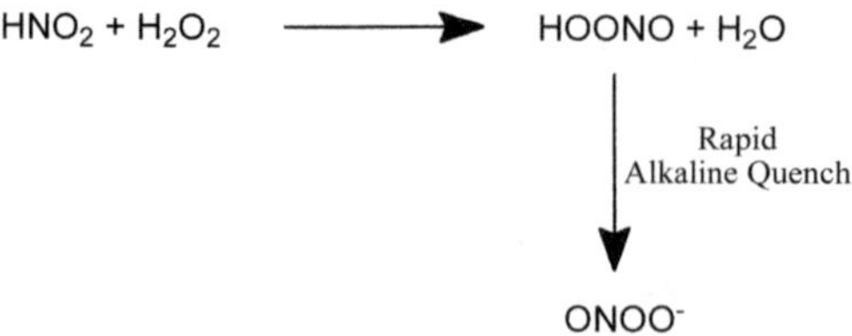

Scheme 3.1   Peroxynitrite synthesis from the reaction of nitrite and acidified H_2O_2 followed by fast quenching in concentrated NaOH solution.

Saha *et al.*[21] further improved the synthesis of peroxynitrite from acidic hydrogen peroxide and sodium nitrite based on earlier procedures. They were interested in optimizing the conditions to maximize the yield of peroxynitrite and to minimize the unreacted H_2O_2 and nitrite in the reaction mixture. They described the synthesis process of peroxynitrite *via* an efficient double mixer with a quenched-flow reactor. Nitrite and acidified H_2O_2 are mixed through the first mixing chamber, then the reaction mixture is run through a short connector to the second mixing chamber, where NaOH is added to quench the reaction. The yellow peroxynitrite is formed and collected through the outlet tube. An optimum flow rate of 162 mL min^{-1} at 20–25 °C was found to produce ~178 mmol L^{-1} of peroxynitrite as determined by measuring the absorbance at 302 nm.[21] With this optimized setting, they achieved an 85–90% yield of peroxynitrite at room temperature with residual nitrite. It was also shown that the conventional use of ice-cold solutions of the reactants and the alkali solutions was no longer necessary if an efficient mixer and appropriate quenching times are used.[21]

3.2.2 Synthesis of Peroxynitrite in a Two-Phase System Using Isoamyl Nitrite and H_2O_2

Uppu and Pryor reported a method for the synthesis of peroxynitrite based on two-phase displacement.[22] The hydroperoxide anion in the alkaline aqueous (pH 12.5–13) phase reacts with isoamyl nitrite in the organic phase for the preparation of up to 1 M of peroxynitrite.[22]

The buffered H_2O_2 solution is stirred vigorously with an equimolar amount 0.02–0.20 mol of isoamyl nitrite for 1–15 h at room temperature. The peroxynitrite product remains in the aqueous phase, and isoamyl alcohol forms in the organic phase along with the unreacted isoamyl nitrite (Scheme 3.2).

The formation of peroxynitrite is monitored using reaction aliquots as a function of reaction time using the peroxynitrite maximum absorbance at 302 nm (ε = 1705 M^{-1} cm^{-1}), as shown in Figure 3.2, from a typical preparation in the authors' laboratory. The aqueous phase contains some isoamyl alcohol and the unreacted H_2O_2, but no isoamyl nitrite.

Removal of isoamyl alcohol, and traces of isoamyl nitrite, is typically accomplished by washing the aqueous phase with dichloromethane and

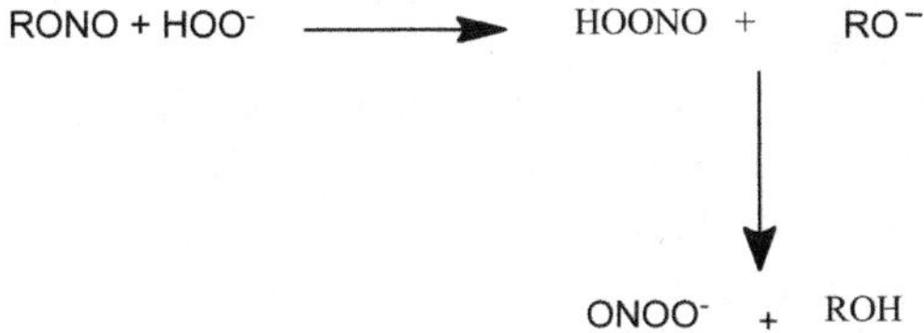

Scheme 3.2 Peroxynitrite synthesis *via* the reaction of hydroperoxide anion and alkyl nitrites.

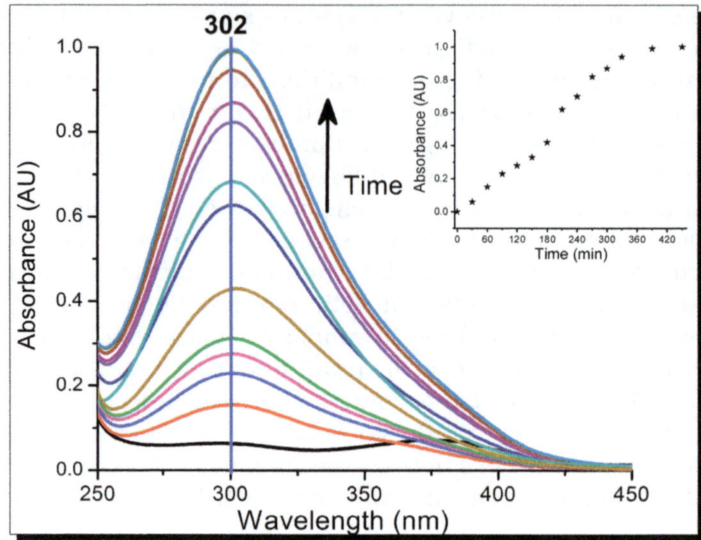

Figure 3.2 The increase in absorbance of peroxynitrite as the reaction between hydroperoxide anion in the alkaline aqueous phase and isoamyl nitrite in the organic phase proceeds. The inset shows the absorbance at 302 nm as a function of time.

chloroform. The aqueous phase is further filtered through a column filled with granular MnO_2 to efficiently remove any unreacted H_2O_2.[22]

This method is reported to give peroxynitrite solutions containing less nitrite when expressed as a fraction of the peroxynitrite formed compared with samples prepared by the ozonation of azide (described in Section 3.2.4), or the reaction of nitrite with acidified H_2O_2, or even the autooxidation of hydroxylamine.[22]

The same class of reaction can also be carried out in one phase using solvents of intermediate polarities such as alcohols.

3.2.3 The Reaction of H_2O_2 and Water Soluble Nitrites such as 2-Ethoxyethyl Nitrite in Basic Medium

The two-phase displacement reaction by hydroperoxide on isoamyl nitrite described in Section 3.2.2 is actually a modification of a prior procedure described by Leis in 1993.[23]

The method also produces stable $ONOO^-$ but in homogeneous alkaline solution as a result of the nucleophilic attack of hydroperoxyl radical (HOO^-) on the nitroso group of alkyl nitrites (Scheme 3.2). Normally, under such conditions, alkaline hydrolysis of the alkyl nitrile is expected to take place. However, the reaction with HOO^- outperforms the simple reaction with hydroxide ion (HO^-). In fact, kinetic studies have shown that the bimolecular rate constant between HOO^- and 2-ethoxyethyl nitrite is $\sim 3 \times 10^3$ higher than that of the reaction with HO^-, which essentially eliminates interference from the competitive process of alkaline hydrolysis of the alkyl nitrite.[23] The reaction is usually carried out with equimolar amounts of H_2O_2 to avoid large amounts of peroxide in the final solution.

3.2.4 The Reaction of Ozone with Azide Ions in the Presence of a Low Concentration of Alkali

Ozonation of an aqueous solution of azide is perhaps one of the oldest methods on record for the preparation of peroxynitrite. This 1929 method was reexamined by Pryor *et al.* in 1995 and was found to provide a great source of peroxynitrite solution that is free from H_2O_2 contaminant. This method offers a way to prepare clean peroxynitrite that is useful for biological studies. The starting concentration of azide dictates the final concentrations of peroxynitrite that is formed. Some azide stays in the final solution, but continuing ozonation towards the end of the conversion tends to lower the final amount.[24] Based on kinetic observations and the fact that nitrous oxide (N_2O) is observed as a byproduct, the mechanism likely involves two sequential reactions. First, ozone reacts with azide to provide N_2O and a postulated nitrogen peroxide (NOO^-) intermediate; the latter further reacts with O_3 to give $ONOO^-$ and dioxygen (O_2; Scheme 3.3).

Apparently, peroxynitrite solutions prepared from ozonation of azide at pH 7.0 do not show detectable levels of H_2O_2, and even with partial consumption, the concentration of peroxynitrite in these solutions is still 200–3500 fold higher than H_2O_2. Unlike other methods of preparation such as the two-phase displacement reaction of hydrogen peroxide anions and alkyl nitrite under alkaline conditions, this method of preparation does not require further removal of excess hydrogen peroxide.[24]

3.2.5 Other Methods of Peroxynitrite Synthesis

Other methods of peroxynitrite synthesis that have been described include the oxidation of hydroxylamine by oxygen in alkaline solution[25] (Scheme 3.4), and the reaction of potassium superoxide with nitric oxide.[26]

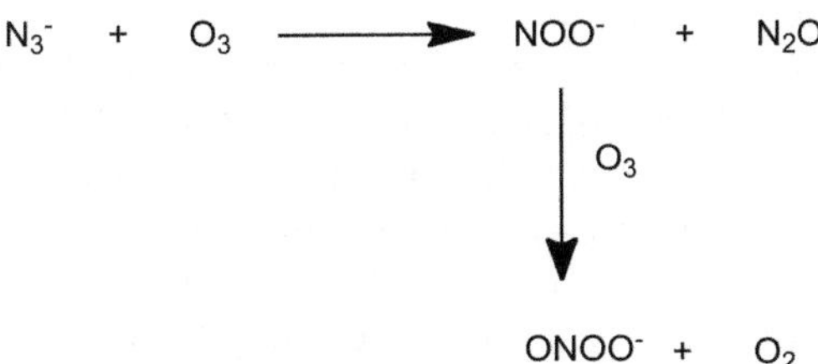

Scheme 3.3 Ozonation of aqueous solutions of azide results in $ONOO^-$ with no detectable levels of H_2O_2.

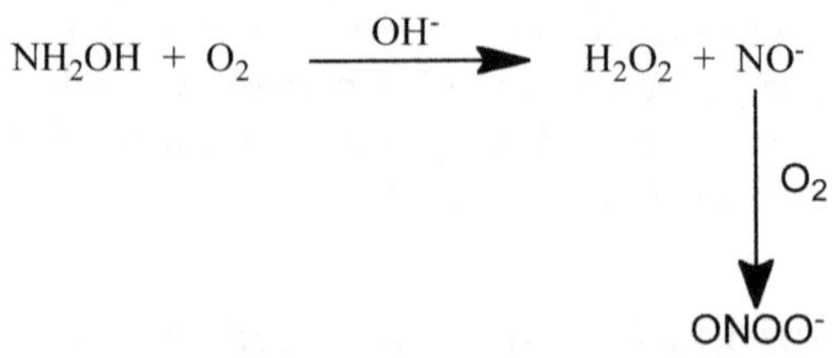

$$NH_2OH + O_2 \xrightarrow{OH^-} H_2O_2 + NO^-$$

$$\downarrow O_2$$

$$ONOO^-$$

Scheme 3.4 ONOO$^-$ synthesis through the oxidation of hydroxylamine under alkaline conditions.

Similarly, a solid-phase synthesis of peroxynitrite by the reaction of solid potassium superoxide dispersed in quartz sand flushed with an inert gas and then exposed to nitric oxide gas has also been described.[26] The corresponding reaction of tetramethyl ammonium superoxide with nitric oxide gas was also reported to yield relatively pure tetramethyl ammonium peroxynitrite ([N(CH$_3$)$_4$]$^+$ [ONOO]$^-$).[27,28] Sturzbecher-Hohne *et al.* described an efficient method based on cation exchange to swap the tetramethyl ammonium cation with Li$^+$ or Na$^+$ ions, which are more relevant to biological investigations.[29] The cation exchange method seems to preserve the purity of peroxynitrite.

Although the bimolecular reaction of *N*-acetyl-*N*-nitrosotryptophan and H$_2$O$_2$ was recently reported to generate peroxynitrite at a wide range of pH values,[30] the system would not be a viable source of peroxynitrite for validation purposes. In fact, the bimolecular reaction is relatively slow and of course yields samples that are not free of contaminants.

3.2.6 Generation of Peroxynitrite from 3-(4-Morpholinyl) Sydnonimine Hydrochloride

3-Morpholinosydnonimine-*N*-ethylcarbamide (SIN-1; Figure 3.2A) belongs to the family of sydnonimines, which were first prepared by Peter Brooks and independently by Masaki Ohta in 1957.[31] These compounds, and particularly the parent compound (molsidomine; Figure 3.2B), have been viewed and used as NO donors in the treatment of angina pectoris. Molsidomine is just the stable prodrug form of SIN-1 that is acylated at the exocyclic nitrogen to stabilize the oxadiazolium ring against hydrolysis. The prodrug is deacylated by host esterases and converted to the active metabolite SIN-1, which then releases its biological activity.

As shown in Figure 3.3A, the decomposition of SIN-1 leads to the release of O$_2^{\cdot-}$ radical and NO, which quickly combine to form ONOO$^-$.[32-34]

Although this method of *in situ* generation of peroxynitrite is becoming widespread because of both ease and convenience, it is not always straightforward under all conditions, particularly in the presence of redox agents. In fact, literature shows that the same molecule (SIN-1) was also labeled as an NO donor[35] or behaved as though it predominantly releases NO.[36] More information about the limitations and known controversies of SIN-1 as a blind source of peroxynitrite are addressed in the next section.

A

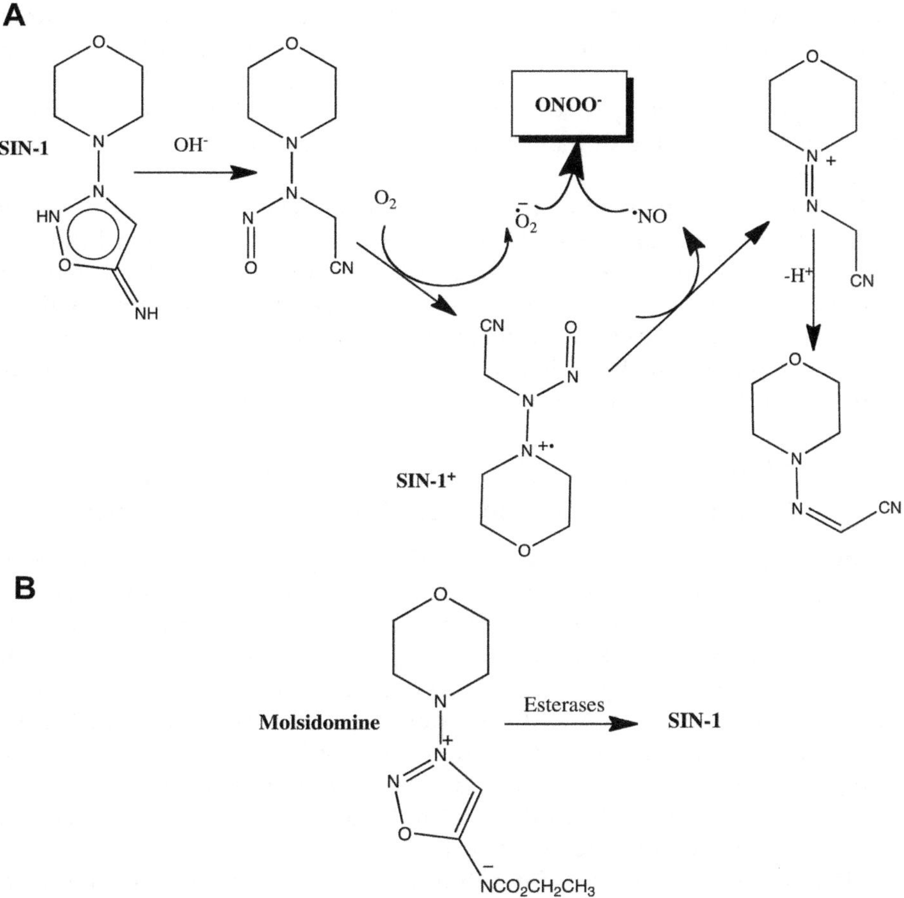

Figure 3.3 (A) Formation of peroxynitrite *via* spontaneous decomposition of SIN-1. At physiologic or alkaline pH, SIN-1 undergoes rapid hydrolysis and opening of the oxadiazolium ring. The ring-opened intermediate reduces molecular oxygen to generate superoxide and the oxidized form, which further releases one equivalent of nitric oxide. The latter quickly combines with superoxide anion to generate peroxynitrite. (B) SIN-1 is the active metabolite of the parent prodrug molsidomine, which undergoes deacetylation to give SIN-1.

3.3   Advantages and Limitations of the Various Peroxynitrite Synthesis Routes

As we briefly surveyed in the previous section, there are several established methods for peroxynitrite synthesis.[37–39] Obviously, the method of preparation and associated impurities play a key role in peroxynitrite quantifications and assay validation. Impurities in peroxynitrite solutions produced by typical methods include nitrite, nitrate, peroxide, and azide. The variation in assays may also come from assumed peroxynitrite levels and assumed purity levels based on side products that are specific for each method of preparation. It is therefore very important to ensure the quality and know the composition of the peroxynitrite solutions used. This may be crucial particularly in attempts to validate methods of determination at very low limits of detections.

The method of peroxynitrite synthesis or peroxynitrite generation may also dictate the conditions under which assays are run, and may limit reliable extensions to settings suitable for measurements under biological conditions. The medium in which (and for which) methods of peroxynitrite determination are developed and validated may also impose its own limitations and constraints. We will give two examples of buffer-based limitations before we address the limitations of synthetic procedures themselves.

3.3.1 Characterization of Peroxynitrite Sensors and Evaluation of Performance: Are Buffers Indifferent?

Before addressing issues that are related to methods of peroxynitrite preparation, it is important to note that in comparing reports about limits of detections and sensitivities of probes and sensors, one has to pay close attention to the nature of the buffers used between reports. In this regard, it has been shown that tris(hydroxymethyl)aminomethane (TRIS) buffer significantly accelerates the decomposition kinetics of peroxynitrite compared with phosphate buffer.[40] Considering these differences, it is conceivable that the effective peroxynitrite concentrations in the two buffer systems, with otherwise identical conditions, will be different, and may indeed affect measured limits of detection and reported sensitivities.

Along the same lines, but for a different reason, tertiary amines, including piperazine-type buffers such as HEPES (4-(2-hydroxyethyl)-1-piperazineethanesulfonic acid), react with and consume peroxynitrite, and generate instead significant amounts of H_2O_2.[41] The group led by Herbert de Groot investigated this problem very closely in order to explain the observed H_2O_2-like cytotoxicity of the murine fibroblast-derived cell line L929 in the presence of a source of peroxynitrite.[41,42] In the presence of HEPES, not only did the H_2O_2 levels and cytotoxicity increase, but also the amount of peroxynitrite (authentic or generated *in situ*) was significantly lower than expected. It turns out that in the presence of HEPES, a peroxynitrite-mediated 1-electron oxidation of the piperazine ring generates the corresponding radical cation, which loses a proton and forms the corresponding α-aminoalkyl radical that transiently reacts with molecular oxygen and ultimately yields a superoxide radical anion. The latter undergoes dismutation to give molecular oxygen and hydrogen peroxide (Figure 3.4). This proposed mechanism is not specific to HEPES but would be operational with other tertiary amines.

The two examples cited above (TRIS and HEPES) highlight the influence of buffer systems and show that the conditions used in the development and validation of methods for peroxynitrite determination should be considered

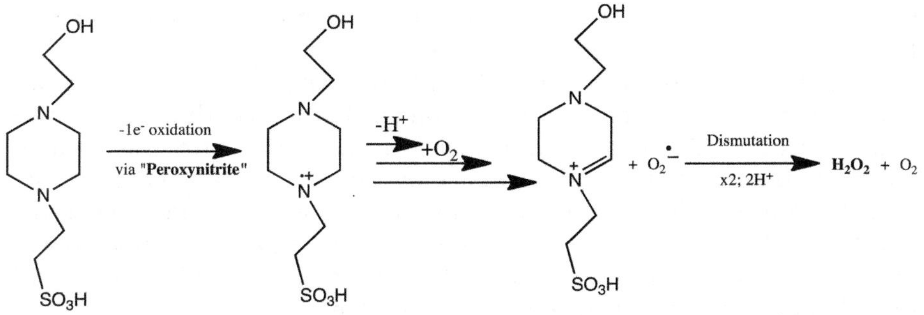

Figure 3.4 Simplified scheme outlining the reaction of an oxidant such as peroxynitrite and piperazine-based buffers such as HEPES. The redox process ultimately consumes peroxynitrite and generates instead H_2O_2.[42]

when one compares reported values of sensitivity and limits of detection of various methods.

3.3.2 Advantages and Limitations of the Various Methods of Peroxynitrite Synthesis and Generation for Sensing Purposes

Let us now turn to the various methods of synthesis of authentic peroxynitrite and their respective limitations in the development and validation of peroxynitrite sensing. The first is the general method of synthesis of peroxynitrite from acidic hydrogen peroxide and nitrite followed by fast base quenching. As we reported in Section 3.2.1, the method seems to be relatively simple and, if the timing of the quenching step is optimized, can yield only residual nitrite in the sample. However, there is variability in yields of peroxynitrite and levels of residual nitrate and nitrite as a result of PON isomerization or decomposition under experimental conditions. Some levels of hydrogen peroxide in the peroxynitrite samples obtained by this method also persist in the final sample.[43] Both nitrite and hydrogen peroxide can affect the validation of peroxynitrite determination, particularly for methods based on electrochemical detection. Hydrogen peroxide can also interfere with fluorescence-based assays. In fact, the dioxaborolane moiety attached to some fluorescent probes for peroxynitrite determination[44] is the same target on other fluorescent probes developed to measure hydrogen peroxide itself.[45] The same is true for dichlorodihydrofluorescein-based probes.[46–48] While excess H_2O_2 can be removed using filtration through MnO_2 or addition of MnO_2 followed by centrifugation, there is a concern that MnO_2 treatment itself may introduce metal ions as contaminants in the peroxynitrite sample. This of course would not be desirable because many metal complexes are known to facilitate the decomposition of peroxynitrite.[49,50] The residual metal ions not only enhance peroxynitrite decomposition,[51] but they can also—conceivably—interfere with catalytic peroxynitrite detection, particularly when using electrochemical sensors.

The same is true for the method of synthesis using the nucleophilic attack of hydroperoxide anion (alkaline H_2O_2) on alkyl nitrites both in one- or two-phase reactions. The method also results in H_2O_2 and nitrite as contaminants despite the post-reaction processing. While the two-phase reaction of hydroperoxide with isoamyl nitrite introduced by Uppu and Pryor results in relatively less nitrite as a fraction of peroxynitrite formed compared with other methods,[38] the hydrogen peroxide and nitrite may still interfere with the validation of methods of peroxynitrite determination.

Peroxynitrite samples that are prepared by ozonation of azide[24] contain variable levels of azide depending on the extent of the ozonation. Amperometric sensors for peroxynitrite based on metalloporphyrin and metallophthalocyanine films have been proposed.[52–58] Conceivably, many anionic ligands, including azide, can bind to the transition metal catalytic center in these systems, and thus present a potential interference with amperometric signals in electrochemical methods. Azide anion is known to bind to various metalloporphyrins and metalloproteins. For instance, it readily binds to iron and manganese porphyrins.[59,60] It also binds to the manganese center in manganese superoxide dismutase (MnSOD),[61] and in fact interferes with its peroxynitrite-driven self-nitration. Similarly, azide binds to iron heme proteins[62–65] as well as to metal-substituted hemoproteins such as manganese myoglobin.[66,67] In addition, membrane sensors based on the use of manganese(III)porphyrin and cobalt(II)phthalocyanine ionophores dispersed in plasticized poly(vinyl chloride) have actually been used in potentiometric determination of azide.[68] Therefore, the synthetic source of exogenous peroxynitrite used in methods of determination and validation of this analyte is critical, and the effects of interferents such as azide, if present, need to be carefully assessed.

While this may not be directly related to validation of methods of determination of peroxynitrite, azide impurities left in prepared samples may also interfere with fluorescence studies that measure the extent of protein nitration after exposure to synthesized samples of peroxynitrite. In fact, sodium azide has been reported as a chemical agent used in fluorescence microscopy to reduce photobleaching of fluorescein isothiocyanate (FITC).[69] On the other hand, FITC-conjugated secondary antibodies have been used as fluorescent probes associated with polyclonal anti-nitrotyrosine antibodies to detect and measure nitrated proteins after exposure to exogenous peroxynitrite.[70] In biological settings, either in cell culture or in animal models, nitration of select proteins mediates cellular responses to peroxynitrite-driven oxidative stress. These select proteins may act as *in situ* peroxynitrite sensors.[71–73] The variability in the levels of azide in the peroxynitrite samples used in validation steps will therefore introduce variability in fluorescence if FITC-based fluorescence is used as the measured signal. This, in turn, may bias conclusions about the levels of peroxynitrite-driven nitration of proteins used as *in situ* biological peroxynitrite sensors.

Sources such as SIN-1 [32–34,41,74] that continuously generate exogenous peroxynitrite have been used to deliver this reactive species at physiologic pH in a way that mimics its *in situ* biological generation. In fact, the bolus addition of authentic aliquots of peroxynitrite from prepared stock solutions is often not considered reflective of the continuous flow of peroxynitrite under physiologic conditions.[75] However, the SIN-1 system is *not* an innocuous source of peroxynitrite, and reports using this system to generate peroxynitrite under physiologic conditions showed conflicting results.[75,76] The multistep chemistry of peroxynitrite generation from SIN-1 may introduce a level of complication in the use of this system to characterize the performance of sensors and probes.

The rate of generation of peroxynitrite from the decomposition of SIN-1 in the presence of oxygen has been estimated at 1 μM min^{-1} from 1 mM SIN-1 solution.[77] The half-life of steady-state production of peroxynitrite was measured in select buffers and was found to range from 14 to 26 min depending on the buffer used. The maximum peroxynitrite levels that can be achieved depend on the concentration of SIN-1.[78] When SIN-1 is used as a precursor of peroxynitrite in the validation of methods of determination of this analyte, experimental details, including concentration of the precursor,[78] levels of dissolved oxygen,[9] incubation time prior to injections, and controls for peroxynitrite sinks such as carbon dioxide,[79] all have to be very carefully described in order to compare performance reports of different sensors and probes. Redox agents and precursors of radical species may interfere with the continuous fluxes of superoxide and nitric oxide, particularly at low levels, which may affect the amount of peroxynitrite that is potentially expected.[36,75,80] Furthermore, while steady-state levels of peroxynitrite from a given *in situ* amount of SIN-1 may be determined, peroxynitrite levels from aliquots transferred from external SIN-1 stock solutions into separate buffers where sensor characterization takes place may not accurately be known. Unless the assumed added concentration levels are carefully confirmed by independent means, the sensor responses observed may not reflect the stated performance. Also, as we discussed earlier, experimental conditions may affect the ultimate levels of peroxynitrite released.[40–42] Assumptions based on stoichiometry alone and the expected spontaneous decomposition yield may be misleading, and may seriously bias the reported sensitivities and limits of detection for sensors and sensing systems that use SIN-1 as a source of peroxynitrite.

References

1. J. O. Edwards and R. C. Plumb, *Prog. Inorg. Chem.*, 1994, **41**, 599.
2. E. Halfpenny and P. L. Robinson, *J. Chem. Soc.*, 1952, 928.
3. E. Halfpenny and P. L. Robinson, *J. Chem. Soc.*, 1952, 939.
4. M. N. Hughes and H. G. Nicklin, *J. Chem. Soc. A*, 1970, 925.
5. M. N. Hughes and H. G. Nicklin, *J. Chem. Soc. A*, 1968, 450.
6. W. G. Keith and R. E. Powell, *J. Chem. Soc. A*, 1969, 90.
7. L. R. Mahoney, *J. Am. Chem. Soc.*, 1970, **92**, 5262.

8. J. S. Beckman, T. W. Beckman, J. Chen, P. A. Marshall and B. A. Freeman, *Proc. Natl. Acad. Sci. U. S. A.*, 1990, **87**, 1620.
9. C. D. Reiter, R. J. Teng and J. S. Beckman, *J. Biol. Chem.*, 2000, **275**, 32460.
10. N. V. Blough and O. C. Zafiriou, *Inorg. Chem. Res. Toxicol.*, 1985, **24**, 3502.
11. P. Pacher, J. S. Beckman and L. Liaudet, *Physiol. Rev.*, 2007, **87**, 315.
12. P. Ascenzi, A. di Masi, C. Sciorati and E. Clementi, *Biofactors*, 2010, **36**, 264.
13. R. Radi, G. Peluffo, M. N. Alvarez, M. Naviliat and A. Cayota, *Free Radical Biol. Med.*, 2001, **30**, 463.
14. R. Kissner, T. Nauser, C. Kurz and W. H. Koppenol, *IUBMB Life*, 2003, **55**, 567.
15. J. S. Beckman, *Nature*, 1990, **345**, 27.
16. A. Denicola, J. M. Souza and R. Radi, *Proc. Natl. Acad. Sci. U. S. A.*, 1998, **95**, 3566.
17. C. Szabo, H. Ischiropoulos and R. Radi, *Nat. Rev. Drug Discovery*, 2007, **6**, 662.
18. J. W. Reed, H. H. Ho and W. L. Jolly, *J. Am. Chem. Soc.*, 1974, **96**, 1248.
19. M. H. Zou and V. Ullrich, *FEBS Lett.*, 1996, **382**, 101.
20. J. W. Reed, H. H. Ho and W. L. Jolly, *J. Am. Chem. Soc.*, 1974, **96**, 1248.
21. A. Saha, S. Goldstein, D. Cabelli and G. Czapski, *Free Radical Biol. Med.*, 1998, **24**, 653.
22. R. M. Uppu and W. A. Pryor, *Anal. Biochem.*, 1996, **236**, 242.
23. J. R. Leis, M. E. Pena and A. Rios, *J. Chem. Soc. Chem. Commun.*, 1993, 1298.
24. W. A. Pryor, R. Cueto, X. Jin, W. H. Koppenol, M. Nguschwemlein, G. L. Squadrito, P. L. Uppu and R. M. Uppu, *Free Radical Biol. Med.*, 1995, **18**, 75.
25. M. N. Hughes and H. G. Nicklin, *J. Am. Chem. Soc.*, 1971, **93**, 164.
26. W. H. Koppenol, R. Kissner and J. S. Beckman, *Methods Enzymol.*, 1996, **269**, 296.
27. P. Latal, R. Kissner, D. S. Bohle and W. H. Koppenol, *Inorg. Chem.*, 2004, **43**, 6519.
28. D. S. Bohle, E. S. Sagan, W. H. Koppenol, and R. Kissner, *Inorganic Syntheses*, Wiley, New York, 2004, vol. 34.
29. M. Sturzbecher-Hohne, R. Kissner, T. Nauser and W. H. Koppenol, *Chem. Res. Toxicol.*, 2008, **21**, 2257.
30. M. Kirsch and M. Lehnig, *Org. Biomol. Chem.*, 2005, **3**, 2085.
31. P. Brookes and J. Walker, *J. Am. Chem. Soc.*, 1957, **79**, 4409.
32. F. Regoli and G. W. Winston, *Toxicol. Appl. Pharmacol.*, 1999, **156**, 96.
33. N. Hogg, V. M. Darley-Usmar, M. T. Wilson and S. Moncada, *Biochem. J.*, 1992, **281**, 419.
34. J. L. Trackey, T. F. Uliasz and S. J. Hewett, *J. Neurochem.*, 2001, **79**, 445.
35. L. Y. Xu, J. S. Yang, H. Link and B. G. Xiao, *J. Immunol.*, 2001, **166**, 5810.
36. S. Alcon, S. Morales, P. J. Camello, J. M. Hemming, L. Jennings, G. M. Mawe and M. J. Pozo, *J. Physiol.*, 2001, **532**, 793.
37. R. M. Uppu, G. L. Squadrito, R. Cueto and W. A. Pryor, *Methods Enzymol.*, 1996, **269**, 285.

38. R. M. Uppu and W. A. Pryor, *Anal. Biochem.*, 1996, **236**, 242.

39. R. M. Uppu, *Anal. Biochem.*, 2006, **354**, 165.

40. C. Molina, R. Kissner and W. H. Koppenol, *Dalton Trans.*, 2013, **42**, 9898.

41. E. E. Lomonosova, M. Kirsch, U. Rauen and H. de Groot, *Free Radical Biol. Med.*, 1998, **24**, 522.

42. M. Kirsch, E. E. Lomonosova, H. G. Korth, R. Sustmann and H. de Groot, *J. Biol. Chem.*, 1998, **273**, 12716.

43. E. Hrabarova, P. Gemeiner and L. Soltes, *Chem. Pap.*, 2007, **61**, 417.

44. F. Yu, P. Song, P. Li, B. Wang and K. Han, *Analyst*, 2012, **137**, 3740.

45. V. S. Lin, B. C. Dickinson and C. J. Chang, *Methods Enzymol.*, 2013, **526**, 19.

46. J. P. Crow, *Nitric Oxide*, 1997, **1**, 145.

47. A. S. Keston and R. Brandt, *Anal. Biochem.*, 1965, **11**, 1.

48. H. Wang and J. A. Joseph, *Free Radical Biol. Med.*, 1999, **27**, 612.

49. W. H. Koppenol, J. J. Moreno, W. A. Pryor, H. Ischiropoulos and J. S. Beckman, *Chem. Res. Toxicol.*, 1992, **5**, 834.

50. M. K. Stern, M. P. Jensen and K. Kramer, *J. Am. Chem. Soc.*, 1996, **118**, 8735.

51. J. T. Groves and S. S. Marla, *J. Am. Chem. Soc.*, 1995, **117**, 9578.

52. J. S. Cortés, S. G. Granados, A. A. Ordaz, J. A. L. Jiménez, S. Griveau and F. Bedioui, *Electroanalysis*, 2007, **19**, 61.

53. J. Xue, X. Ying, J. Chen, Y. Xian and L. Jin, *Anal. Chem.*, 2000, **72**, 5313.

54. R. Kubant, C. Malinski, A. Burewicz and T. Malinski, *Electroanalysis*, 2006, **18**, 410.

55. S. F. Peteu, B. Saleem, M. M. Gunesekera, P. Peiris, O. A. Sicuia and M. Bayachou, in *Oxidative Stress: Diagnostics, Prevention, and Therapy*, ed. S. H. M. Andreescu, American Chemical Society, 2011, vol. 1083, p. 311.

56. S. Peteu, P. Peiris, E. Gebremichael and M. Bayachou, *Biosens. Bioelectron.*, 2010, **25**, 1914.

57. R. Oprea, S. F. Peteu, P. Subramanian, W. Qi, E. Pichonat, H. Happy, M. Bayachou, R. Boukherroub and S. Szunerits, *Analyst*, 2013, **138**, 4345.

58. S. F. Peteu, T. Bose and M. Bayachou, *Anal. Chim. Acta*, 2013, **780**, 81.

59. W. R. Scheidt and C. A. Reed, *Chem. Rev.*, 1981, **81**, 543.

60. V. Day, B. Stults, E. Tasset, R. Day and R. Marianelli, *J. Am. Chem. Soc.*, 1974, **96**, 2650.

61. N. B. Surmeli, N. K. Litterman, A. F. Miller and J. T. Groves, *J. Am. Chem. Soc.*, 2010, **132**, 17174.

62. D. C. Blumenthal and R. J. Kassner, *J. Biol. Chem.*, 1979, **254**, 9617.

63. T. Jacobson, J. Williamson, A. Wasilewski, J. Felesik, L. B. Vitello and J. E. Erman, *Arch. Biochem. Biophys.*, 2004, **422**, 125.

64. R. Maurus, R. Bogumil, N. T. Nguyen, A. G. Mauk and G. Brayer, *Biochem. J.*, 1998, **332**(1), 67.

65. M. Bayachou, L. Elkbir and P. J. Farmer, *Inorg. Chem.*, 2000, **39**, 289.

66. N.-T. Yu and M. Tsubaki, *Biochemistry*, 1980, **19**, 4647.

67. R. Lin, C. E. Immoos and P. J. Farmer, *JBIC, J. Biol. Inorg. Chem.*, 2000, **5**, 738.

68. S. S. M. Hassan, A. E. Kelany and S. S. Al-Mehrezi, *Electroanalysis*, 2008, **20**, 438.

69. G. Bock, M. Hilchenbach, K. Schauenstein and G. Wick, *J. Histochem. Cytochem.*, 1985, **33**, 699.

70. T. Grune, I. E. Blasig, N. Sitte, B. Roloff, R. Haseloff and K. J. Davies, *J. Biol. Chem.*, 1998, **273**, 10857.

71. C. Lindemann, N. Lupilova, A. Muller, B. Warscheid, H. E. Meyer, K. Kuhlmann, M. Eisenacher and L. I. Leichert, *J. Biol. Chem.*, 2013, **288**, 19698.

72. D. Martins, I. Bakas, K. McIntosh and A. M. English, *Free Radical Biol. Med.*, 2015, **85**, 138.

73. H. J. Forman, F. Ursini and M. Maiorino, *J. Mol. Cell. Cardiol.*, 2014, **73**, 2.

74. R. Floris, S. R. Piersma, G. Yang, P. Jones and R. Wever, *Eur. J. Biochem.*, 1993, **215**, 767.

75. S. McLean, L. A. Bowman and R. K. Poole, *Microbiology*, 2010, **156**, 3556.

76. K. Konishi, N. Watanabe and T. Arai, *Nitric Oxide*, 2009, **20**, 270.

77. N. Kuzkaya, N. Weissmann, D. G. Harrison and S. Dikalov, *Biochem. Pharmacol.*, 2005, **70**, 343.

78. F. J. Martin-Romero, Y. Gutierrez-Martin, F. Henao and C. Gutierrez-Merino, *J. Fluoresc.*, 2004, **14**, 17.

79. M. K. Hulvey, C. N. Frankenfeld and S. M. Lunte, *Anal. Chem.*, 2010, **82**, 1608.

80. S. Goldstein and G. Czapski, *Chem. Res. Toxicol.*, 2000, **13**, 736.

Peroxynitrite-Sensitive Electrochemically Active Matrices

SABINE SZUNERITS*[a], SERBAN PETEU[b],
AND RABAH BOUKHERROUB[a]

[a]Institut d'Electronique, de Microélectronique et de Nanotechnologie,
CNRS UMR8520, Avenue Poincaré – CS 60069, 59652 Villeneuve d'Ascq,
France; [b]Chemical Engineering and Materials Science, Michigan State
University, 3523 Engineering Building, East Lansing, Michigan 48824,
United States
*E-mail: sabine.szunerits@iri.univ-lille1.fr

4.1 Introduction: Challenges for the Formation of Selective and Sensitive Matrices for Peroxynitrite Detection

Over the years, a variety of approaches for the sensitive and selective detection of peroxynitrite have been developed. Most of the techniques for assaying peroxynitrite are, however, indirect methods, relying on measurements of secondary species. The widely used biochemical assays are thus based on the nitration of tyrosine by peroxynitrite resulting in 3-nitrotyrosine, which is then detected by immunochemical or chromatographic techniques.[1–3] The oxidation of chromophores, fluorescent and chemiluminescent probes by

RSC Detection Science Series No. 7
Peroxynitrite Detection in Biological Media: Challenges and Advances
Edited by Serban F. Peteu, Sabine Szunerits, and Mekki Bayachou

Published by the Royal Society of Chemistry, www.rsc.org

peroxynitrite, routinely used for *in vivo* peroxynitrite analysis under physiological conditions, is another indirect detection approach.[4-7] The most notable drawback of these methods is that selectivity towards peroxynitrite alone is often not ensured as other reactive oxygen and nitrogen species can contribute to the results of the assays.

Recently, much effort has been put into the development of fluorescent probes capable of reacting *exclusively* with peroxynitrite (for more details see Chapters 10 and 11 of this book).[7] Such probes are, unfortunately, not commercially available and require advanced skills in organic chemistry for their synthesis. It has also been speculated that introduction of any organic probe (a fluorescent ligand in this case) into a biological environment can create unwanted alterations of cell functions. Electrochemical sensing strategies have consequently been seen as a viable alternative when it comes to real-time, label-free and direct quantification of peroxynitrite. The problems to overcome here were related to fouling of the sensor interface, moderate sensitivity and low selectivity. The use of adapted surface chemistry approaches, making the electrode more selective and in addition less prone to nonspecific adsorption events, is one of the strategies employed to overcome these shortcomings. This chapter will look at the different approaches currently available and in use. Surprisingly, only a limited number of research groups have attempted to fabricate peroxynitrite-selective matrices (Figure 4.1). While platinized carbon micro- and nano-electrodes were extensively used by Amatore *et al.*[8-12] and showed that the direct oxidation of peroxynitrite takes place at 0.27 V *vs.* saturated calomel electrode (SCE; for more details see Chapter 6 of this book), this chapter exclusively investigates electrodes coated with polymers, metallophthalocyanine (MPc) films and modified reduced graphene oxide (rGO) nanostructures. This chapter might motivate researchers to implement their most recent knowledge in this area and help advance the field.

4.2   Polymeric Films

The use of electrodes modified with polymeric films is a widely developed and used approach for the selective detection of biologically relevant species. Such an approach has also become popular for peroxynitrite. Some of the first examples date back to the beginning of the 21st century,[13-15] and are based on polymeric films of phthalocyanine[13,16] and porphyrin[14] derivatives, notably hemin.

4.2.1   Metallophthalocyanine Films (MPCs)

MPcs are transition metal complexes that have a number of characteristic properties that contribute in a major way to their extraordinary versatility. One research field where MPcs have found a wide range of applications is the development of electrochemical sensors, where MPcs are employed as electrocatalysts. Electrodes modified with phthalocyanine-based macrocycles have shown to be outstanding sensors for a variety of analytes, including oxygen,[17] hydrogen peroxide,[18] nitrite,[19-21] nitrate[22] and nitric oxide (NO).[23] Most

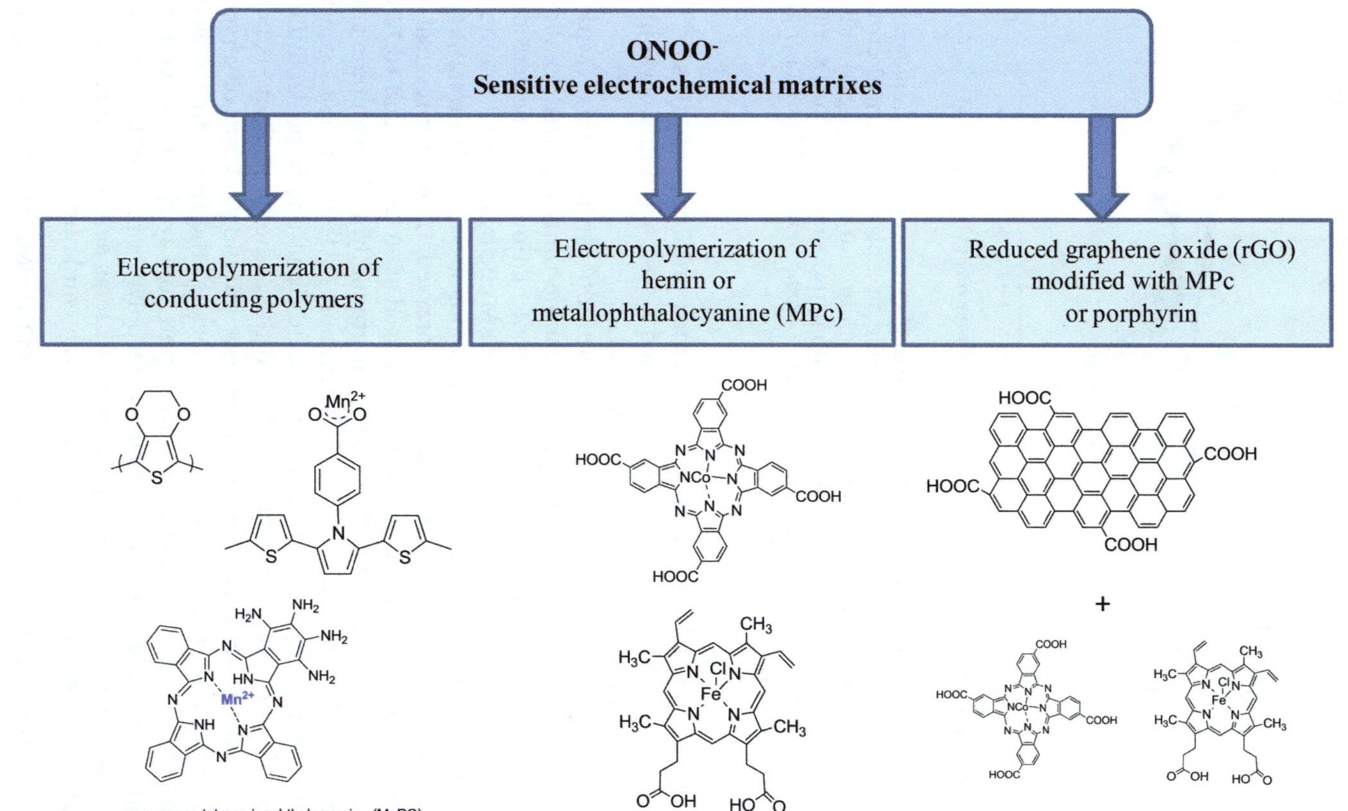

Figure 4.1 Different electrochemical active matrices for peroxynitrite detection.

(A)

(B)

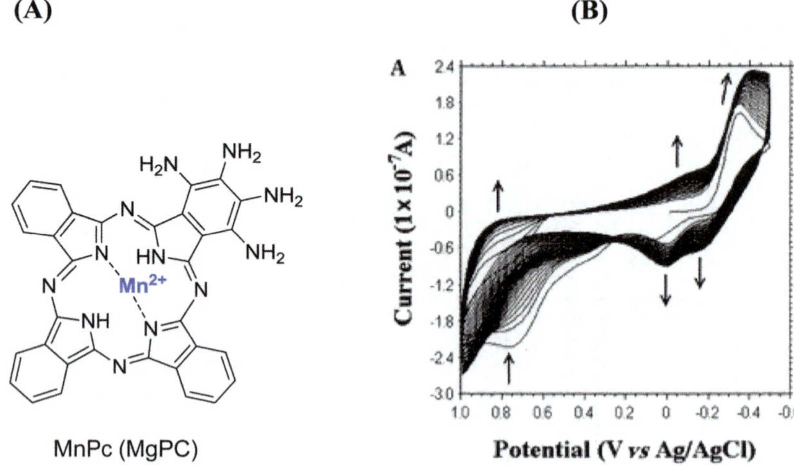

Figure 4.2 (A) Chemical structure of MnPc. (B) Consecutive cyclic voltammograms of MnPc (5 mM) in dimethyl sulfoxide on a platinum microelectrode (15 μm diameter) at a potential of −0.5 to +1.0 V (*vs.* Ag/AgCl) with a scan rate of 100 mV s^{-1}. Reprinted with permission from J. Xue, X. Ying, J. Chen, Y. Xian, L. Jin and J. Jin, *Anal. Chem.*, 2000, **72**, 5313. Copyright 2000 American Chemical Society.

of the applications rely critically upon the redox properties of MPcs.[21,23–27] Due to their macrocyclic nature with extended π-systems, phthalocyanines are capable of undergoing fast redox processes.

Manganese(II)-coordinated tetraaminophthalocyanine films (MnPc; Figure 4.2A) were one of the first phthalocyanine films reported to allow the detection of peroxynitrite.[15] Modification of platinum microelectrodes was achieved through electropolymerization of MnPc by potential scanning between −0.5 and +1 V at a scan rate of 100 mV s^{-1} for 30 cycles (Figure 4.2B). On the first scan, a cathodic peak at −0.35 V was observed. Upon scan reversal to 1.0 V, three significant anodic peaks were observed at −0.23, +0.1 and +0.75 V. The redox waves at 0.1 and 0.75 V disappear gradually and the anodic current related to the oxidation of the phthalocyanine cycle around 0.8 V gently increases. On further potential scanning, two additional anodic peaks shifted to potential values of −0.16 and +0.02 V and the cathodic peak at −0.35 V shifted negatively. Another broad cathodic wave around 0.0 V was observed. The current of both redox waves increased because the redox processes of Mn$^{II/III}$ and Mn$^{III/IV}$ increased with continued potential cycling. Cyclic voltammetry suggests the formation of a redox-active polymer film on the electrode surface that reacts catalytically with peroxynitrite as described in eqn (4.1), where peroxynitrite is oxidized to nitric dioxygen and nitrite.[28]

$$Poly(TAPc)Mn^{(III)}(H_2O) + O=N-O-O^- \rightarrow [poly(TAPc)Mn^{(III)}-OONO]^- + H_2O$$

$$2[poly(TAPc)Mn^{(III)}-OONO]^- \rightarrow 2poly(TAPc)Mn^{(IV)}=O + NO_2^- + NO_2$$

$$poly(TAPc)Mn^{(IV)}=O + 2e + 2H^+ \rightarrow poly(TAPc)Mn^{(III)}(H_2O) \tag{4.1}$$

Under optimal conditions, this sensor exhibited a detection limit for peroxynitrite of 18 nM with a sensitivity of 2.4×10^{-3} nA nM^{-1}. However, the origin of the signals attributed to peroxynitrite is questionable under the chosen conditions.[29] Bedioui and co-workers reconsidered this peroxynitrite sensor in alkaline solutions in order to ensure the performance of the poly-tetraaminophthalocyanine manganese(III) [poly-Mn(TAPc)] modified electrode and reported a 5 μM detection limit with a sensitivity of 14.6 nA mM^{-1}.[16]

4.2.2 Porphyrins

Porphyrins are structurally related to other macrocycles, both featuring four pyrrole-like subunits linked to from a 16-membered ring. Metalloporphyrins of iron and manganese have been shown to react directly with peroxynitrite in bimolecular reactions with the transition metal centers in one or two electron redox processes. One of the most widely used porphyrins is hemin (Figure 4.3A), an iron protoporphyrin shown to have peroxidase-like activity[30] with the Fe^{III}/Fe^{II} redox center being the electrocatalytic site for the reduction of peroxynitrite.[30,31] Carbon fiber electrodes (CFEs) modified with electro-polymerized hemin (Figure 4.3B) have been shown to mediate the oxidation of peroxynitrite (Figure 4.3C).[32] In a typical cyclic voltammetry response of the electro-polymerized hemin film added to peroxynitrite (0.45 mM, pH 10.5), the hemin-modified electrode shows an oxidation peak at +1.07 V *vs.* Ag/AgCl. This peak is neither observed for the bare glassy carbon electrode (GCE) nor for the protoporphyrin modified GCE, confirming the role of the iron redox center (Figure 4.3D). In fact, in the same conditions, the direct oxidation of peroxynitrite on GCE is rather sluggish, with a less well defined peak, and occurs at potentials above +1350 mV *vs.* Ag/AgCl.[32] A series of controls examined the effect of nitrite or nitrate on the voltammetric waves. Thus, the response of hemin-modified electrodes to 1 mM nitrite and 1 mM nitrate was compared with the typical response observed with 400 μM peroxynitrite (Figure 4.3E). The peak current recorded with the 400 μM peroxynitrite is over seven times bigger than that obtained with 1 mM nitrite. Furthermore, the response to 1 mM nitrate is relatively silent. These results suggest selectivity against nitrite or nitrate.[32]

4.2.3 Conducting Polymers

The electrochemical polymerization of thiophene and its derivatives is well documented and widely used for the construction of biosensors. The advantage of using conducting polymers is that very thin films (nanometers thick) can be prepared reproducibly, ensuring a rapid and stable response of the electrochemical sensor to the analyte. Organic functional groups such as amine or carboxylic acid can be used for further anchoring of analyte-specific ligands. This approach was used by Koh *et al.* who modified platinum microelectrodes with a conducting polymer formed electrochemically from [(2,5-di(2-thienyl)-1*H*-pyrrole)-1-(*p*-benzoic acid)] (Figure 4.4A).[33] The polymer backbone provides binding sites for Mn^{2+} ions, which interact with

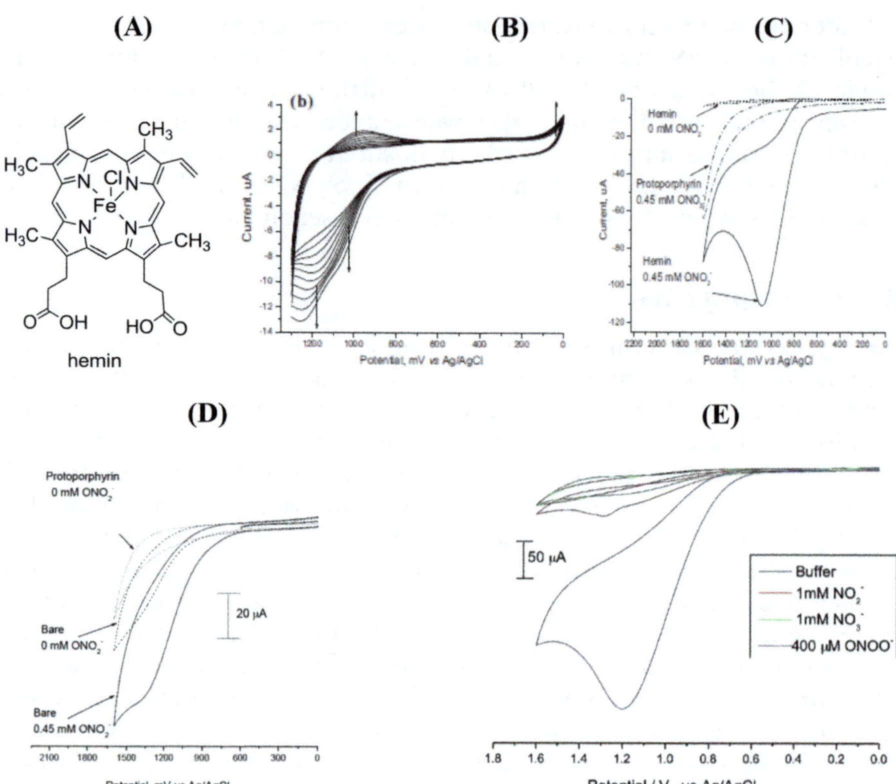

Figure 4.3   (A) Chemical structure of hemin. (B) Cyclic voltammograms of the electropolymerization of hemin in dichloromethane (0.1 M TBABF$_4$). (C) Response of hemin and protoporphyrin (no iron center) modified electrodes to the addition of peroxynitrite. (D) Cyclic voltammograms of a bare GCE and a protoporphyrin modified GCE towards peroxynitrite. (E) Cyclic voltammetry of hemin/GCE in pH 10.5 3-(cyclohexylamino)-1-propanesulfonic acid (CAPS) buffer in the presence of 400 μM peroxynitrite, 1 nM nitrite and 1 mM nitrate; scan rate = 150 mV s^{-1}. Reprinted from ref. 58: *Anal. Chim. Acta*, **780**, 81, S. F. Peteu, T. Bose and M. Bayachou, Polymerized hemin as an electrocatalytic platform for peroxynitrite's oxidation and detection, copyright 2013 with permission from Elsevier.

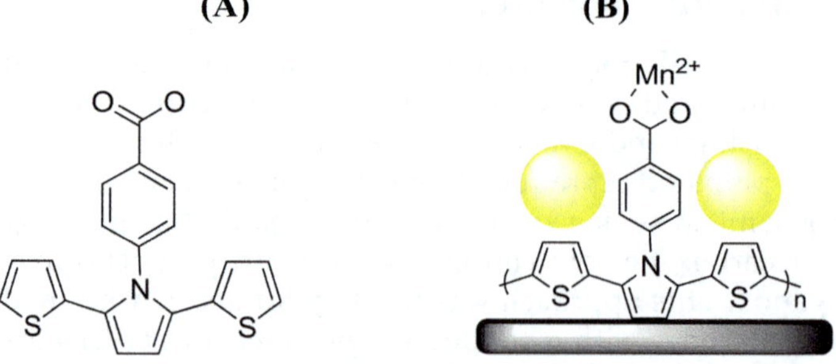

Figure 4.4   (A) Chemical structure of (2,5-di(2-thienyl)-1*H*-pyrrole)-1-(*p*-benzoic acid). (B) Electrode architecture for peroxynitrite detection.

peroxynitrite by being oxidized to Mn^{3+} (Figure 4.4B). The amperometric calibration showed a dynamic range from 2×10^{-8} to 3×10^{-5} M with a detection limit of 1.9 nM.

The other conducting polymer matrix investigated was polyethylenedioxythiophene (PEDOT). PEDOT has received a significant amount of attention in recent years as a redox active, low potential and highly stable conducting polymer. In a follow-up work by Peteu *et al.* on the use of hemin-modified electrodes, a PEDOT/hemin matrix was used for the electrocatalytic oxidation of peroxynitrite (Figure 4.5A). Figure 4.5B displays the scanning electron microscopy (SEM) image of CFEs after modification with PEDOT/hemin, revealing a 3D branched multi-globular surface with features 100–300 nm in size.[34] The energy dispersive spectrum of the same PEDOT/hemin matrix showed a sulfur peak (S = 16.04 wt%), characteristic of PEDOT, and an iron peak (Fe = 2.30 wt%), present in hemin, but

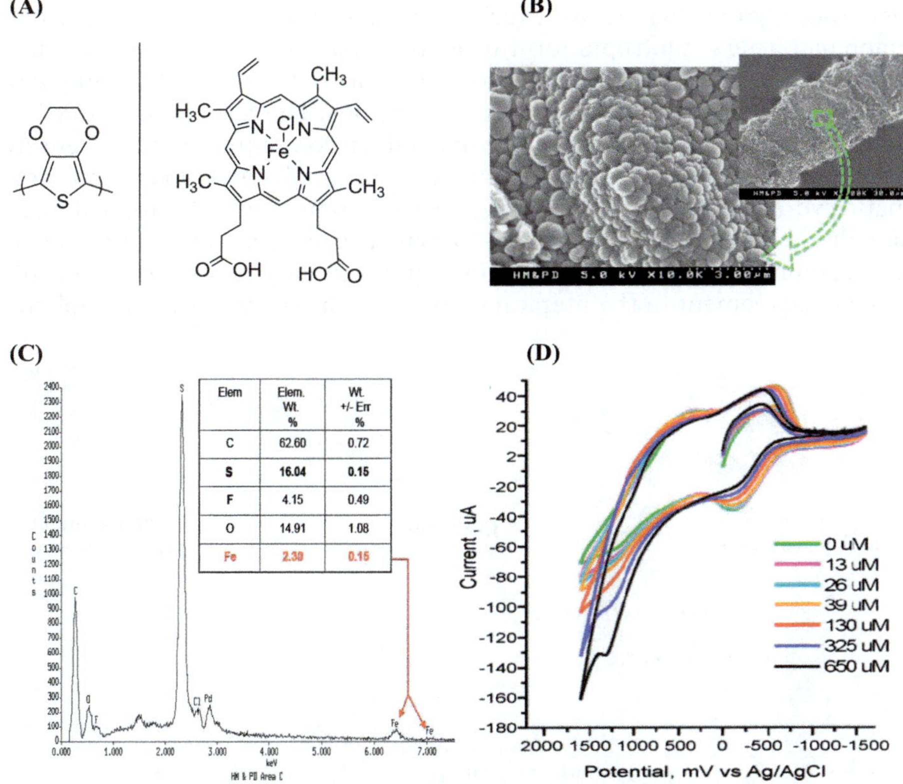

Figure 4.5 (A) Structures of PEDOT and hemin. (B) SEM image of a CFE electrochemically modified with PEDOT/hemin. (C) Energy dispersive spectrum of PEDOT/hemin. (D) Cyclic voltammograms of the electrodes upon addition of peroxynitrite at pH 10.5. Reprinted from ref. 34: *Biosens. Bioelectron.*, **25**, 1914, S. Peteu, P. Peiris, E. Gebremichael and M. Bayachou, Nanostructured poly(3,4-ethylenedioxythiophene)-metalloporphyrin films: Improved catalytic detection of peroxynitrite, copyright 2010 with permission from Elsevier.

absent in protoporphyrin. The electrochemical response of this interface to the addition of peroxynitrite resulted in an increase of the redox current at 0 V, assigned to PEDOT, and 1.25 V characteristic of hemin with a sensitivity of 13 nA μM^{-1} and a detection limit of 200 nM at the more anodic potential (Figure 4.5D).[34] These results supported the "synergy" hypothesis between the conductive PEDOT matrix and hemin electrocatalyst. The high sensitivity is believed to be a result of the morphology of the interface, which is highly nanostructured and porous.

4.3   Graphene-Based Matrices

Carbon materials have been widely used in both analytical and industrial electrochemistry, out-performing traditional noble metal electrodes in many areas. This stems largely from their carbon structural polymorphism, chemical stability, low cost, wide potential window, rich surface chemistry and electro-catalytic activities for a variety of redox reactions.[35] More recently, graphene, a new allotropic form of carbon, has started to dominate and replace other carbon based nanomaterials as its potential for designing low cost, sensitive and selective sensors is enormous. The term graphene refers to a planar monolayer sheet of sp^2 bonded carbon atoms that are densely packed into a 2D honeycomb lattice. It is essentially a very large polyaromatic hydrocarbon and the building block for all fullerene allotropic dimensionalities. Thus, in addition to its 2D planar form, graphene can be rolled into carbon nanotubes or stacked into graphite (Figure 4.6). There is still some disagreement in the literature regarding the point at which graphene

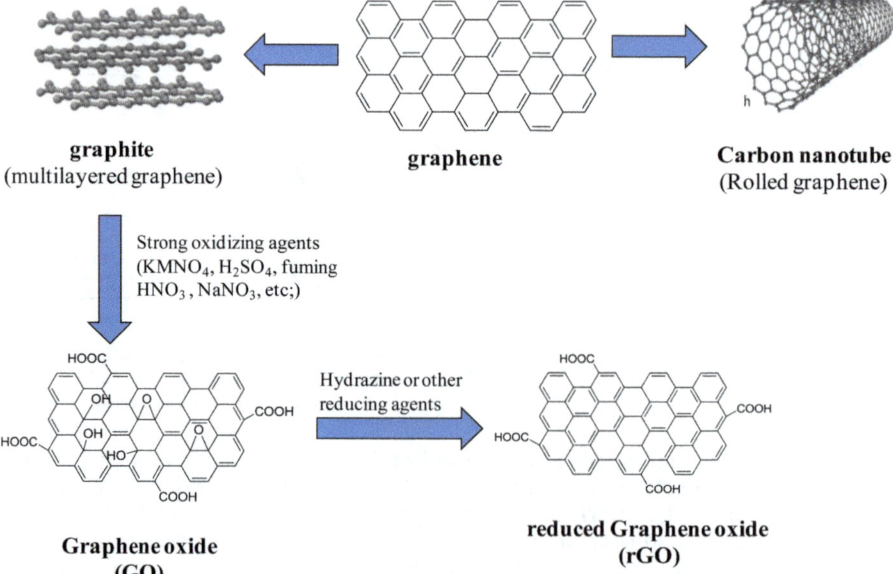

Figure 4.6   Structure of graphene and its derivatives.

becomes graphite. Geim and Novoselov have reported that structures composed of up to around ten graphitic sheets have electronic properties sufficiently different from bulk graphite to be classified as graphene.[36] Pumera recommended that structures consisting of up to 100 layers should be thought of as graphene.[37]

This 2D structure offers a unique environment for electron transport with fast heterogeneous electron transfer at the edges.[38] The electrochemical properties are similar to those of carbon nanotubes, with one huge advantage of not containing any other heterogeneous materials (metal contamination). Carbon nanotubes are typically grown from carbon-containing gases with the use of catalytic nanoparticles, which remain in the nanostructures even after extensive purification procedures.[39] Such residual metallic impurities can dominate the electrochemistry of carbon nanotubes and thus represent a problem for the construction of reliable sensors and energy devices.[40,41] In addition, these metallic impurities possess toxicological hazards within biological samples even for concentrations as low as 50 ppm.[41]

For the integration of graphene into electrochemical devices it is essential to have a simple, reproducible and controllable technique to produce large-area graphene sheets on a large scale. The use of chemically derived rGO rather than graphene appears to be a promising alternative (Figure 4.6). It is based on the chemical oxidation of graphite to graphene oxide (GO) *via* one of the three principle methods developed by Brodie,[42] Staudenmeier,[43] and Hummers.[44] The Hummers method is probably the most widely used and involves graphite soaking in a solution of sulfuric acid and potassium permanganate (Figure 4.6).[44] Stirring or sonication of the resulting graphite oxide is then performed to obtain single layers of GO. GO consists of oxidized graphene sheets with the basal planes decorated mostly with epoxide and hydroxyl groups, in addition to carbonyl and carboxyl groups located preferentially at the edges.[45,46] These oxygen functions render GO layers hydrophilic and water molecules can readily intercalate into the interlayers. As GO is insulating, it is necessary to restore its conductivity for electrochemical applications by reducing it to rGO using different reducing agents.[47]

The interest in using rGO for the detection of peroxynitrite is linked to the fact that organic molecules such as hemin and other porphyrin species interact strongly with rGO through π–π stacking. Their electron rich character also makes them ideal candidates for the reduction of GO to rGO. Indeed, electron donating molecules such as dopamine,[48] tetrathiafulvalene,[49,50] 4-aminophenylbenzoic acid,[51] or tyrosine[52] have shown their ability to work as reducing agents for GO, while being incorporated *via* π–π stacking interactions into the rGO matrix. We have recently used this approach for the construction of a peroxynitrite sensitive matrix (Figure 4.7A). A rGO matrix with incorporated hemin can be easily obtained by mixing an aqueous solution of GO with hemin and sonicating the suspension for 5 h at room temperature. The incorporation of hemin in the rGO matrix is evidenced by the presence of the Fe2p band (2.3%) at 710.16 eV (Fe2p$_{3/2}$) and 723.16 eV (Fe2p$_{3/2}$) and of the N1s band (8.3%) at 398 eV (Figure 4.7B). The rGO/hemin modified electrode shows an irreversible

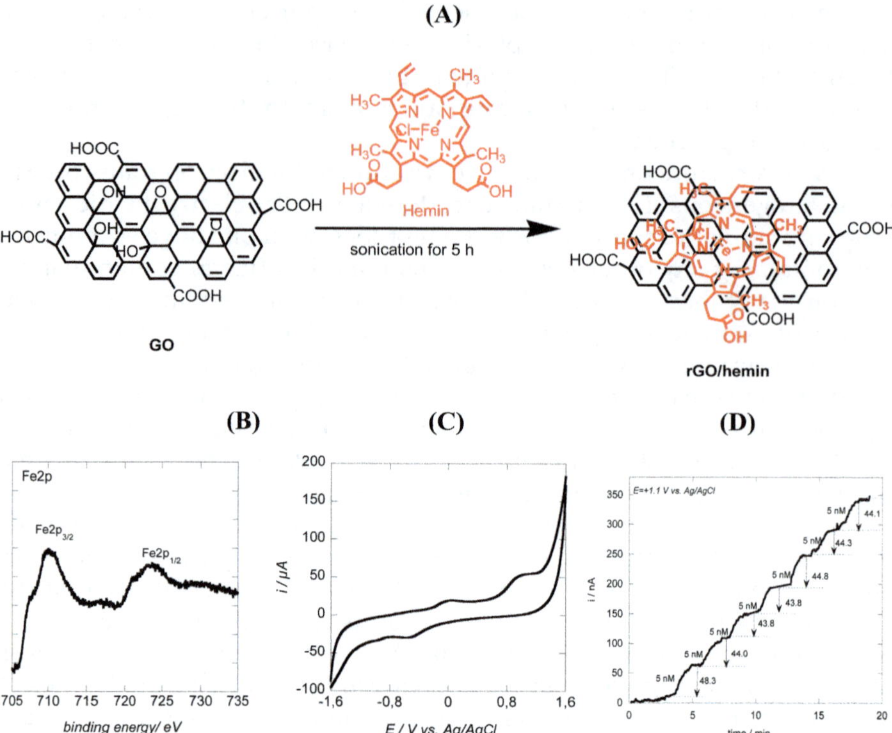

Figure 4.7   (A) Preparation of rGO/hemin matrix. (B) High resolution Fe2p XPS spectrum of rGO/hemin composite. (C) Cyclic voltammograms of rGO/hemin modified GCE under N_2-saturated phosphate-buffered saline buffer (0.1 M, pH 7.4); scan rate: 100 mV s^{-1}. (D) Amperometric response curve obtained at +1.1 V *vs.* Ag/AgCl polarized interface upon subsequent addition of peroxynitrite. Reproduced from ref. 54 with permission from The Royal Society of Chemistry.

wave at about E = +0.95 V *vs.* Ag/AgCl assigned to ring oxidation of the pyrrole macrocycle in hemin to its radical cation and a reversible redox couple at −0.3 V of the Fe^{3+}/Fe^{2+} hemin redox center. Application of a potential of 1.17 V *vs.* Ag/AgCl results in oxidation of the Fe^{3+} hemin center to the high valent iron form (*e.g.* iron oxo intermediate [$Fe^{4+}=O$]), which in the presence of ONOO$^-$ is reduced back to Fe^{3+} for further turnover. Indeed, subsequent additions of peroxynitrite at this potential result in oxidation current increases (Figure 4.7C). This interface exhibited an electrocatalytic reaction towards peroxynitrite at a potential of +1.1 V and from chronoamperometric response curves one can see that the current response scales linearly with the concentration of peroxynitrite. A detection limit of 5 ± 0.5 nM and sensitivity of 7.5 nA nM^{-1} were recorded on such an interface.[54]

A similar strategy was recently investigated using cobalt phthalocyanine (CoPc) complexes to reduce GO to rGO and as electrocatalysts for peroxynitrite

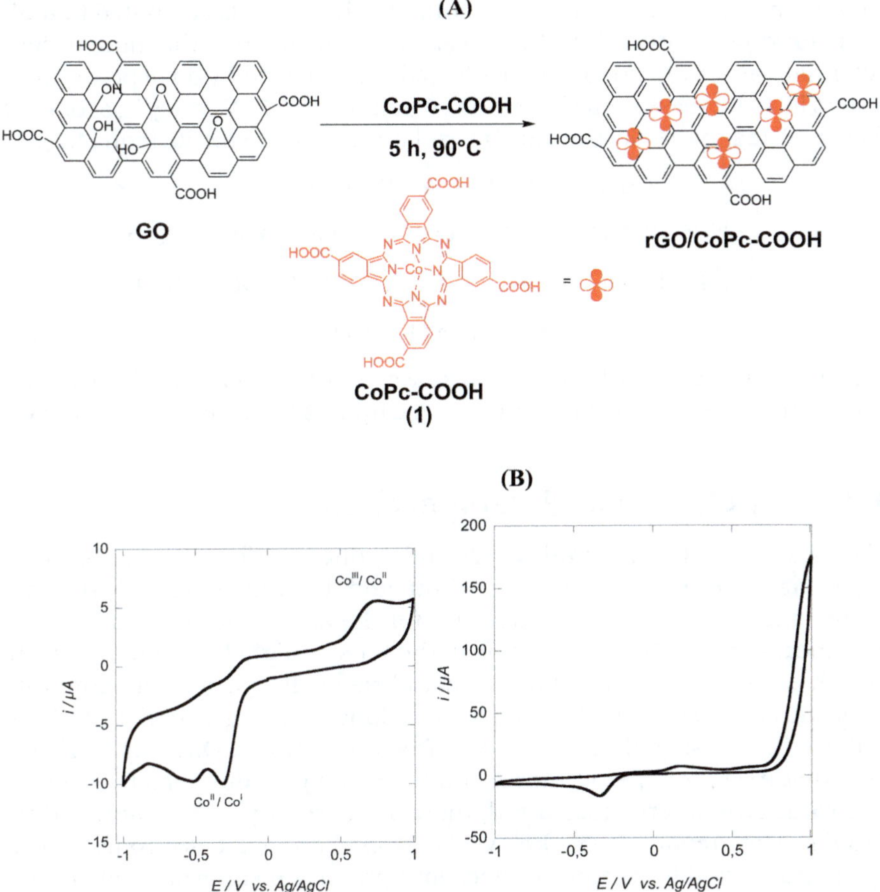

Figure 4.8   (A) Synthesis of rGO/CoPc–COOH. (B) Cyclic voltammograms recorded on GCE modified by drop-casting rGO/CoPc–COOH in 3-(cyclohexyl-amino)-1-propanesulfonic acid (CAPS) buffer (pH 10) in the absence (left) and presence (right) of 15 nM peroxynitrite; scan rate: 100 mV s^{-1}. Reproduced from ref. 58 with permission from the Royal Society of Chemistry.

(Figure 4.8). One major limitation of electrodes modified by physically adsorbed CoPc is that the complex detaches from the electrode surface with time and the complex has relatively low conductivity. Host matrices such as graphene are thus required to form stable electrochemically active electrodes.[55–57] As unmodified phthalocyanine derivatives exhibit a high aggregation tendency that makes them insoluble in almost all organic solvents and in aqueous solutions, we used a CoPc–tetracarboxylic acid (COOH) complex (Figure 4.8), which is soluble in dimethylformamide (DMF).

The electrochemical behavior of rGO/CoPc–COOH modified GCE in DMF shows an irreversible redox band at 0.72 V *vs.* Ag/AgCl attributed to the Co^{III}/ Co^{II} redox couple, and two redox bands at −0.31 V and −0.53 V *vs.* Ag/AgCl are seen on the cathodic scan (Figure 4.8B). The first band is attributed to multi-layer stacking of CoPc–COOH complex, while the band at the more negative potential was attributed to the Co^{II}/Co^{I} redox couple. Addition of peroxynitrite results in an increase in the oxidation current from 0.72 V *vs.* Ag/AgCl onwards that is related to the electrocatalytic oxidation of peroxynitrite (eqn (4.2)):

$$Co^{II}Pc\text{–}COOH \rightarrow Co^{III}Pc\text{–}COOH + e^-$$

$$Co^{III}Pc\text{–}COOH + O{=}N\text{–}O\text{–}O^- \rightarrow Co^{II}Pc\text{–}COOH\text{–}OONO^\circ$$

$$Co^{II}Pc\text{–}COOH\text{–}OONO^\circ + e^- \rightarrow Co^{III}Pc\text{–}COOH + ONO_2^\circ$$

$$2ONO_2^\circ \rightarrow O_2 + NO_2 \tag{4.2}$$

The sensitivity towards peroxynitrite was ~11.5 ± 1.0 nA nM^{-1} with a limit of detection of ~1.7 nM and a linear range up to 20 nM with good selectivity.

4.4 Conclusion and Perspectives

Chemically modified electrodes have undergone significant progress, mainly regarding lowering the detection limit and increasing selectivity. These results open an interesting perspective for the use of such electrodes in real biological samples. Miniaturization of the sensor might be an important step for achieving this goal. It is known that ultra-small sensors can afford rapid responses and better analyte sensitivity along with high spatial resolution and use in non-stirred solutions or soft solid matrices. Ongoing work aims to miniaturize these peroxynitrite and hydrogen peroxide sensitive/selective electrocatalytic interfaces. Indeed, there are currently no miniaturized rGO based sensors reported. The knowledge gained on macroelectrodes can easily be transferred to microelectrodes and (micro)sensor arrays able to measure nitric oxide, superoxide, and peroxynitrite simultaneously.

While significant progress has been made in the development of peroxynitrite-sensitive methods, the quantitative determination of peroxynitrite continues to be a challenging task. One of the inherent difficulties that has not been entirely overcome concerns the accurate production of the true *in vivo* kinetics of peroxynitrite in model experiments. The use of rGO matrices, graphene modified with catalytic molecules towards peroxynitrite, might be one way to go in the future for *in vivo* applications. Questions such as stability of the peroxynitrite-sensitive matrix and cytotoxicity of such matrices are still not answered and certainly need further investigations.

The development of methods for real-time *in vivo* monitoring of peroxynitrite in very complex media or for clinical tests remains a significant challenge and could benefit largely from a multidisciplinary and meticulous approach through the application of recent advances in nanotechnology, physics, materials science, chemical engineering, and biology.

Meanwhile, the quantification of peroxynitrite in *in vivo* relevant conditions still provides ample room for work and advances. Beside such technological advances, there is still scope for the development of selective and sensitive peroxynitrite matrices that would allow a simple method to make peroxynitrite sensors. In addition, the efforts to develop individually addressable microelectrode arrays are expected to be further refined. These methods could not only prove useful for the detection of early disease states, but could also allow for a screening-type analysis of potential signal transduction pathways in cells.

Acknowledgements

R. B. and S. S. gratefully acknowledge financial support from the Centre National de Recherche Scientifique (CNRS), the University Lille 1 and Nord Pas de Calais region. S. S thanks the Institut Universitaire de France (IUF) for financial support. S. F. P. acknowledges Agentia Nationala de Cercetare Stiintifica (MEN-ANCS-UEFISCDI) for the PN-II project 184/2011. All authors acknowledge support *via* the BHC Bilateral Partenariat project 28684VM.

References

1. R. Radi, G. Peluffo, M. N. Alvarez, M. Naviliat and A. Cayota, *Free Radical Biol. Med.*, 2001, **30**, 463–488.
2. M. M. Tarpey and I. Fridovich, *Circ. Res.*, 2001, **89**, 224.
3. M. R. Radabaugh, O. V. Nemirovsky, T. P. Misko, P. Aggarwal and W. R. Mathews, *Anal. Biochem.*, 2008, **380**, 68.
4. J. P. Crow, *Nitric Oxide*, 1997, **1**, 145.
5. N. A. Sieracki, B. N. Gantner, M. Mao, J. H. Horner, R. D. Ye, A. B. Malik, M. E. Newcomb and M. G. Bonini, *Free Radical Biol. Med.*, 2013, **61**, 40.
6. J. Glebska and W. H. Koppenol, *Free Radical Biol. Med.*, 2003, **35**, 676.
7. D. Yang, H. L. Wang, Z. N. Sun, N. W. Chung and J. G. Shen, *J. Am. Chem. Soc.*, 2006, **128**, 6004.
8. C. Amatore, S. Arbault, C. Bouton, K. Coffi, J. C. Drapier, H. Ghandour and Y. Tong, *ChemBioChem*, 2006, **7**, 653–661.
9. C. Amatore, S. Arbault, D. Bruce, P. De Oliveira, M. Erard and M. Vuillaume, *Faraday Discuss.*, 2000, **116**, 319.
10. C. Amatore, S. Arbault, D. Bruce, P. De Oliveira, M. Erard and M. Vuillaume, *Chem.–Eur. J.*, 2001, **7**, 4171.
11. C. Amatore, S. Arbault, Y. Chen, C. Crozatier and I. Tapsoba, *Lab Chip*, 2007, **2**, 233.
12. C. Amatore, S. Arbault, M. Guille and F. Lemaitre, *Chem. Rev.*, 2008, **108**, 2585.
13. J. Xue, X. Ying, J. Chen, Y. Xian, L. Jin and J. Jin, *Anal. Chem.*, 2000, **72**, 5313.
14. R. Kubant, C. Malinski, A. Burewicz and T. Malinski, *Electroanalysis*, 2006, **18**, 410.
15. J. S. Cortes, S. G. Granados, A. A. Ordaz, J. A. L. Jimenez, S. Griveau and F. Bedioui, *Electroanalysis*, 2007, **19**, 61.

16. J. Sandoval Cortes, S. Gutierrez Grabados, A. Alatorre, J. A. Lopez Jiminez, S. Griveau and F. Bedioui, *Electroanalysis*, 2007, **19**, 61.

17. A. Sivanesan and J. S. Abraham, *Electrochim. Acta*, 2008, **53**, 6629.

18. P. Mashazi, C. Togo, J. Limson and T. Nyokong, *J. Porphyrins Phthalocyanines*, 2010, **14**, 252.

19. N. Nombona, P. Tau, N. Sehlotho and T. Nyokong, *Electrochim. Acta*, 2008, **53**, 3139.

20. B. O. Agboola, K. I. Ozoemena and T. Nyokong, *Electrochim. Acta*, 2006, **51**, 6470.

21. C. A. Caro, F. Bedioui and J. H. Zagal, *Electrochim. Acta*, 2002, **47**, 1489.

22. M. Shibata and N. Furuya, *Electrochim. Acta*, 2003, **48**, 3953.

23. C. M. Yap, G. Q. Xu and S. G. Ang, *Anal. Chem.*, 2013, **85**, 107.

24. B. O. Agboola, K. I. Ozoemena and T. Nyokong, *Electrochim. Acta*, 2006, **51**, 6470.

25. J. S. Cortes, S. G. Granados, A. A. Ordaz, J. A. L. Jimenez, S. Griveau and F. Bedioui, *Electroanalysis*, 2007, **19**, 61.

26. T. Malinski and Z. Taha, *Nature*, 1992, **358**, 676.

27. Y. H. Tse, P. Janda, H. Lain and A. Lever, *Anal. Chem.*, 1995, **67**, 981.

28. S. S. Marla, J. Lee and J. T. Groves, *Proc. Natl. Acad. Sci. U. S. A.*, 1997, **94**, 14243.

29. F. Bedioui, D. Quninton, S. Griveau and T. Nyokong, *Phys. Chem. Chem. Phys.*, 2010, **12**, 9976.

30. Y. Guo, L. Deng, J. Li, S. Guo, E. Wang and S. Dong, *ACS Nano*, 2011, **5**, 1282.

31. G. P. Zhang and K. Dasgupta, *Anal. Chem.*, 1992, **64**, 517.

32. S. F. Peteu, T. Bose and M. Bayachou, *Anal. Chim. Acta*, 2013, **780**, 81.

33. W. C. A. Koh, J. I. Son, E. S. Choe and Y. B. Shim, *Anal. Chem.*, 2010, **82**, 10075.

34. S. Peteu, P. Peiris, E. Gebremichael and M. Bayachou, *Biosens. Bioelectron.*, 2010, **25**, 1914–1921.

35. R. L. McCrerry, *Chem. Rev.*, 2008, **108**, 2646.

36. A. K. Geim and K. S. Novoselov, *Nat. Mater.*, 2007, **6**, 183.

37. M. Pumera, *Chem. Soc. Rev.*, 2010, **39**, 4146.

38. F. Chen and N. Tao, *Acc. Chem. Res.*, 2009, **42**, 429.

39. M. Pumera, *Langmuir*, 2007, **23**, 6453.

40. C. E. Banks, A. Crossley, C. Salter, S. J. Wilkins and R. G. Compton, *Angew. Chem., Int. Ed.*, 2006, **45**, 2533.

41. M. Pumera and Y. Miyahara, *Nanoscale*, 2009, **1**, 260.

42. B. C. Brodie, *Ann. Chim. Phys.*, 1860, **59**, 466–472.

43. L. Staudenmaier, *Ber. Dtsch. Chem. Ges.*, 1898, **31**, 1481.

44. W. S. Hummers and J. R. E. Offerman, *J. Am. Chem. Soc.*, 1958, **80**, 1339.

45. H. He, J. Klinowski, M. Forster and A. Lerf, *Chem. Phys. Lett.*, 1998, **287**, 53.

46. A. Lerf, H. He, M. Forster and J. Klinowski, *J. Phys. Chem. B*, 1998, **102**, 4477.

47. O. C. Compton, B. Jain, D. A. Dikin, A. Abouimrane, K. Amine and S. T. Nguyen, *ACS Nano*, 2011, **5**, 4380.
48. I. Kaminska, A. Barras, Y. Coffinier, W. Lisowski, J. Niedziolka-Jonsson, P. Woisel, J. Lyskawa, M. Opallo, A. Siriwardena, R. Boukherroub and S. Szunerits, *ACS Appl. Mater. Interfaces*, 2012, **4**, 5386.
49. I. Kaminska, M. R. Das, Y. Coffinier, J. Niedziolka-Jonsson, J. Sobczak, P. Woisel, J. Lyskawa, M. Opallo, R. Boukherroub and S. Szunerits, *ACS Appl. Mater. Interfaces*, 2012, **4**, 1016.
50. I. Kaminska, M. R. Das, Y. Coffinier, J. Niedziolka-Jonsson, P. Woisel, M. Opallo, S. Szunerits and R. Boukherroub, *Chem. Commun.*, 2012, **48**, 1221.
51. Q. Wang, I. Kaminsak, J. Niedziolka-Jonsson, M. Opallo, L. Musen, R. Boukherroub and S. Szunerits, *Biosens. Bioelectron.*, 2013, **50**, 331.
52. Q. Wang, M. Li, S. Szunerits and R. Boukherroub, *Electroanalysis*, 2014, **26**, 156.
53. Z.-X. Liang, H.-Y. Song and S.-J. Lia, *J. Phys. Chem. C*, 2011, **115**, 2604.
54. R. Oprea, S. F. Peteu, P. Subramanian, W. Qi, E. Pichonat, H. Happy, M. Bayachou, R. Boukherroub and S. Szunerits, *Analyst*, 2013.
55. L. Cui, L. Chen, M. Xu, H. Su and S. Ai, *Anal. Chim. Acta*, 2012, **712**, 64.
56. L. Cui, T. Pu and X. He, *Electrochim. Acta*, 2013, **88**, 559.
57. H. Hosseini, M. Mahyari, A. Bagheri and A. Shaabani, *Biosens. Bioelectron.*, 2014, **52**, 136.
58. I. S. Hosu, Q. A. Wang, A. Vasilescu, S. F. Peteu, V. Raditoiu, S. Railian, V. Zaitsev, K. Turcheniuk, Q. Wang, M. Li, R. Boukherroub and S. Szunerits, *RSC Adv.*, 2015, **5**, 1474.

Electrochemical Detection of Peroxynitrite in Biological Solutions: Challenges and Perspectives

SOPHIE GRIVEAU[a,b,c,d] AND FETHI BEDIOUI*[a,b,c,d]

[a]PSL Research University, Chimie ParisTech, Unité de Technologies Chimiques et Biologiques pour la Santé, Paris, France; [b]CNRS, Unité de Technologies Chimiques et Biologiques pour la Santé UMR 8258, Paris, France; [c]Université Paris Descartes, Unité de Technologies Chimiques et Biologiques pour la Santé, Paris, France; [d]INSERM, Unité de Technologies Chimiques et Biologiques pour la Santé (N°1022), Paris, France
*E-mail: fethi.bedioui@chimie-paristech.fr

5.1 Challenges in the Detection of Peroxynitrite in Biological Conditions

In biological systems, peroxynitrite (PON) is the product of a very fast reaction between two radicals, namely nitric oxide (NO) and superoxide ($O_2^{\bullet-}$):[1]

$$NO + O_2^- = ONOO^- \ (k = 4 - 16 \times 10^9 \ M^{-1} \ s^{-1})$$

From a physico-chemical point of view, PON possesses both redox and acido basic properties: it is a strong biological oxidant and is the conjugated base of

RSC Detection Science Series No. 7
Peroxynitrite Detection in Biological Media: Challenges and Advances
Edited by Serban F. Peteu, Sabine Szunerits, and Mekki Bayachou

Published by the Royal Society of Chemistry, www.rsc.org

peroxynitrous acid [ONOOH; pK_a(ONOOH/ONOO$^-$) = 6.8]. Both ONOO$^-$ and ONOOH have high and complex chemical reactivity in biological systems[2] (Scheme 5.1). In biological systems, PON mainly reacts with carbon dioxide, thiols, metalloproteins and glutathione peroxidase, or undergoes hemolytic cleavage. PON is thus highly reactive and possesses a short half-life (<1 s)

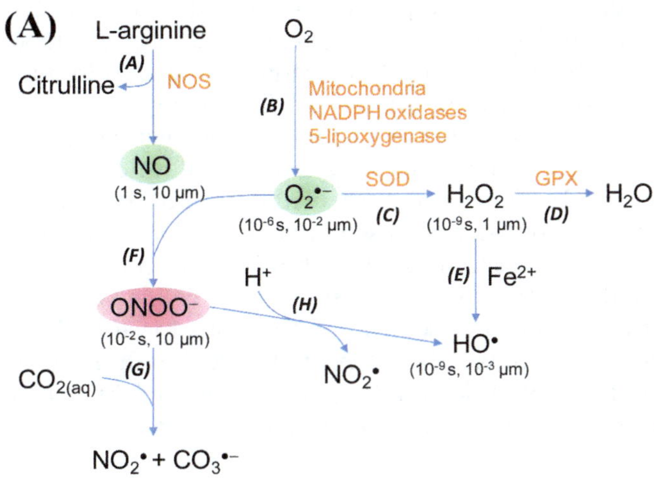

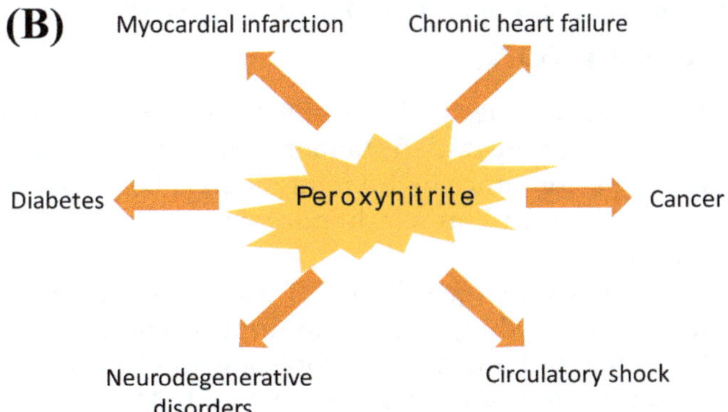

Scheme 5.1 (A) Scheme of the main ROS and RNS found in cells and their reaction pathways. (a) NO is formed upon oxidation of L-arginine by NO-synthases (NOS). (b) O$_2$$^{•-}$ is generated mainly in mitonchondria or by NADPH oxidases and 5-lipoxygenase, and (c) then transformed into H$_2$O$_2$ by superoxide dismutase (SOD). (d) H$_2$O$_2$ is reduced in water by glutathione peroxidase (GPX). (e) Reactive hydroxyl radicals are produced *via* a reaction between H$_2$O$_2$ and Fe^{2+} following a Fenton-type reaction. (f) Through the reaction between NO with O$_2$$^{•-}$, ONOO$^-$ is generated. (g) PON reacts quickly with CO$_2$ leading to nitrite and carbonate radicals while (h) ONOOH decomposes to nitrite and hydroxyl radicals. The estimated half-lives and mean travelled distances before decay by chemical reaction are provided in parentheses. Adapted from Erdmann *et al.*[4] (B) Main pathophysiological effects of PON. Adapted from Pacher *et al.*[5]

in vivo at physiological pH. Despite its short half-life, PON is able to diffuse outside the cell within a distance equivalent to one to two cell diameters and induce damage to surrounding cells.[3] The role, main reaction pathways and biological actions of PON have been reported and can be found in several reviews.[2,4,5] PON mainly acts as a nitrooxidative agent, and interacts (either by itself or *via* its degradation products such as $NO_2^{\bullet}$) with DNA, lipids and proteins, leading to several injuries summarized in Scheme 5.1B.[1,2,4,5] Thus there is a need for its accurate and specific detection *in vitro* and *in vivo* in order to have a better insight into its variations in level in human diseases and to be able to propose appropriate medical therapies.

The ideal method to detect and quantify PON *in vivo* should thus be sensitive, selective, real time, minimally invasive to avoid perturbation of the studied biological samples and direct (without the need for labeling or chemical probing). One of the most used approaches for PON detection in cells consists of using fluorescent or chemiluminescent probes, which may lack specificity and/or disturb the redox state of the cell.[3,6] We describe and discuss here the significant research contributions towards the development of electrochemical miniaturized sensors for PON with particular emphasis on the importance of understanding the reactivity of this molecule and the potential interference when developing such sensors.

5.2   Challenges in the Preparation and Storage of PON

Several protocols exist for the synthesis of PON solutions or PON-based salt that can be used for the preparation standard solutions of PON for the evaluation of a sensor's performance in different media. The preparation of these standard solutions is of the utmost importance and necessitates detailed and meticulous protocols.

A simple way to prepare large volumes of stock solutions of PON consists of reacting nitrite and hydrogen peroxide in acid conditions followed by quenching.[7–9] The method proposed by Robinson and Beckman led to highly concentrated alkaline PON solutions (180–190 mM). However, the solutions contain unreacted hydrogen peroxide and nitrite, and also nitrate. Hydrogen peroxide is eliminated from the solutions by the addition of a manganese dioxide (MnO_2) catalyst. The levels of nitrite and nitrate should be carefully quantified before use of the PON solution, since nitrite can interfere with some of the electrochemical detection techniques (as detailed below). Thus accurate analytical controls should be performed to ensure the purity of the stock PON solution being used. Another method for the preparation of stock solutions of concentrated PON (~30–80 mM) is the ozonation of slightly alkaline azide ion solutions.[10] This allows the PON solutions being prepared to be free from hydrogen peroxide, nitrite or other potential interfering species. However, it necessitates the use of ozone, which is known to be a strong irritant for the lungs and eyes. Whatever the method of preparation, PON solutions have to be kept at low temperatures (<−20 °C) and the initial concentration of PON (as well as contaminants) should be controlled spectrophotometrically at 302 nm (ε = 1700 L cm^{-1} mol^{-1}) at the beginning of each experiment.[11,12]

Addition of PON aliquots to mimic a "pulse" or "bolus" of PON to biological systems can be achieved by using PON-releasing molecules (or PON donors) such as 3-morpholinosydnonimine-*N*-ethylcarbamide (SIN-1). SIN-1 spontaneously decomposes in aqueous solution at neutral pH by releasing NO and O_2^-, thus generating PON with a rate of decomposition dependent on time. The maximal flux of PON generation is obtained after 20–30 min and ranges from 1.2 to 3.6% of added SIN-1 per minute (for a SIN-1 concentration of 0.1 mM).[13,14] The actual concentration of PON at a given time will be dependent upon the kinetics of its rate of formation by SIN-1 and its rate of decomposition in a given milieu. It is worth noting that it is not possible to obtain a steady concentration of PON using SIN-1 as a donor since the rate of formation changes with time, unless sequential SIN-1 injection pulses are used, as proposed by Martin-Romero and co-workers.[14] Finally, PON concentration in solutions is generally controlled by ultraviolet (UV)-visible spectrophotometric measurements at 302 nm. However, SIN-1 presents two bands of absorption, one of which has a maximum located at 296 nm. Therefore, the spectrophotometric signal measured at 302 nm is most likely not free from interference coming from remaining SIN-1 and from its products of degradation such as SIN-1C.[15]

Recently, photocontrollable PON donor molecules have been synthesized (P-NAP), allowing the generation of PON under UV-A irradiation (330–380 nm), leading to photodecomposition.[16] Due to the recent development of these compounds, no electrochemical sensor has yet been calibrated with solutions of such photocontrollable compounds.

5.3 Main Electrochemical Approaches for the Detection of PON in Biological Solutions

5.3.1 Electrochemical Strategies for the Detection of PON

The electrochemical detection of PON constitutes an attractive strategy for its direct, real-time, *in situ* and *in vivo* detection, generally coupled to the use of micro- and nanometric electrodes. From a general point of view, miniaturized electrodes can be positioned in the vicinity of cells to perform measurements at the unique cell level with optimal detection of released (or consumed) species thanks to the minimized distance between the receptor (the electrode) and the generator (the cell). Only a limited number of electrochemical methods have been described in the literature compared with wider reported techniques such as the use of fluorescence. As illustrated in Figure 5.1, these strategies are either based on direct oxidation or reduction or on electrocatalytic approaches (as described below). It should be noted that most of these detection methods have been detailed in several reviews, so only the basic principles of detection will be reported here.[15,17]

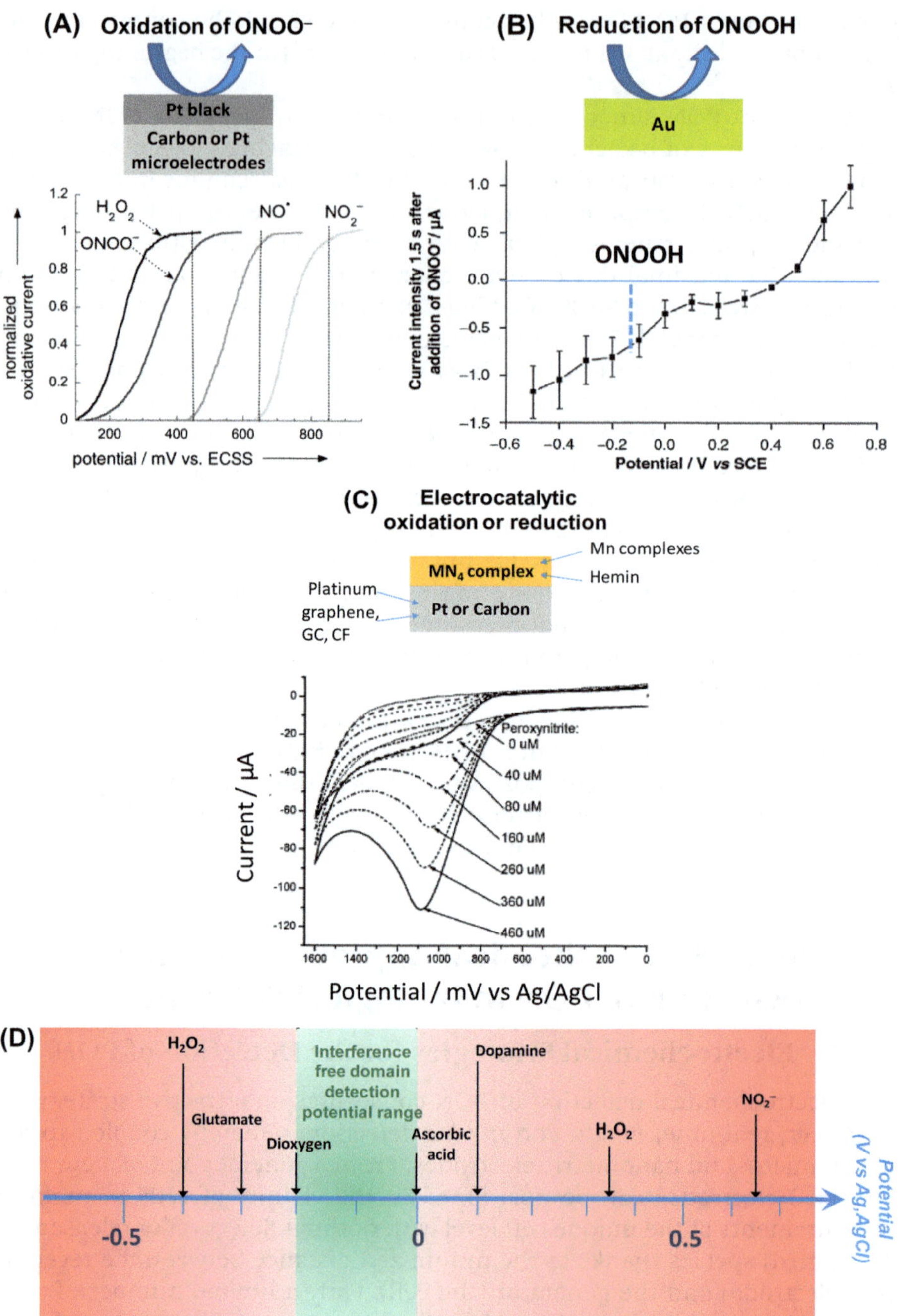

Figure 5.1 The main electrochemical strategies for PON sensing and voltammograms obtained at these electrodes. (A) Detection at a platinized (Pt) electrode. Reprinted with permission from C. Amatore, S. Arbault, C. Bouton, K. Coffi, J.-C. Drapier, H. Ghandour and Y. Tong, *ChemBioChem*,

5.3.1.1   Electrochemical Detection of PON at Metallic Electrodes

Amatore and co-workers have developed an electroanalytical methodology for the amperometric detection of PON released by biological cells by inducing direct electrochemical oxidation at carbon microfiber or micro-wire platinum electrodes covered by a thin platinum black coating.[18-21] In fact, the oxidation potential of the main reactive oxygen species (ROS; O_2^- and H_2O_2) and reactive nitrogen species (RNS; NO and $ONOO^-$) was shown to be in a potential range where the detection of these species suffers from interference. As illustrated in Figure 5.1A, the resolution between two oxidation waves is large enough (half-wave potential separation ~0.2 V, except for H_2O_2 and $ONOO^-$) that it is possible to make amperometric measurements at several potentials where either two (H_2O_2 and $ONOO^-$ at lower potential) or all species (at higher potential) are detected. The obtained signals can be deconvoluted using an algorithm to calculate each analyte flux. The originality of this approach is thus linked to the use of several working potentials to detect selectively ROS and RNS at miniaturized platinized electrodes. This electroanalytical method is very attractive due to its quick response time, the relative simplicity of electrode preparation, its high sensitivity and the quantitative input of several short-lived analytes simultaneously. It was applied to the study of ROS and RNS release by different biological systems, as explained below.

Amatore's group has also developed microfluidic chips with integrated microelectrodes to analyze simultaneously H_2O_2, $ONOO^-$, NO and NO_2^- using platinized microelectrodes (as shown on Figure 5.2).[22,23] The microelectrodes possess high sensitivity and a good detection limit for PON (40 nM at pH 8.4). This approach has recently been applied to the detection of ROS and RNS released by macrophages.[24]

Another approach using microchips was proposed by Hulvey *et al.*, by developing microfluidic-based electrophoresis for the separation of PON from other electroactive species of the sample, and PON amperometric detection at bare Pd electrodes (aligned in an end-channel configuration).[25] The strategy was applied to the analysis of artificial samples at pH 11 using either PON standards or SIN-1 as a PON donor molecule. The amperometric detection was conducted at several potentials and a comparison with current ratios was used to identify the nature of each peak. PON was detected without

---

7, 653–661. Copyright © 2006 WILEY-VCH Verlag GmbH & Co. KGaA, Weinheim. (B) Detection at a bare electrode. Reprinted from *Electrochemistry Communications*, vol. 12, D. Quinton, S. Griveau, F. Bedioui, Electrochemical approach to detect the presence of peroxynitrite in aerobic neutral solution, 1446–1449, Copyright (2010) with permission from Elsevier. (C) Detection at MN_4 modified electrodes. Reprinted from *Analytica Chimica Acta*, vol. 780, S. F. Peteu, T. Bose, M. Bayachou, Polymerized hemin as an electrocatalytic platform for peroxynitrite's oxidation and detection, 81–88, Copyright (2013) with permission from Elsevier. (D) The detection potential range available for electrochemical detection of PON at pH 7. The arrows represent the foot of the wave potential for the interfering species considered.

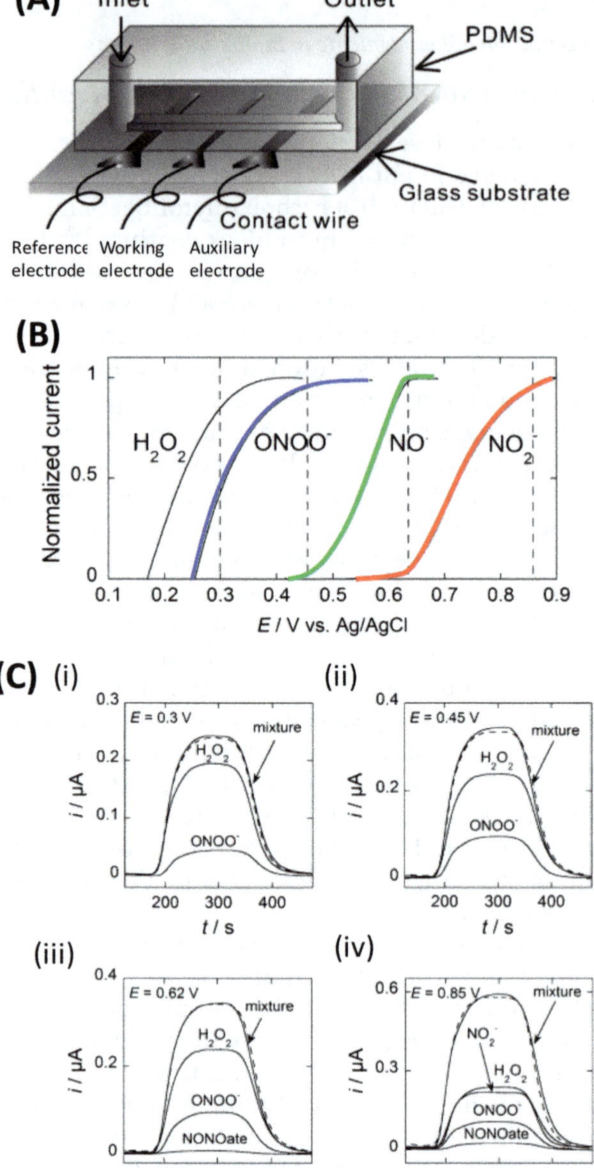

Figure 5.2 Microchips with integrated platinized working microelectrode for the simultaneous amperometric detection of ROS and RNS *in vitro*. (A) Scheme of the microchannel with an integrated sensing platinized electrode. PDMS: poly(dimethylsiloxane). (B) Position of the individual oxidation waves for main ROS and RNS at the platinized platinum microelectrode. The dashed line indicates the optimal potential for sensitivity and selectivity by amperometry. (C) Amperometric detection of a mixture of H_2O_2, $ONOO^-$, NO and NO_2^- (0.1 mM each) at several potentials at pH 8.4. Reprinted from *Electrochimica Acta*, vol. 144, Y. Li, C. Sella, F. Lemaître, M. Guille-Collignon, L. Thouin and C. Amatore, Electrochemical Detection of Nitric Oxide and Peroxynitrite Anion in Microchannels at Highly Sensitive Platinum-Black Coated Electrodes, Application to ROS and RNS Mixtures prior to Biological Investigations, 111–118, 2014 with permission from Elsevier.

significant degradation. The comparison of microchip performance with capillary electrophoresis showed shorter separation times (~25 s) and higher temporal resolution. At this stage, this approach was not applied to the study of biological samples. For real biological sample analysis, the presence of other "unknown" species would lead to optimizations for separation and the short PON half-life at physiological pH should be carefully considered.

Building on the studies by Kurz *et al.* related to the reduction of PON in acidic solutions,[26] Quinton *et al.* have recently explored the detection of PON in aerobic neutral solutions (pH = 7.1) by analyzing the direct electrochemical reduction of its conjugated acid at a rotating millimetric bare gold electrode at −0.1 V *vs.* saturated calomel electrode (SCE), as shown in Figure 5.1B.[27] Since the use of bare gold electrodes may give rise to interference from other electroactive molecules (NO_2^- H_2O_2, ascorbic acid, dopamine and glutamate), the authors optimized the operating potential for the detection of PON by performing a precise evaluation of amperometric behavior. The optimized working potential for amperometry was −0.1 V *vs.* SCE, leading to the selective amperometric detection of PON (Figure 5.1B). The sensitivity of the gold electrode towards the reduction of ONOOH was not accessible due to the evanescent nature of PON at pH 7.1.

Direct electrochemical detection of PON was also reported by Compton *et al.*, who evaluated its electrochemical reduction at mercury film electrodes in alkaline solutions (pH 9–13) by voltammetry.[28] The presence of PON was detected *via* a cathodic inverted peak observed at ~0.1 V *vs.* Ag/AgCl and the mechanism was studied in detail. The observed peak current varied linearly with PON level in the range 10 to 100 µM; no selectivity study was performed at this stage. This approach would have limited application for PON detection in real biological matrices due to the low response at a high scan rate, the potential interference with other biological species such as dioxygen and the toxicity of mercury.

5.3.1.2 Electrocatalytic Detection of PON at Chemically Modified Electrodes

Several studies report on the use of chemically modified electrodes with MN_4 complexes such as metalloporphyrins and metallophthalocyanines, or other complexes for the electrocatalytic detection of PON. Such macrocycles have already been adapted for the modification of the electrode surface for the electrochemical detection of other RNS such as NO. The first strategy was proposed by Xue *et al.* concerning the modification of a carbon fiber (CF) electrode surface with a thin electropolymerized film of manganese(ii) tetraamino phthalocyanine (poly-MnTAPc) and poly(4-vinylpyridine).[29] The reported data make the participation of PON itself in the electroanalytical measurements taken when using this modified electrode questionable. Indeed, the key data were obtained by amperometry, and show a plateau-like shape without any decay, which is expected due to the very short half-life of PON. These data suggest that PON remains stable at pH 7.4 for more than 400 seconds, which contradicts all the previously reported data about the stability of PON at physiological pH. Later, Kubant *et al.* also reported on the use of CF modified with a thin polymerized film of manganese(III)-[2,2]*para*cyclophenylporphyrin.[30] However, in both approaches, the conditions of calibration were not detailed enough. Therefore, the calibration of the poly(MnTAPc)-modified platinum microelectrode in alkaline solutions was re-investigated by Cortes *et al.*, using conditions where PON is stable enough to ensure the accuracy of the evaluation of the performance of the electrode.[31] The detection of PON was observed at −0.5 V *vs.* Ag/AgCl (in anaerobic alkaline solutions) and the performance evaluated. Such a detection potential avoids amperometric signals from oxidizing interfering compounds but not the reduction of dioxygen. Thus, this approach was not fully satisfactory for *in vitro* or *in vivo* detection.

Koh *et al.* reported an attempt to prepare an electrochemical ultramicrosensor for the detection of PON in physiological solution [phosphate-buffered saline (PBS) at pH 7.4] using a platinum microelectrode modified with an electropolymerized manganese-[2,5-di-(2-thienyl)-1*H*-pyrrole)-1-(*p*-benzoicacid)] (Mn-pDPB) complex, gold nanoparticles and polyethyleneimmine.[32] However, missing information, inconsistencies and contradictions cast doubt on the origin of the measured electrochemical signals, and thus the performance of the proposed method and the conclusion of the work. This work is introduced to alert the readers to some missing details.[33]

Peteu *et al.* have studied the modification of carbon-based electrodes [glassy carbon (GC) and CF] with iron protoporphyrin IX (hemin). The electrode was modified either by trapping the metalloporphyrin in a nanostructured electrodeposited coating of poly(3,4-ethylenedioxythiophene) (PEDOT) or by direct hemin electrodeposition.[34,35] The electrocatalytic oxidation of PON was observed at ~1.2 V *vs.* Ag/AgCl at pH 10.5 at the hemin–PEDOT-modified electrode.[34] The authors reported a synergetic effect between PEDOT and hemin in the catalysis[17] and they later re-investigated electrocatalysis at electrodes modified with hemin only, for which the catalytic oxidation occurred at ~1.0 V *vs.* Ag/AgCl at pH 10.5 (Figure 5.1C). The catalytic effect was attributed to the formation of an iron-oxo intermediate that acts as mediator of PON oxidation. Lower current intensities were obtained for PON oxidation at hemin compared with hemin–PEDOT electrodes for the same PON level,[35] which could be attributed to the larger surface:volume ratio of the latter. The selectivity of detection at the hemin-modified electrode was analyzed against NO, nitrite and nitrate, despite the fact that nitrate is not electrochemically active. This analysis should have been completed with other interfering species to confirm the selectivity of the proposed electrodes. It should also have been performed at lower pH values.

Later, Peteu's group studied the catalysis of PON oxidation at physiological pH at GC electrodes modified with reduced graphene oxide functionalized with hemin (GC/graphene/hemin).[36] The study was conducted using SIN-1 as a PON donor at physiological pH. A large oxidation current starting at ~0.9 V *vs.* Ag/AgCl was observed at the GC/graphene/hemin electrode, and chronoamperometry was conducted at +1.1 V *vs.* Ag/AgCl to assess the sensitivity of the electrode at pH 7.4 (Table 5.1). It is notable that the concentration of PON in solution was deduced from UV-visible measurements at 302 nm. However,

Table 5.1 Main electrochemical sensing approaches developed for the detection of ONOO$^-$ in biological samples.[a]

Sensor type	Dimensions	Detection potential (*vs.* Ag/AgCl)	Medium and biological sample	Linear dynamic range, LOD and sensitivity	Ref.
Platinized carbon microfiber	10 µm diameter	+750 mV	Unique human fibroblast cell in PBS	Specific amperometric technique providing selectivity *via* the chosen applied potential	18–21
	10 µm diameter	+450 mV	Macrophage RAW 264.7 in PBS		
Platinized platinum nano- and microelectrode	120 nm diameter	+850 mV	Nanoelectrode inside a single macrophage RAW 264.7 (in PBS) and microelectrode positioned ~5 µm above the same macrophage	Detection of ROS and RNS	37
Platinized platinum microband electrode	200 µm width and 200 µm length	+450 mV	Alkaline solution (pH 11.1) No biological sample studied	Linear range 100 nM to 5 mM LOD = 40 nM Sensitivity = 2.64 A M^{-1} cm^{-2}	23
Gold micro- and millimetric electrode	3 mm diameter 50 µm diameter	−500 mV −500 mV	PBS (pH 7.4)	Selectivity provided by the working potential (against H$_2$O$_2$, nitrite, O$_2$, DA and AA)	27 41
Carbon microfiber/ poly-MnTAPc/ poly(4-vinylpyridine)	7 µm diameter	+200 mV (DPV)	Single myocardial cell	Selectivity against H$_2$O$_2$, NO, nitrite, O$_2$, O$_2^{\cdot-}$, DA, AA, L-arg, GSH and GSSH	29
Carbon microfiber/polymeric film of Mn(III)-[2,2]*paracy*-clophenyl porphyrin	Not reported	−350 mV	HUVEC cells *In vivo*: implantation in femoral artery	LOD ≈ 1 nM	30
Platinum microelectrode/ poly-MnTAPc	25 µm diameter	−500 mV	Alkaline solution (pH 12) No biological sample studied	Linear range 12–500 µM LOD = 5000 nM Sensitivity = 14.6 nA mM^{-1}	31

(*continued*)

Table 5.1 (*continued*)

Sensor type	Dimensions	Detection potential (*vs.* Ag/AgCl)	Medium and biological sample	Linear dynamic range, LOD and sensitivity	Ref.
Platinum microelectrode/ manganese-[poly-2,5-di-(2-thienyl)-1*H*-pyrrole)-1-(*p*-benzoicacid)]/ polyethyleneimmine	100 μm diameter	+200 mV	PBS Spiked rat plasma with PON Rat glioma YPEN-1 cells in HBSS buffer	Selectivity tested against NO, H_2O_2, $O_2^{\cdot-}$, serotonin, AA, DA, bilirubin and O_2	32
Carbon microfiber/(PEDOT/ hemin)	30 μM diameter	+750 mV	CAPS buffer (pH 10.5) No biological sample studied	LOD = 200 nM Sensitivity = 13 nA μM^{-1}	34
Carbon microfiber/polymerized hemin	30 μm diameter	+700 mV	CAPS buffer (pH = 10.5) No biological sample studied	LOD = 5 nM Sensitivity = 7.5 nA nM^{-1}	35
Glassy carbon/reduced graphene oxide/hemin	5 mm diameter	+1100 mV	pH 7.4	LOD = 11 nM Sensitivity = 5 nA nM^{-1}	36
Carbon nanofiber/ Mn(III) [2,2] paracyl-cophenyl-porphyrin/ poly(4-vinylpyridine)	300 nm diameter	−300 mV	HUVEC cells in MCBD-131 medium	Linear range = 20 nM to 1 μM LOD = 1 nM	39

[a]Abbreviations: AA: ascorbic acid; CAPS: 3-(cyclohexylamino)-1-propanesulfonic acid; DA: dopamine; DPV: differential pulse voltammetry; GSH: reduced glutathione; GSSH: oxidized glutathione; HBSS: Hank's Balanced Salt Solution; L-arg: L-arginine; LOD: limit of detection; PBS: phosphate buffer saline.

as explained above, such analyses are not free from interference. In addition, to confirm the nature of the electrochemical signal attributed to the electrocatalytic oxidation of PON, a study of the evolution of the voltammograms as a function of elapsed time after addition of SIN-1 in the medium should have been used to demonstrate this hypothesis.

5.3.1.3   Summary of Electrochemical Approaches

Table 5.1 summarizes the various electrodes conceived for the detection of PON. Most of the developed sensors are amperometric, the detection being performed at a given applied potential. However, few studies have reported or discussed the selectivity of the proposed sensors, which should have been assessed because of the high operating potentials.[19,27] Additionally, due to the short half-life of PON at physiological pH, the main performance of the sensors has been reported in mild alkaline or alkaline solutions. Users should be aware of this shortcut since both sensitivity and selectivity may be affected by the pH of the medium in which the sensors will be used. In order to gain strength, the proposed methods for PON detection should take into account the analysis of the selectivity of the sensors against the main known interfering species in a given biological environment, as illustrated in Figure 5.1D.

Furthermore, it should be noted that in some of the reported electrochemical methods, the experimental details regarding the preparation of the PON stock solutions are occasionally missing or not comprehensive enough. These experimental data are, however, key for the justification and validation of the performance of the sensors. Indeed, as indicated above, the detection of PON in neutral solution cannot give rise to stationary current in long time scales (in amperometry) due to its short half-life (~1 s at pH 7.4), so reported electrochemical data should be considered with care.

5.3.2   Significant Examples of *In Vivo* Electrochemical Sensing of PON

Amatore and co-workers have worked intensively on the elaboration of platinized ultramicroelectrodes (UMEs) and electroanalytical methodology to simultaneously detect reactive RNS and ROS released at the single cell level, as indicated above. The authors developed a configuration similar to an "artificial synapse" where the UME is positioned in close proximity to the cell under investigation (distance <5 μm). Next, the cell being studied is stimulated either by mechanical or chemical stress and the fluxes of endogenously released reactive species are analyzed through successive short time frame measurements at several potentials corresponding to the detection of the reactive species (multiple potential step chronoamperometry).[21] Using this smart strategy, the authors have explored several biological systems: human fibroblast, skin cells and murine macrophages. They have, for example, reported for the first time the production of PON by immunostimulated macrophages and the evolution of the ratio between H_2O_2, NO and PON as a function of time after stimulation.[21] Recently, intracellular measurements of RNS and ROS, taken by inserting a nanoelectrode inside a macrophage cell, were reported for the first time by the same group.[37] Similarly to previous studies on UME, the nanoelectrode (~120–180 nm diameter platinum disk) was platinized with caution to obtain the optimal sensitive layer.[38] The nanoelectrode was introduced into one macrophage and amperometry was performed to monitor ROS and RNS inside the cell. In order to study concomitantly intra- and extracellular bursts of species, a platinized UME was positioned above the cell to detect extracellular released species (Figure 5.3A). Due to the high working potential (+0.85 V *vs.* Ag/AgCl), the obtained signals were representative of the totality of ROS and RNS collected intra- and extracellularly (Figure 5.3C). The position of individual voltammetric waves of ROS and RNS may vary with the properties of the platinized layer. This can be seen by comparing the responses of micrometric and nanometric electrodes (Figures 5.1A and 5.3B), so the position of the voltammetric wave for each analyte has to be examined for each prepared electrode. In addition, the position of the electrochemical wave may also vary with pH, so it would be interesting to measure simultaneously the pH outside and inside the cell.

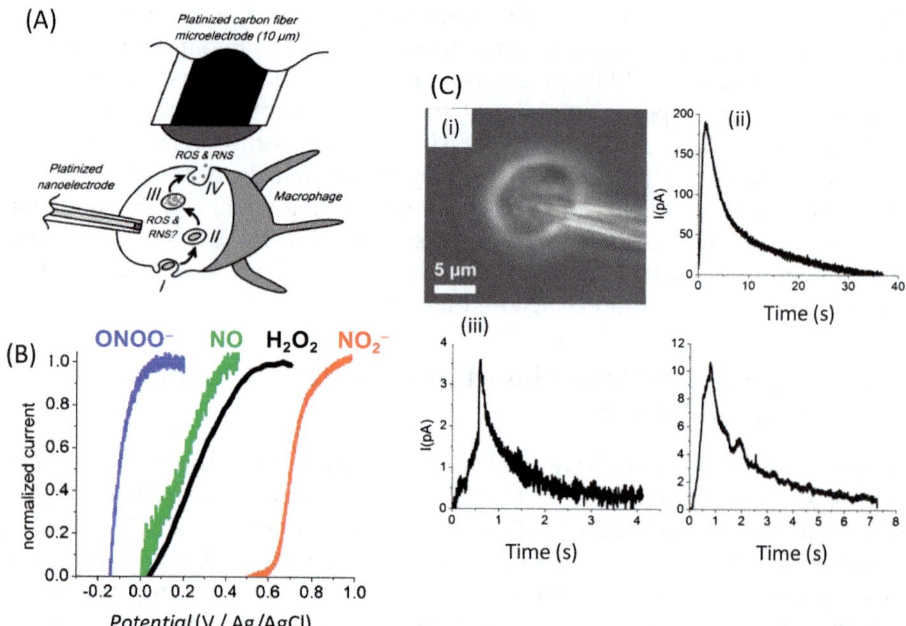

Figure 5.3   Detection of ROS and RNS by platinized nano- and microelectrodes positioned inside and in the vicinity of a single macrophage, respectively. (A) Schematic view of the sensors positioned inside and outside the macrophage. (B) *In vitro* analysis of the main ROS and RNS, giving the detection potentials for each analyte. (C) (i) Photograph of the nanoelectrode implanted into a single macrophage, (ii) amperometric monitoring at +850 mV *vs.* Ag/AgCl with the platinized microelectrode, and (iii) with the implanted platinized nanosensor. Reprinted from Y. Wang, J.-M. Noël, J. Velmurugan, W. Nogala, M. V. Mirkin, C. Lu, M. G. Collignon, F. Lemaître and C. Amatore, *Proc. Natl. Acad. Sci.*, 2012, **109**, 11534–11539, with permission.

Malinski's group has reported on *in vivo* simultaneous detection of PON and its precursors, NO and O_2^-, through the use of nanosensors positioned above a single endothelial cell (cell–sensor distance ~5 µm or less) as well as inside the wall of the femoral artery of rats. This was aimed at analyzing the PON:NO balance in different models of vascular diseases (ischemia/reperfusion, dysfunctional endothelium) in order to use this ratio as a marker of cell function or dysfunction.[30,39] Concurrent measurements of NO, O_2^- and PON were performed with electrochemical nanosensors combined into one working unit (Figure 5.4A). Each nanosensor was conceived to selectively detect a given analyte (PON, NO and O_2^-; Table 5.1). Illustrative curves obtained

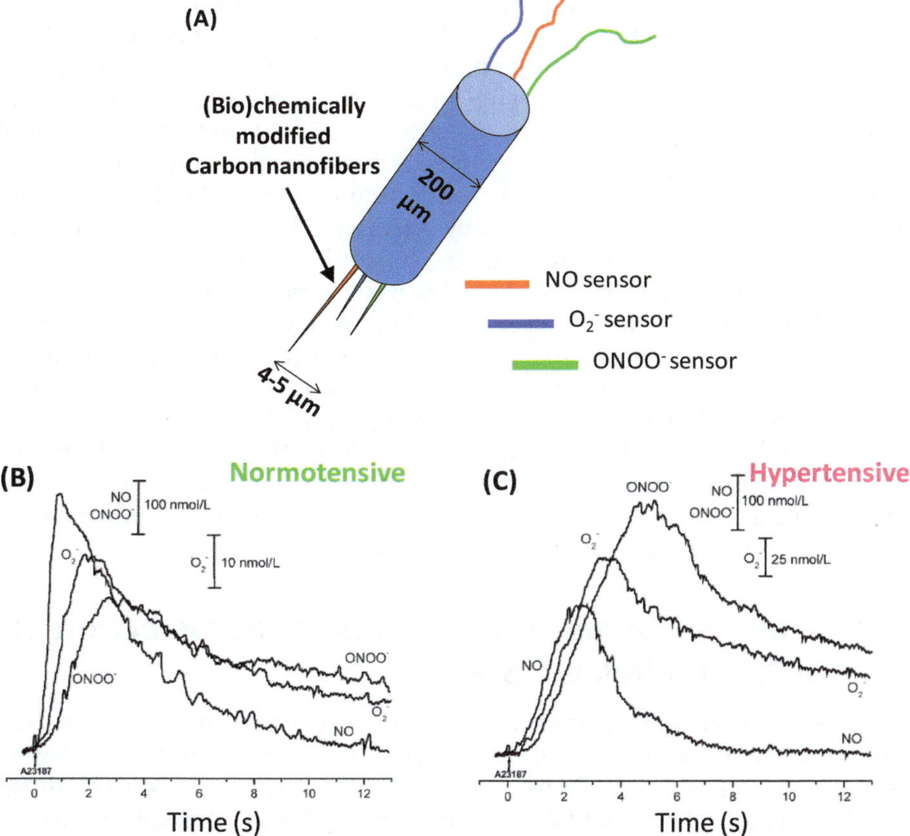

Figure 5.4 PON/nitric oxide balance in an ischemia/reperfusion injury nanomedical approach. Concurrent measurements of NO, $O_2^{\cdot-}$ and $ONOO^-$ were performed with electrochemical nanosensors combined into one working unit. (A) Scheme of the three sensors' configuration for the simultaneous detection of NO, $O_2^{\cdot-}$ and $ONOO^-$. (B) Simultaneous chronoamperograms of NO, $O_2^{\cdot-}$ and $ONOO^-$ near the surface of a single endothelial cell (aorta) of normotensive rats and (C) same as (B) for hypertensive rats. Stimulation of NO, $O_2^{\cdot-}$ and $ONOO^-$ releases a calcium ionophore. Reprinted with permisson from R. Kubant, C. Malinski, A. Burewicz, T. Malinski, Electroanalysis, 2006, **18**, 410–416. Copyright © 2006 WILEY-VCH Verlag GmbH & Co. KGaA, Weinheim.

above single aortic endothelial cells of normo- (WKY) and hypertensive (SHR) rats are shown in Figure 5.4B. The curve shapes for WKY and SHR cells demonstrate differences in the kinetics of production of each analyte as well as in their level of production between cells. It should be noted that very few experimental details are given concerning the fabrication of the sensors, their surface modification and calibration, or the specificity of each sensor, meaning that it could be very difficult to reproduce these sensors and accurately analyze the obtained results.

Bedioui's group has recently reported on the development of new assembled arrays of UMEs for the simultaneous and real-time sensing of NO and PON released by cultured cells.[27,40,41] The use of arrays was able to improve the sensitivity of detection of both species, which are known to be produced at low basal concentrations in real biological systems. The detection of both species was conducted by amperometry: NO was detected at the modified electrode array (at +0.8 V *vs.* Ag/AgCl); and PON was detected at the bare gold electrode array (at −0.1 V *vs.* Ag/AgCl; Figure 5.5A and B). The proof of concept was developed by mimicking PON production through the reaction between exogenous NO (released by a NO-donor molecule, diazeniumdiolate) and endogenous O_2^- (released by activated HL60 cells). The formation of PON was followed in real-time at both NO and PON sensing arrays (Figure 5.5C, reduction signal for PON). The observed signal at the PON sensors was attributed to PON reduction since no variations in current were observed in the presence of excess superoxide dismutase, which induces quick removal of O_2^-. This device shows the ability to detect both RNS species in real time without crosstalk between the two sensing arrays. The developed devices could be adapted for the extracellular detection RNS in other cell lines or perfused slices of tissue. They constitute an attractive alternative to study the key roles of these two RNS where a chemical imbalance is implicated in several pathologies and physiological disorders.

5.4   Conclusion and Perspectives in the Detection of PON and Related Species

The electrochemical strategies developed for PON detection as well as their application to the study of *in vitro* and *in vivo* biological systems are reviewed. The advantages of electrochemical detection are mainly related to its direct (no labeling), real-time and *in situ* monitoring of PON through the use of miniaturized electrodes located in the vicinity of one (single electrode) or several (electrode arrays) biological cells. Using this approach, the detection of PON was demonstrated under several physiological and pathophysiological conditions. However, due to the reactive nature of PON, it is difficult to calibrate the electrochemical sensors at physiological pH due to the very rapid evolution of PON's concentration with time. Furthermore, the electrochemical detection of PON is generally not free from interference from other electroactive species present in the media, meaning that an accurate study of the sensor's selectivity should always be conducted, whatever the sensor, to

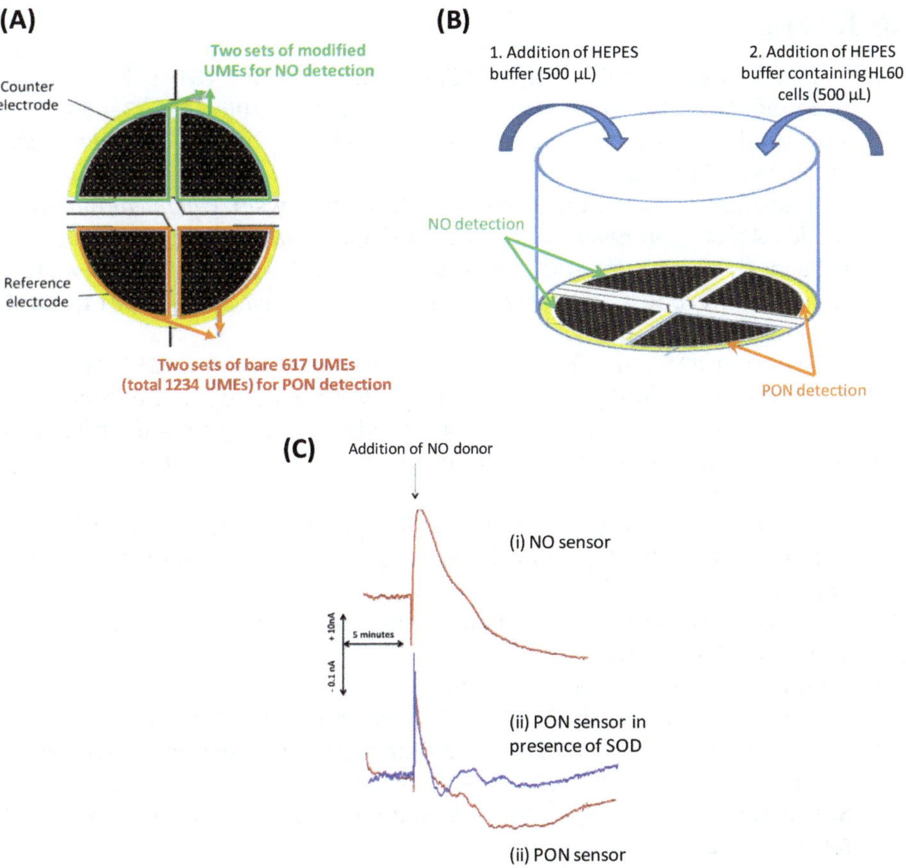

Figure 5.5 Simultaneous monitoring of NO and PON at ultramicroelectrode (UME) arrays. PON is formed by the reaction between $O_2^{\cdot-}$ produced by DMSO-treated HL60 cells and NO (released from a NO donor species). (A) Configuration of the array of gold ultramicrosensors: two sets of chemically modified electrodes are dedicated to NO sensing (oxidation at +800 mV *vs.* Ag/AgCl) while the two bare others sense PON (reduction at 100 mV *vs.* Ag/AgCl). (C) Real-time and simultaneous amperometric detection of (i) NO and (ii) PON produced by the reaction between NO and $O_2^{\cdot-}$ generated by activated HL60 cells (iii) control measurement at PON sensor in presence of SOD. Adapted from ref. 40 and Reprinted from *Electrochimica Acta*, vol. 140, L. T. O. Thi Kim, V. Escriou, S. Griveau, A. Girard, L. Griscom, F. Razan and F. Bedioui, Array of ultramicroelectrodes for the simultaneous detection of nitric oxide and peroxynitrite in biological systems, 33–36, Copyright (2014) with permission from Elsevier.

guarantee its performance. In the future, in order to have a deeper analysis of PON production/consumption in biological systems, it would be interesting to couple electrochemical detection to another technique, such as fluorescence, to simultaneously monitor electrochemical and fluorescence events and thus obtain more information about intra- and extra-cellular behavior of PON with enhanced sensitivity and selectivity.

References

1. G. Ferrer-Sueta and R. Radi, *ACS Chem. Biol.*, 2009, **4**, 161–177.
2. C. Szabó, H. Ischiropoulos and R. Radi, *Nat. Rev.*, 2007, **6**, 662–680.
3. S. Carballal, S. Bartesaghi and R. Radi, *Biochim. Biophys. Acta, Gen. Subj.*, 2014, **1840**, 768–780.
4. V. A. Erdmann, W. T. Markiewicz and J. Barciszewski, *Chemical Biology of Nucleic Acids: Fundamentals and Clinical Application*, Springer, 2014.
5. P. Pacher, S. Beckman and L. Liaudet, *Physiol. Rev.*, 2007, **87**, 315–424.
6. C. C. Winterbourn, *Biochim. Biophys. Acta, Gen. Subj.*, 2014, **1840**, 730–738.
7. A. Baeyer and V. Villiger, *Ber. Dtsch. Chem. Ges.*, 1901, **34**, 755–763.
8. K. M. Robinson and J. S. Beckman, *Synthesis of peroxyntirite from nitrite and hydrogen peroxide*, pp. 207–214, Elsevier, San Deigo, California, 2005.
9. A. Saha, S. Goldstein, D. Cabelli and G. Czapski, *Free Radical Biol. Med.*, 1998, **24**, 653–659.
10. W. A. Pryor, R. Cueto, X. Jin, W. H. Koppenol, M. Ngu-Schwemlein, G. L. Squadrito, P. L. Uppu and R. M. Uppu, *Free Radical Biol. Med.*, 1995, **18**, 75.
11. D. S. Bohle, B. Hansert, S. C. Paulson and B. D. Smith, *J. Am. Chem. Soc.*, 1994, **116**, 7423–7424.
12. P. Griess, *Chem. Ber.*, 1879, **12**, 426.
13. N. Ashki, K. C. Hayes and F. Bao, *Neuroscience*, 2008, **156**, 107–117.
14. F. J. Martin-Romero, Y. Gutierrez-Martin, F. Henao and C. Gutierrez-Merino, *J. Fluoresc.*, 2004, **14**, 17–23.
15. S. F. Peteu, R. Boukherroub and S. Szunerits, *Biosens. Bioelectron.*, 2014, **58**, 359–373.
16. N. Ieda, H. Nakagawa, T. Peng, D. Yang, T. Suzuki and N. Miyata, *J. Am. Chem. Soc.*, 2012, **134**, 2563–2568.
17. S. F. Peteu and S. Szunerits, *Peroxynitrite Electrochemical Quantification: Recent Advances and Challenges*, The Royal Society of Chemistry, 2013.
18. C. Amatore, S. Arbault, D. Bruce, P. DeOliveira, M. Erard and M. Vuillaume, *Faraday Discuss.*, 2000, **116**, 319.
19. C. Amatore, S. Arbault, D. Bruce, P. de Oliveira, M. Erard and M. Vuillaume, *Chem.–Eur. J.*, 2001, **7**, 4171–4179.
20. C. Amatore, S. Arbault, C. Bouton, K. Coffi, J.-C. Drapier, H. Ghandour and Y. Tong, *ChemBioChem*, 2006, **7**, 653–661.
21. C. Amatore, S. Arbault and A. C. W. Koh, *Anal. Chem.*, 2010, **82**, 1411–1419.
22. Y. Li, C. Sella, F. Lemaître, M. Guille Collignon, L. Thouin and C. Amatore, *Electroanalysis*, 2013, **25**, 895–902.
23. Y. Li, C. Sella, F. Lemaitre, M. Guille-Collignon, L. Thouin and C. Amatore, *Electrochim. Acta*, 2014, **144**, 111–118.
24. Y. Li, *High Performance Electrochemical Detection of Reactive Oxygen/Nitrogen Species inside Microfluidic Devices. Application for Monitoring Oxidative Stress from Living Cells*, PhD Thesis, Université Pierre et Marie Curie, Paris, France, 2014.

25. M. K. Hulvey, C. N. Frankenfeld and S. M. Lunte, *Anal. Chem.*, 2010, **82**, 1608–1611.
26. C. Kurz, X. Zeng, S. Hannemann, R. Kissner and W. H. Koppenol, *J. Phys. Chem. A*, 2005, 965.
27. D. Quinton, S. Griveau and F. Bedioui, *Electrochem. Commun.*, 2010, **12**, 1446–1449.
28. E. A. Zakharova, T. A. Yurmazova, B. F. Nazarov, G. G. Wildgoose and R. G. Compton, *New J. Chem.*, 2007, **31**, 394–400.
29. J. Xue, X. Ying, J. Chen, Y. Xian, L. Jin and J. Jin, *Anal. Chem.*, 2000, 72, 5313–5321.
30. R. Kubant, C. Malinski, A. Burewicz and T. Malinski, *Electroanalysis*, 2006, **18**, 410–416.
31. J. S. Cortes, S. G. Granados, A. A. Ordaz, J. A. L. Jimenez, S. Griveau and F. Bedioui, *Electroanalysis*, 2007, **19**, 61–64.
32. W. C. A. Koh, J. I. Son, E. S. Choe and Y.-B. Shim, *Anal. Chem.*, 2010, **82**, 10075–10082.
33. F. Bedioui, S. Griveau and D. Quinton, *Anal. Chem.*, 2011, **83**, 5463–5464.
34. S. Peteu, P. Peiris, E. Gebremichael and M. Bayachou, *Biosens. Bioelectron.*, 2010, **25**, 1914–1921.
35. S. F. Peteu, T. Bose and M. Bayachou, *Anal. Chim. Acta*, 2013, **780**, 81–88.
36. R. Oprea, S. F. Peteu, P. Subramanian, W. Qi, E. Pichonat, H. Happy, M. Bayachou, R. Boukherroub and S. Szunerits, *Analyst*, 2013, **138**, 4345–4352.
37. Y. Wang, J.-M. Noël, J. Velmurugan, W. Nogala, M. V. Mirkin, C. Lu, M. Guille Collignon, F. Lemaître and C. Amatore, *Proc. Natl. Acad. Sci.*, 2012, **109**, 11534–11539.
38. J. Velmurugan and M. V. Mirkin, *ChemPhysChem*, 2010, **11**, 3011–3017.
39. A. Burewicz, H. Dawoud, L.-L. Jiang and T. Malinski, *Am. J. Anal. Chem.*, 2013, **4**, 30–36.
40. D. Quinton, A. Girard, L. T. T. Kim, V. Raimbault, L. Griscom, F. Razan, S. Griveau and F. Bedioui, *Lab Chip*, 2011, **11**, 1342–1350.
41. L. T. O. Thi Kim, V. Escriou, S. Griveau, A. Girard, L. Griscom, F. Razan and F. Bedioui, *Electrochim. Acta*, 2014, **140**, 33–36.

Real Time Monitoring of Peroxynitrite by Stimulation of Macrophages with Ultramicroelectrodes

CHRISTIAN AMATORE[*a], MANON GUILLE-COLLIGNON[a], AND FRÉDÉRIC LEMAÎTRE[a]

[a]Ecole Normale Supérieure, Paris, France
*E-mail: christian.amatore@ens.fr

6.1 Introduction

Oxidative stress is a hazardous situation mainly related to a side route of dioxygen metabolism, which leads to oxygen reduction into superoxide anion ($O_2^{\cdot-}$) probably as a result of the NADPH oxidase (NOX) type enzymes (eqn (6.1)). Moreover, nitric oxide (NO) can also be produced from the activation of the NO synthases (NOS) (eqn (6.2)).

$$2O_2 + NADPH \rightarrow 2O_2^- + NADP^+ + H^+ \tag{6.1}$$

$$\text{L-arginine} + 2O_2 + 2NADPH + 2H^+ \rightarrow NO + \text{citrulline} + 2NADP^+ + 2H_2O \tag{6.2}$$

Indeed, superoxide and NO are relatively weak oxidants by themselves and are also involved, if their production is controlled, in signaling pathways.

RSC Detection Science Series No. 7
Peroxynitrite Detection in Biological Media: Challenges and Advances
Edited by Serban F. Peteu, Sabine Szunerits, and Mekki Bayachou
© The Royal Society of Chemistry 2016
Published by the Royal Society of Chemistry, www.rsc.org

However, both primary species can rapidly evolve into more oxidant and deleterious species. Superoxide can thus disproportionate into hydrogen peroxide (H_2O_2) through a catalyzed reaction by superoxide dismutase enzymes (SOD; with a reactivity constant $k = 2 \times 10^9$ mol^{-1} L s^{-1} at 25 °C and pH 7.4). H_2O_2 can then lead to hydroxyl radical (OH$^{\bullet}$) formation. Subsequently, the reaction between $O_2^{\bullet-}$ and NO can lead to the harmful peroxynitrite anion (ONOO$^-$) and its derivatives by a diffusion-limited reaction ($k \approx 10^{10}$ mol^{-1} L s^{-1}). These reactive oxygen species (ROS) and reactive nitrogen species (RNS) are able to oxidize cell membranes, proteins and DNA (Figure 6.1). Fortunately, there are many efficient protection pathways (antioxidants or metalloenzymes) that can limit and control ROS/RNS production. However, when such protection pathways are overwhelmed, the oxidant forces exceed the antioxidant defenses and induce what is called an "oxidative stress situation" that is suspected to be involved in many pathologies (*e.g.* aging, Parkinson's and Alzheimer's diseases, and cancers). Among all of the ROS/RNS, peroxynitrite appears as a central species because its formation rate is about three to five times faster than superoxide disproportionation by SOD.[1,2] Although very unstable at physiological pH ($t_{1/2} \approx 1$ s),[3] peroxynitrite is the source of many oxidative, nitrating, and nitrosating processes. In this way, peroxynitrite may decompose through the formation of its conjugated acid ONOOH ($pK_a \approx 6.8$), which can then evolve into very reactive radicals NO$_2^{\bullet}$ and OH$^{\bullet}$. Additionally, peroxynitrite can also react with carbon dioxide, leading to the formation of the short-term intermediate ONOOCO$_2^-$, which gives rise to the formation of the powerful nitrating and carboxylating radicals NO$_2^{\bullet}$ and CO$_3^{\bullet-}$.

Due to its great instability,[4] real time detection of peroxynitrite remains challenging because it requires a fast response time from the analytical method in addition to a good selectivity. This is why, as recently reported in the excellent review of Peteu *et al.*, the many analytical developments that have been achieved for the quantification of peroxynitrite are notably based on the use of optical (*e.g.* fluorescence and chemiluminescence) and electrochemical techniques.[2]

Optical detection of peroxynitrite is beyond the focus of the present chapter and we will instead emphasize its electrochemical monitoring. As stated

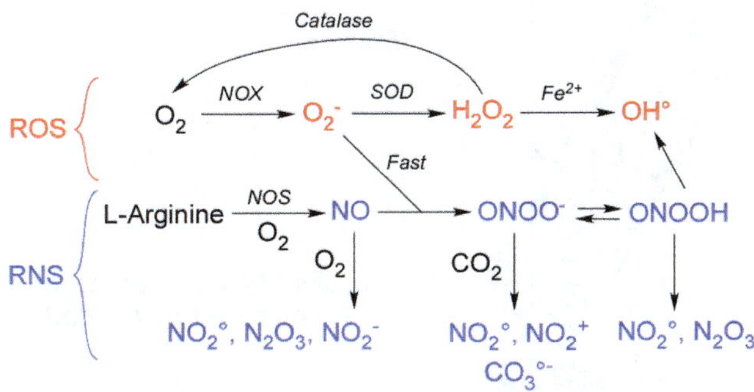

Figure 6.1   Schematic representation of ROS/RNS production *via* the side routes of oxygen metabolism.

above, peroxynitrite is an important but not primary species generated by an oxidative stress situation. Electrochemical detection and quantification of such a species thus imposes specific and contrasting conditions. On the one hand, the electrochemical sensor obviously needs to be *sensitive* and *selective* towards the analytical target, *i.e.* ONOO⁻. On the other hand, monitoring peroxynitrite alone does not provide enough information in a biological context. It is thus advisable to consider a sensor prone that can report altogether on *primary species* and *by-products* to better understand how ONOO⁻ was produced in order to successfully monitor its dynamic flux. Such a compromise with regard to selectivity obviously determines the nature of the working electrode and particularly its surface modifications. Simultaneously, the electrode surface should be modified to allow *real time* detection, *i.e.*, not reducing the response time of the sensor too much. In addition, the electrode material/surface modifications are not the unique analytical keystone. The great instability of peroxynitrite means that the electrochemical probe must be positioned in close vicinity, or at least, *as close as possible* to the release site. Finally, analysis of peroxynitrite production should be done with biologically relevant models, that is, with living cells and possibly at the single cell level. All these requirements are difficult to reconcile but are globally combined concomitantly in the case of the analytical "artificial synapse" configuration using platinized carbon (C-Pt) ultramicroelectrodes (UMEs).

6.2 Electrochemical Detection Using C-Pt UMEs

6.2.1 The "Artificial Synapse" Configuration

The "artificial synapse" configuration corresponds to a peculiar experimental situation for which the release of given molecules by a single emitting cell is electrochemically monitored. In practice, an UME is positioned in the close vicinity, *i.e.*, a few hundred nanometers above the investigated single cell (Figure 6.2). If molecules released by the cell are electroactive,

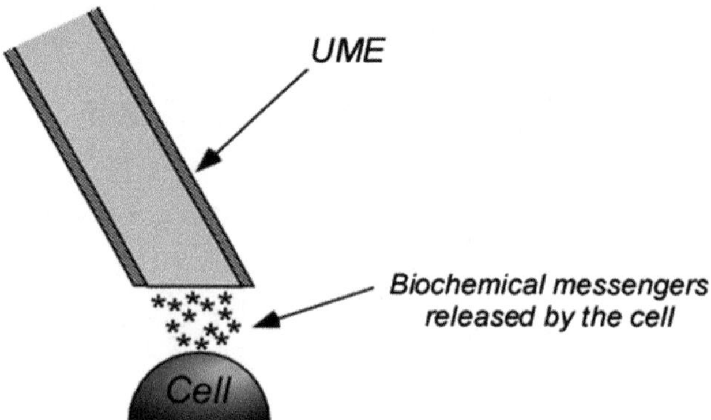

Figure 6.2   Schematic representation of the "artificial synapse" configuration.

electrochemical methods [*e.g.* chronoamperometry, and cyclic voltammetry (CV)] can thus be applied to allow one to quantify the kinetics of release and/or the amount of species emitted. Indeed, such a situation mimics the chemical synapse involved during neural communication and for which an emitting neuron releases neurotransmitters at the synaptic knob of its axon in the close vicinity of the dendrite of the receiving neuron. As described below, the "artificial synapse" configuration takes full advantage of the chemical synapse but also of the UMEs themselves.

6.2.1.1 Analytical Properties of UMEs

Electrochemical analysis at the single cell suggests the need to work with an electrochemical sensor whose size directly corresponds to the usual micrometric cell dimensions. This is why UMEs are well adapted for investigations at the single cell level. What is an UME? The precise definition is not elementary and one can consider that *an UME is an electrode that reaches spherical steady-state diffusion in usual electrochemical experiments*. It has to be emphasized that such a definition should not be related to the electrode dimensions. Yet, in practice, only electrodes for which *at least one dimension is less than a few microns* can be considered as UMEs. In order to better understand this notion, it is necessary to make comparisons between millimetric and micrometric electrodes (see a summary in Table 6.1).

In this manner, for *usual millimetric* electrodes within a solution containing a species "R" and possibly oxidized at the electrode surface (R = O + ne^-) in a diffusion controlled way, diffusion mainly proceeds along the perpendicular axis to the electrode surface because the micrometric thickness of the diffusion layer is significantly lower than the electrode size. In this case, the diffusion layer thickness [$\sim(Dt)^{1/2}$ where D is the diffusion coefficient of the electroactive species] increases as a function of time but will never exceed the micrometric range due to the convection-free domain (a few hundred micrometers) and the relatively short duration of experiments. A linear diffusion thus applies at the surface of a millimetric electrode. Conversely, if considering micrometric electrodes, the diffusion layer still increases as a function of time but can quickly reach the electrode dimensions while it remains contained within the convection free-domain. In this case, the radial diffusion cannot be neglected, thus leading to a convergent diffusional regime, *i.e.* steady-state spherical diffusion with a constant diffusion layer. As a consequence, spherical diffusion can only be reached for an electrode that has at least one dimension comparable to (or smaller than) the micrometric diffusion layer dimensions. Thus, UMEs are practically (sub)micrometric in size in usual electrochemical experiments.

In order to evaluate the analytical properties of UMEs, analytical properties of millimetric and micrometric electrodes are compared in Table 6.2.

Let us consider a chronoamperometry experiment, *i.e.* the working electrode (disk radius r_0) potential is kept constant (*vs.* a reference electrode) to perform R oxidation. For millimetric electrodes, the faradaic current (i_F)

Table 6.1 Comparison of milli- and micrometric electrodes (disk radius r_0) in a solution containing a species "R" ($D = 10^{-9}$ m^2 s^{-1}) and oxidized at the electrode surface ($R = O + ne^-$) in a diffusion controlled way. The working electrode potential is maintained constant (*vs.* a reference electrode) at the appropriate value to perform R oxidation (chronoamperometry).

	Millimeter sized electrode ($r_0 = 1$ mm)	Micrometer sized electrode ($r_0 = 1$ µm)
Expected switching time between linear and spherical diffusion	$t = \dfrac{(r_0)^2}{D} = 1000\,\text{s}$	$t = \dfrac{(r_0)^2}{D} = 1\,\text{ms}$
Expected diffusion layer thickness (δ_{diff}) for a spherical diffusion *vs.* convection-free domain thickness (δ_{conv})	$\delta_{\text{diff}} \sim \text{mm} \gg \delta_{\text{conv}} \sim 100\ \mu\text{m}$	$\delta_{\text{diff}} \sim \mu\text{m} < \delta_{\text{conv}} \sim 100\ \mu\text{m}$
Conclusion	Spherical diffusion can never be reached: only a linear diffusion regime applies	Linear diffusion is only transient. A spherical diffusion regime is quickly reached

Table 6.2 Analytical properties of a usual millimetric electrode and a UME (disk radius r_0) in a solution containing a species "R" (concentration C_R) and oxidized at the electrode surface ($R = O + ne^-$) in a diffusion controlled way. The working electrode potential is maintained constant (*vs.* a reference electrode) at the appropriate value to perform R oxidation (chronoamperometry).

	Millimeter sized electrode ($r_0 = 1$ mm)	UME ($r_0 = 1$ μm)
Diffusion layer thickness	$\delta_{\text{diff}} = \sqrt{\pi D t}$	$\delta_{\text{diff}} = \dfrac{\pi r_0}{4}$
Faradaic current	$i = \dfrac{nF\pi(r_0)^2 DC_R}{\sqrt{\pi D t}}$	$i = 4nFDC_R r_0$
Time constant	$\tau = RC \propto r_0$	
Ohmic drop	$Ri \propto r_0$	$Ri \neq f(r_0)$
S/N	$S/N \neq f(r_0)$	$S/N \propto 1/r_0$

follows the law of Cottrell. It is thus proportional to the electrode surface, *i.e.* r_0^2 and decreases with time ($\propto t^{-1/2}$) in relation to the linear diffusion regime. Conversely, for an UME, the spherical diffusional regime leads to a steady state (with a constant diffusion layer) and a constant faradaic current that only depends on r_0. Therefore, such differences drastically modify the electroanalytical properties of the working electrodes. First of all, the undesired capacitive current i_C (related to the double layer formed at the electrode–solution interface) is lower with an UME since it is governed by an exponential decay whose time constant $\tau = RC$ (R being the resistance of the solution and C the electrical capacitance) depends on r_0. As a consequence, the electroanalytical *sensitivity* of an UME is particularly high since the signal-to-noise ratio (S/N *i.e.* i_F/i_C) for an UME is proportional to $1/r_0$. Finally, UMEs represent a convenient tool to limit the problems related to the *ohmic drop* (Ri), which can prevent the control of the working electrode potential *vs.* that of the reference electrode. Indeed, for a millimetric electrode, the ohmic drop directly depends on the electrode radius and decreases when the electrode dimensions are reduced. For an UME, such an ohmic drop value is minimized and constant. The usual counter electrode needed for limiting the ohmic drop with a millimetric electrode is thus no longer required for an UME and a two-electrode configuration may be used, thus facilitating biological experiments at the single cell level.

6.2.1.2 *Intrinsic Advantages of the "Artificial Synapse" Configuration*

The "artificial synapse" configuration takes great benefits from the analytical properties of UMEs, but offers other major advantages related to the experimental conditions themselves. Indeed, in such a configuration, the UME is positioned at a submicrometric distance from the single cell being investigated. Therefore, the volume of the cleft in which the molecules are released will be restricted to a few femtoliters. It thus means that even minute amounts of electroactive species released will lead to significant local concentration variations that are particularly well adapted to electroanalysis since the faradaic current recorded is proportional to the concentration of the electroactive species and not to the amount.

Finally, the "artificial synapse" configuration is appropriate for detecting unstable species because a minimized electrode–cell distance helps the quantitative collection of the electroactive molecules released within the cleft, thus limiting their decomposition before reaching the electrode surface.

6.2.1.3 Electrochemical Detection of Oxidative Stress: What Type of Technique and Which Electrode Materials Should be Used?

As described above, UMEs involved in the "artificial synapse" configuration allow one to monitor minute amounts of species released by a single living cell. It has to be emphasized that the spherical steady-state diffusion regime is rapidly reached (*ca.* milliseconds). Therefore, concentration changes in the milliseconds range can be measured with an excellent S/N. As such, exocytosis has been widely investigated using carbon fiber UMEs in this experimental configuration.[1,5–7]

Nevertheless, such an ability to resolve fast kinetics of cell secretion has to be tempered in the context of oxidative stress. Indeed, contrary to exocytotic release for which a unique kind of electroactive species (*e.g.*, catecholamines) can be considered in a given biological context, many more species are believed to be involved during an oxidative stress situation than just $ONOO^-$ alone. Beyond the fact that the potential window is restricted at potentials positive to oxygen reduction, various electrode materials and surface modifications can be used to selectively detect the species of interest. As suggested above, $ONOO^-$ should not be detected alone because it is not a primary species of oxidative stress. Such a compromise between moderate selectivity and high sensitivity can be obtained if considering carbon fiber platinized UMEs that are able to detect four species involved in an oxidative stress situation: H_2O_2, $ONOO^-$ and NO, which have been described above, and nitrite (NO_2^-), which is not a deleterious RNS but is a decomposition product of NO and $ONOO^-$.

Carbon is a suitable material for cellular detection in biological media because it is inexpensive, robust, easily hand-made and provides a stable noise background. However, H_2O_2 and NO, two species belonging, respectively, to the group of ROS and RNS, are not easily oxidizable at bare carbon surfaces. Appropriate surface modifications are, therefore, required. In this way, carbon fiber surfaces can be platinized by reducing hydrogen hexachloroplatinate in the presence of lead acetate. Thus, using this type of dendritic platinum (Pt) deposit, a stable surface was created, as well as a high number of electroactive sites (which prevents the passivation phenomena) and lower applied potentials for hydrogen peroxide, peroxynitrite, NO and NO_2^- (compared with a bare carbon surface). It has to be noted that such an additional Pt layer will increase the time constant of the UME from 1 ms (for a usual 10 µm diameter carbon fiber) to around 100 ms. Temporal resolution remains adapted to real time analysis because the ROS/RNS delay of release for the cell models described here ranges from several dozens of seconds to up to an hour.

Beyond the nature of the UME surface and the species monitored in the "artificial synapse" configuration, the question remains regarding the most appropriate electroanalytical technique between amperometry and CV.

In CV, the potential between the working electrode and the reference electrode is usually applied in a voltage ramp, the corresponding current being concomitantly measured. The cyclic voltammogram (position and shape of the peaks) characterizes a given electroactive species and is a potent method to identify and quantify the species analyzed. Nevertheless, if dynamic changes of the electroactive species concentration occur during the experiment (which is the case for a cell in the artificial synapse configuration), many cyclic voltammograms should be recorded consecutively at relatively high scan rates. However, even at UMEs, capacitive currents at high scan rates significantly alter the S/N and have to be subtracted after the recordings have been made in a data treatment process. While CV is an important technique involved in electrochemical detection of exocytosis, it cannot be achieved in the context of ROS/RNS detection. Indeed, for electroanalyses with not enough stable capacitive currents and/or relatively slow electron transfer processes—as is typically the case for detection of H_2O_2, $ONOO^-$, NO and NO_2^- at C-Pt UMEs—CV has too many drawbacks to be considered.

In amperometry, the potential of the working electrode is held at a constant value that corresponds to the oxidization potential of the species analyzed (since the reduction window is precluded to avoid the signal caused by oxygen reduction) and the corresponding current is recorded as a function of time. Thus, the capacitive charging current is reduced by applying the potential value long enough before the cell analysis. The current corresponding to the electroactive species is thus not altered, leading to a high S/N.

Therefore, because the potential applied is constant such that no charging current is involved, the temporal resolution is better (submillisecond) in amperometry than for CV, which is a great advantage for single cell electroanalysis. Additionally, since the current is proportional to the concentration (see above), it gives direct access to the dynamics and to the amount of species involved during cell release. Indeed, knowing the number "n" of exchanged electrons in the oxidation process and Faraday's law ($Q = nNF$, where F is Faraday's constant and N is the species amount), the $I = f(t)$ curve can be converted into fluxes Φ ($= I/nF$) or into the amount of matter, N ($N = Q/nF$, where Q is the area below the $I(t)$ curve). Despite the fact that amperometry is not able to distinguish different oxidizable species, such an analytical tool is preferred to monitor real time ROS/RNS release, and particularly peroxynitrite production, by a single cell in the "artificial synapse" configuration. In this method, all species that could be oxidized at the held potential value have to be known beforehand. Therefore, different potential values will have to be considered in order to separate the peroxynitrite release from the other ROS/RNS, and also to estimate the contribution of H_2O_2, NO and NO_2^- to better analyze the involved enzymatic systems during an oxidative stress situation.

6.2.2 Identification of Peroxynitrite Among the ROS/RNS Detected on Single Cells with C-Pt Fiber UME

Early measurements of oxidative stress were carried out by our group in 1995 when an UME was used to probe the response of a biological cell without electrical or chemical stimulations.[8] In this work, C-Pt fiber microelectrodes were used to monitor the dynamics of electroactive species contained inside cultured simian virus 40 human fibroblasts, known to possess deficient catalase activity (the enzyme that catalyzes the disproportionation of H_2O_2 into water and oxygen), and thus supposed to produce larger amounts of H_2O_2. *In all of the following work on oxidative stress detection published by our group, the same type of carbon UME with a modified surface made of black Pt was employed.* In this early work, two series of experiments aimed at stimulating oxidative stress responses were performed. (The authors remarked that, in any case, the amount of H_2O_2 released *naturally* by cells was undetectable with their electrodes.) One method involved positioning a microelectrode (set by amperometry to an appropriate potential of +600 mV *vs.* Ag/AgCl) over the living cell, locally rupturing the cell membrane by forcing the electrode tip into the cell *via* micromanipulator control, and measuring the transient current corresponding to the complete electrolysis of electroactive species inside the cell. The second series involved puncturing the cell membrane with a closed-tip pulled glass capillary. In this arrangement, the H_2O_2 flux leaking out of the cell through the hole created was measured with an electrode (10 or 20 μm diameter) positioned 4 μm above the cell in the "artificial synapse" configuration described previously. The selectivity of the amperometric measurements was demonstrated through the addition of known H_2O_2 scavengers to the media that bathed the cells. The great reduction or abolition of the amperometric signal when either of these specific scavengers was added suggested that H_2O_2 was the primary substance detected. The use of specific enzymes (catalase) or substrates (*o*-dianisidine) enabled the authors to determine that an important proportion of the detected current corresponded to the oxidation of H_2O_2. This work formed the basis for what was called "physical stimulation" (operated with a full glass microcapillary with *a* <0.5 μm diameter tip) from 1995, and "biochemical stimulation" from 2006 onwards *vide infra*. Here, it is important to emphasize that the oxidative stress measured is provoked by the capillary intrusion, which depolarizes the cell membrane and thus allows the entry of calcium, and this rise in concentration is what the enzymes responsible for oxidative stress are dependent on. Thus, this is not a basal oxidative stress measurement but an active process, as will be shown below. In other words, the H_2O_2 detected here represents an active biological response and its production from the cell is a result of its aggression and injury. Finally, in this work, only H_2O_2 was cited and detected *a priori* as a mediator of oxidative stress. In 1997, Arbault *et al.* measured, with the same established technique, the effects of phenylarsine oxide (PAO; considered to be a potent inhibitor of the NOX enzyme) on human cell lines containing the HIV-1 genome. They showed that PAO, also described as an inhibitor of tyrosine phosphatase activity, inhibited H_2O_2 release from human peripheral blood mononuclear cells triggered by electrochemistry. This work was the first application of the above-cited method using C-Pt fiber UMEs to detect oxidative stress species such as H_2O_2 in deficient and HIV-1-infected cells.[9]

The two following pieces of works from our group started to identify one of the enzyme actors (NOX) of oxidative stress. Our work used the new technique involving C-Pt fiber UMEs for electrochemical detection and "physical stimulation" to mimic a particle intrusion within the cell and was achieved using the full glass microcapillary controlled by micromanipulators. First, in carcinogenesis, Arbault *et al.* activated NOX in human fibroblasts by mechanical intrusion of a single cell with an UME. At this point, it should be emphasized that this study was carried out in real time and at the single cell level (human fibroblasts, studied one by one). The mechanical intrusion induced the emission of large quantities (femtomoles) of H_2O_2 by a unique single cell within a short time frame (less than half a second). Moreover, possible implications of NOX enzymes (which produce $O_2^{\bullet-}$, which quickly dismutates into H_2O_2 and O_2) in H_2O_2 production were deduced by measuring the effects of NOX inhibitors (PAO).[10] Second, in the *Journal of Virology* the authors confirmed the role of membrane-bound NOX in cellular responses to immune activation and they analyzed in real time, lymphocyte per lymphocyte, the T-cell response following activation in relation to the redox state.[11]

In 2000, despite the apparent ubiquity and importance of the phenomenon called "oxidative stress" or "oxidative burst", there was almost no explanation in the literature about the intensity and duration of these "ROS and RNS cocktails", or their nature or chemical composition. However, to identify the exact composition of the oxidative bursts there were two issues.

The first issue is cellular variability: in the experiments with fibroblasts reported here, the variability of current intensities at each potential required the averaging of 25–40 individual cell responses to produce small enough deviations to observe meaningful results at each potential. Besides, transient techniques, such as CV (at 100 V s^{-1}), to investigate the nature of the species were impossible because of the capacitive current changes induced and the slow scans (a few volts per second or less) were not meaningful because the scan duration was comparable to the signal half-width. Owing to these intrinsic constraints, electrochemical detection and analysis of oxidative bursts were carried out at constant potentials above 0.275 V *vs.* silver–silver chloride electrode (SSCE; to avoid electrochemical consumption of oxygen above the cell), step by step, every 5 mV up to 0.850 V *vs.*

SSCE in order to reconstruct the corresponding voltammogram and thus identify the species in the bursts. This necessitated the averaging of 25–40 individual cell responses at a given electrode potential depending on the cell culture examined (in this case fibroblasts). Interestingly, the different $I^{burst} = f(t)$ curves or "amperometric spikes" obtained, when normalized by their maximum intensity $(I^{burst})_{max}$, showed a similar time course. The overall reconstructed curve of the $(I^{burst})_{max}$ as a function of the detection potential demonstrates the presence of a least three electrochemical waves labeled I–III. Furthermore, independent *in vitro* voltammograms with the same UMEs on four different synthetic solutions of H_2O_2, $ONOO^-$, NO and NO_2^- allowed electrochemical identification of the species detected in the cellular bursts. Wave I was attributed to accidental superimposition of those of H_2O_2 ($E_{1/2} = 0.250 \pm 0.005$ V *vs.* SSCE; two-electron oxidation) and $ONOO^-$ ($E_{1/2} = 0.350 \pm 0.005$ V *vs.* SSCE; one-electron oxidation), wave II corresponded to those for NO ($E_{1/2} = 0.555 \pm 0.005$ V *vs.* SSCE; one-electron oxidation), and wave III to those for NO_2^- ($E_{1/2} = 0.730 \pm 0.005$ V *vs.* SSCE; two-electron oxidation). Indeed, at the surface of C-Pt fiber microelectrodes, H_2O_2 and $ONOO^-$ are simultaneously oxidized at the potential values of 300 and 450 mV *vs.* SSCE, whereas NO is oxidized at 650 mV *vs.* SSCE and NO_2^- is oxidized at 850 mV *vs.* SSCE (Figure 6.3). The amount of released H_2O_2 and $ONOO^-$ is determined by comparing the currents observed at 300 and 450 mV *vs.* SSCE since each species contributes in different proportions at each of these potentials. However, before assigning definitively $ONOO^-$ to wave I, its electrochemical signature was obtained. *In vitro* voltammetric measurements were performed

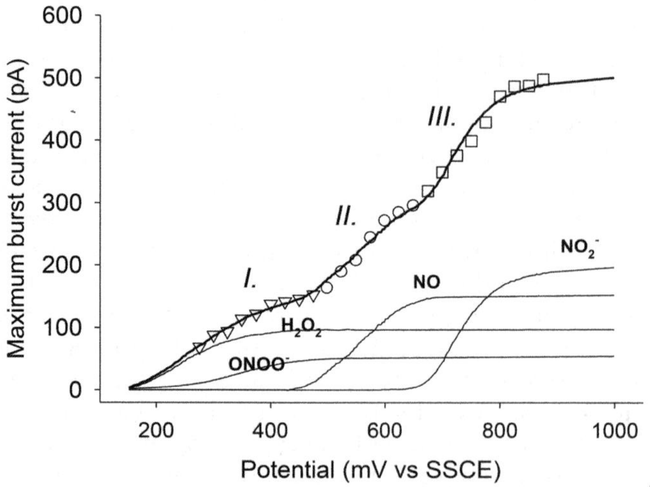

Figure 6.3 Reconstructed steady-state voltammograms from the maximum intensity obtained for each burst at a given potential value [waves I. (triangles), II. (circles) and III. (squares)] and from the arithmetic addition of the individual steady-state voltammograms of the four species: H_2O_2, $ONOO^-$, NO and NO_2^-. Adapted from Amatore *et al.*[13]

in moderately basic aqueous media of peroxynitrite, produced by ozonation of slightly alkaline azide solutions, in which the $ONOO^-$ concentration did not reasonably evolve ($t_{1/2}$ ranging from a few minutes at pH 9 to an hour at pH 12). A strict correlation with the absorbance of the respective solution measured at λ_{max} = 302 nm, which is characteristic of $ONOO^-$, was observed.[12] The electro-oxidation reported was obtained on the same C-Pt fiber UME used on the fibroblasts and corresponds to the following equation:

$$ONOO^- - e^- \rightarrow ONO_2^{\bullet} \qquad (6.3)$$

The electrochemical parameters (E° = 0.27 V *vs.* SSCE, α_{ox} = 0.45, k^{el}_{ox} = 10^{-3} cm s^{-1}) have been determined, as well as an estimate ($t_{1/2} \approx 0.1$ s) for the lifetime of the electro-generated $ONOO^{\bullet}$ radical in phosphate-buffered saline (PBS).[13]

After this important electrochemical identification step, the spatial origin of the oxidative bursts was established to consist mostly of a spherical diffusion wave emitted by the punctured hole made by the physical stimulation of the glass microcapillary at the cell membrane. Spatial features of the oxidative bursts from fibroblasts were investigated later, in 2008, and step-by-step geometrical mapping was used to show that the emission of ROS and RNS occurred from a disk surface of the membrane limited to ~15 μm in radius centered at the puncture hole created by the capillary with a 1 μm tip diameter.[14]

All four species identified above (H_2O_2, $ONOO^-$, NO and NO_2^-) derive from the spontaneous evolution of cross-reactions of the initial tandem production of NO and $O_2^{\bullet-}$, as described in the introduction. Indeed, H_2O_2 results from the spontaneous evolution of $O_2^{\bullet-}$; similarly, $ONOO^-$ is formed by the diffusion-limited coupling of NO and $O_2^{\bullet-}$; and NO_2^- results from the spontaneous decomposition of $ONOO^-$ (Figure 6.4). Therefore, the individual

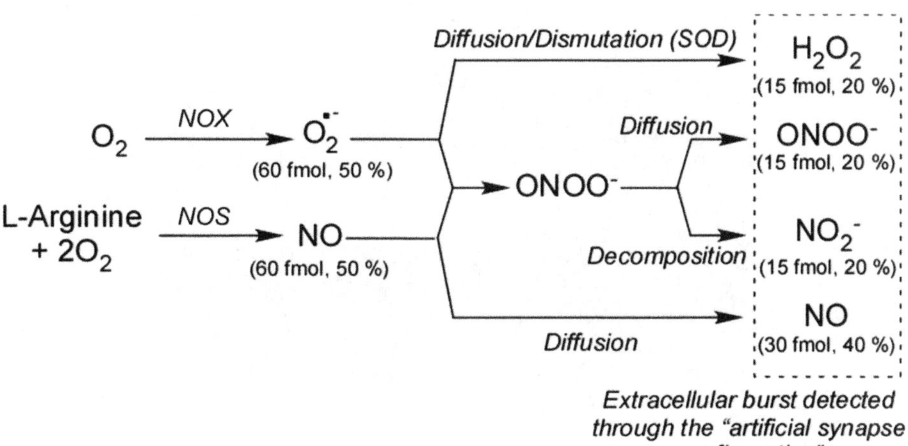

Figure 6.4 Reaction scheme describing the behavior of ROS and RNS during an oxidative burst when a single human fibroblast is physically stimulated.[15,16]

quantities of each species indirectly reflect the release of equimolar amounts of superoxide and NO by the fibroblast cell (60 fmol of each).

The simultaneous detection of $ONOO^-$, NO and H_2O_2 is proof that the NO and $O_2^{\cdot-}$ sources are actually different and located in different cellular compartments approximately 700 nm apart (Figure 6.5). At the same time, the central involvement of NOX and NOS in the oxidative burst was proved by studying the effects of catalase or peroxidases (scavengers of H_2O_2), as well as N-ethylmaleimide (NEM; inhibitor of NOX) and N^G-monomethyl-L-arginine (NMMA; competitive inhibitor specific to NOS).[15,16]

It is important to note that, in 2007, Bedioui and coworkers presented an electropolymerized manganese (Mn) tetraaminophthalocyanine thin film C-Pt UME for the electrochemical detection of $ONOO^-$ in solution. The authors demonstrated a real- time calibration curve of the amperometric determination of a stable $ONOO^-$ aqueous solution at pH 10.2 with a sensitivity of 14.6 nA mM^{-1} and detection limit of 5 µM.[17]

In conclusion, it was established by our group in the early 2000's that:

(i) The use of a C-Pt fiber UME in an "artificial synapse" configuration above a cell, together with micromanipulator-controlled stimulation using a capillary to puncture the cell membrane, triggers oxidative stress (or "oxidative bursts") in different cells types (originally fibroblasts).

(ii) These ROS and RNS bursts are not composed of species present below the cell membrane ready to be delivered upon stimulation, but are

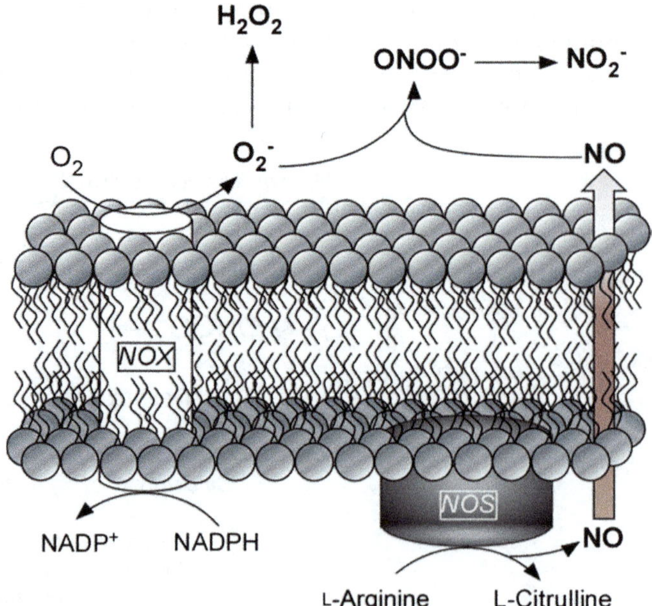

Figure 6.5 Simplified scheme of ROS/RNS production through the NOX and NOS enzymes in the cell membrane. Adapted from Verchier *et al.*[23]

      actually active secretions triggered by the physical stimulation, as shown by the intensities and duration of the oxidative spikes monitored.

(iii) It is possible to identify the exact nature and composition of the burst, that is to say, the four species NO_2^-, NO, H_2O_2 and $ONOO^-$, which can be identified through reconstructed voltammograms (by amperometry step by step every 5 mV) obtained *in vivo* on a single cell (direct voltammetry being impossible because of the too high capacitive current offered by the platinized black Pt modified surface of the UME).

(iv) Two enzymes, NOS and NOX, are implicated in the production of the original primary species of the oxidative burst, NO and $O_2^{\cdot-}$, respectively.

(v) The advantages of this method operating at the single cell level are that measurements are taken in real time, it is sensitive, the fast response time is adapted to the kinetics of the oxidative phenomenon, and it is selective for the four aforementioned species. On the other hand, the single cell level has the drawbacks of long procedures and the selectivity is offered by the potential.

(vi) $ONOO^-$, a very unstable species, can be identified and monitored directly with the desired sensitivity on a single living cell with C-Pt fiber UMEs provided that one uses the "artificial synapse" configuration (because of its half-life of less than a second at 25 °C) and a carbon UME modified with an electrodeposition of black Pt.

Such results, especially taking into account the rather short activation time (initiated by the physical stimulation of less than one tenth of a second), appear to be extremely important owing to the likely involvement of peroxynitrite in several human pathological conditions.

6.2.3 Implications of Peroxynitrite in Numerous Biological Mechanisms as Detected by Electrochemistry

In this section, we would like to emphasize different studies in which the biological role of peroxynitrite species is highlighted during the burst of other ROS and RNS. We also wish to shine more light onto important discoveries regarding the biological synthesis of $ONOO^-$ that were made possible by treatment with the enzyme inhibitors NOS and NOX, and development of different types of stimulation methods: either physical stimulation with the plain glass capillary method described above or biochemical means with interferon-γ (Int-γ)/lipopolysaccharide (LPS) for inducible NO synthase (iNOS) activation and also possibly phorbol 12-myristate 13-acetate (PMA) for NADPH activation.[1,18,19]

6.2.3.1 Physical Stimulation

In all of the following examples, physical stimulation was used to initiate ROS and RNS secretion in the "artificial synapse" configuration and their secretion was measured with C-Pt fiber UMEs. In 2004, our group investigated oxidative stress in cancer prone xeroderma pigmentosum (XP) fibroblasts. Indeed, sun exposure is clearly implicated in premature skin ageing and neoplastic development. These features are exacerbated in patients with XP, a hereditary disease associated at the cellular level with DNA repair defects and low catalase enzyme activity. The implications of oxidative stress in the defects and cancer proneness of XP skin cells (keratinocytes, fibroblasts) are multiple and remain unclear. They were investigated in our study at the level of a single fibroblast by the electrochemical method already described above, *i.e.*, using a C-Pt fiber UME in the "artificial synapse" configuration and physical stimulation with a capillary. The oxidative bursts produced by fibroblasts from normal strains were compared with those of fibroblasts from abnormal ones. All abnormal strains provided responses of higher amplitude and duration than controls. The nature of these species was investigated and it was revealed that cancer prone XP fibroblasts produced higher amounts of $O_2^{\cdot-}$ and H_2O_2, and lower amounts of NO and $ONOO^-$ than normal fibroblasts. Yet the kinetics of release of H_2O_2 are directly related to those of $O_2^{\cdot-}$, while the kinetics of $ONOO^-$ are a function of the kinetics of both $O_2^{\cdot-}$ and NO release. A faster decrease in NO production than $O_2^{\cdot-}$ was observed in abnormal fibroblasts, with $ONOO^-$ kinetics following NO kinetics, suggesting that $ONOO^-$ was probably generated in shorter time periods and then in lower quantities than H_2O_2. However, this explanation does not sufficiently account for the observed 50% decrease in $ONOO^-$. Thus, the authors suggest that other conditions govern the production of peroxynitrite; for example, the production of NO by iNOS in conjunction with or as a replacement of constitutive NO synthase (cNOS). The simultaneous production of H_2O_2 and $ONOO^-$ further implies that the cellular location of NO generating systems may vary so that diffusion of NO towards $O_2^{\cdot-}$ generating sites takes more time when the paths leading to cross-coupling are longer. This would lead to lower quantities of peroxynitrite because of the competitive disproportionation of $O_2^{\cdot-}$ into H_2O_2 *vs.* its reaction with NO. These hypotheses will be tested and confirmed in later publications by the group.[20]

The group of Amatore published a paper in 2006 in which they monitored in real time the release of ROS and RNS by a single macrophage stimulated by mechanical depolarization of its membrane. Macrophages are key cells of the immune system and during phagocytosis they engulf a foreign bacterium, virus or particle into a vacuole, the phagosome, wherein oxidants are produced to neutralize and decompose the threatening element. In the work by Amatore and colleagues, RAW 264.7 macrophages were shown to secrete $ONOO^-$ (9 fmol), NO (14 fmol) and NO_2^- (6 fmol), while H_2O_2 (5 fmol) was produced at lower levels. Evaluation of the original $O_2^{\cdot-}$ and NO production by macrophages was calculated using two equations:[21]

$$\left(\Phi_{O_2^{\cdot-}}\right)^{prod} = 2\left(\Phi_{H_2O_2}\right)^{mes} + \left(\Phi_{ONOO^-}\right)^{mes} + \left(\Phi_{NO_2^-}\right)^{mes} \qquad (6.4)$$

$$\left(\Phi_{NO}\right)^{prod} = \left(\Phi_{NO}\right)^{mes} + \left(\Phi_{ONOO^-}\right)^{mes} + \left(\Phi_{NO_2^-}\right)^{mes} \qquad (6.5)$$

Under the experimental conditions, macrophages were not preactivated by immune factors but through sudden physical depolarization of their membrane. In this case, the observed NO production necessarily arises from a constitutive pool of NO synthase (NOS) enzymes (also named eNOS for endothelial NOS or NOS-III type). This line of evidence, in terms of the kinetics of activation of the cell response, is supported by the known effect of NOS inhibitors. Therefore, the ability of RAW 264.7 macrophages to produce $ONOO^-$ was clearly established in this work. In a subsequent study, the oxidative response of macrophages was examined under conditions of immunological activation in order to offer a comprehensive view of the dynamics of $O_2^{\cdot-}$ and NO derivatives during phagocytosis (*vide infra*).[22] PLB-985 cells, originating from a myeloid cell line prone to differentiate into neutrophils or monocyte-like phagocytes, were investigated with real time selective electrochemical detection of each ROS or RNS using a C-Pt fiber UME and physical stimulation. The results showed that there was high NOS activity in these cells (means of 14 fmol of NO *vs.* 7 fmol of $O_2^{\cdot-}$ for the reconstructed primary fluxes for individual cells) and, even more surprisingly, NOS was able to provide NOX activity in cells where *NOX2* was knocked out.

Indeed, deactivation of the enzyme NOX leads to annihilation of the formation pathway of NO_2^- and peroxynitrite because they arise from the reaction of NO and $O_2^{\cdot-}$. However, in this study, superoxide was still produced on cells where the NOX system was knocked out. This was confirmed using inhibitors of cNOS and iNOS by, respectively, L-NAME and 1400W.[23] A study in 2008 showed that vitamin C, known as ascorbic acid, acts as an antioxidant agent in these PLB-985 cells but as a pro-oxidant in RAW macrophages. Incubation of each cell type with ascorbic acid appears to have an effect mainly on the production of primary RNS, that is to say NO, and only a small effect on peroxynitrite production.[24] Antioxidant and pro-oxidant activities of β-lapachone and α-lapachone were also investigated in macrophages. Once again, the technique allowed the detection of $ONOO^-$ and the reconstitution of the original $O_2^{\cdot-}$ and NO fluxes. Different mechanisms involving opposite reactivities of quinones in living cells mainly depending on incubation times and the quinone formula (pharmacologically active or inactive isomer forms of lapachones) were proposed.[25] Another similar work implicated the azido moiety in oxidative stress when studying the pro-oxidant properties of azidothymidine (AZT; the first drug approved for HIV treatment) and other thymidine analogs in macrophages. Owing to the presence of peroxynitrite detected with the electroanalytical method, nitrative and nitrosative properties were suggested for one of the thymidine analogs.[26] Another study examined the oxidative stress responses of MG63 osteosarcoma cells and showed, on average per single cell, a prominent release of RNS (5 fmol of NO_2^-, 6 fmol of $ONOO^-$ and 17 fmol of NO) with a small quantity of H_2O_2 detected (2 fmol). This work confirms, *via* the important $NO/O_2^{\cdot-}$ and NO/H_2O_2 ratios, that the ability of the osteosarcoma cells to form malignant bone was related to a high level of secretion of NO and a low level of $O_2^{\cdot-}$.[27] Finally, quantitative analysis of ROS and RNS production in breast cancer cell lines incubated with ferrocifens (an original class of ferrocifen-type breast cancer drug) demonstrated that peroxynitrite plays a role in this context, in particular for MDA-MB-231 cells after incubation with the ferrocifen Fc-OH-Tam (corresponding to the structural combination of hydroxytamoxifen and ferrocene).[28]

All of these results shine more light on the role of peroxynitrite (detected by amperometry with physical stimulation using a plain glass capillary and a carbon fiber platinized UME in the "artificial synapse" configuration) in various physiological mechanisms, diseases and biochemical studies of antioxidant molecules including, respectively, phagocytosis, HIV, breast, skin or bone cancers, lapachones and the anti- or pro-oxidant effects of vitamin C, as well as the cooperative roles of NOS and NOX, the two enzymes responsible for, *in fine*, the formation of $ONOO^-$ from NO and $O_2^{\cdot-}$.

6.2.3.2   Biochemical Stimulation

In a study by Nathan and colleagues, biochemical stimulation was employed to elicit secretion of ROS and RNS. Notably, macrophage cells were activated by Int-γ and LPS in order to induce expression of iNOS (NOS-II type), which has been shown to produce large amounts of NO over a long time period.[29] Indeed, Int-γ and LPS are required for synergistic activation of the iNOS promoter. The reactive mixture released by single cells was analyzed, in real time, by amperometry at C-Pt microelectrodes. In 2008, the sequence of work on peroxynitrite detection in macrophages involved in phagocytosis changed the method of stimulation to these biochemical means rather than the physical stimulation described above. All of the electrochemical considerations reported in Section 6.2.2 led the authors to propose a system of four equations with four unknowns to determine the contribution of each of the four oxidative stress species detected ($ONOO^-$, H_2O_2, NO and NO_2^-) using the measurements of the currents at the four potentials (300, 450, 650 and 850 mV *vs.* SSCE):

$$I^{850\,\text{mV}} = I_{H_2O_2} + I_{ONOO^-} + I_{NO} + I_{NO} + I_{NO_2^-} \tag{6.6}$$

$$I^{650\,\text{mV}} = I_{H_2O_2} + I_{ONOO^-} + I_{NO} \tag{6.7}$$

$$I^{450\,\text{mV}} = 0.99I_{H_2O_2} + 0.90I_{ONOO^-} \tag{6.8}$$

$$I^{300\,\text{mV}} = 0.85I_{H_2O_2} + 0.29I_{ONOO^-} \tag{6.9}$$

This system of linear equations is readily solved to provide the individual current contributions of each of the four individual species for carbon platinized fiber UME with a charge of black Pt at the tip of 5 μC. The emission fluxes of each species (Φ_{species}) can then be calculated from their respective current intensities obtained above, by using Faraday's equation:

$$\Phi_{\text{species}} = \frac{I_{\text{species}}}{n_{\text{species}} \times F} \tag{6.10}$$

where $n_{species}$ is the number of electrons per molecule exchanged for oxidation and F is the Faraday constant. Moreover, the current may be time integrated to provide the overall detected charge over any time interval. Thus, direct evidence for the formation of $ONOO^-$ in stimulated macrophages has been presented. Measurements of the ROS and RNS released by a single biochemically stimulated macrophage were obtained and compared with those from physical stimulation (Table 6.3).

In this case, amperometry offered the possibility of real time detection with a much better time resolution than the fluorometric method, for example. Indeed, the amperometric responses presented two distinct behaviors of release: weak broad secretions and sharp amperometric spikes (Figure 6.6).

Interestingly, the large and wide peaks could feature the quasi-continuous detection of ROS or RNS, such as NO or $ONOO^-$, which are known to freely diffuse across cellular membranes,[30–32] while the keen ones could suggest cellular events such as exocytosis of phagolysosomes.[33,34] In the same work,

Table 6.3 ROS and RNS released by a single stimulated RAW 264.7 macrophage depending on the stimulation (biochemical *vs.* physical).

ROS/RNS stimuli	H_2O_2 (fmol)	$ONOO^-$ (fmol)	NO (fmol)	NO_2^- (fmol)
Int-γ/LPS biochemical stimulation	0	7.5 ± 0.9	8.9 ± 1.9	4.0 ± 1.3
Physical stimulation	5 ± 1	9 ± 1	14 ± 2	6 ± 1

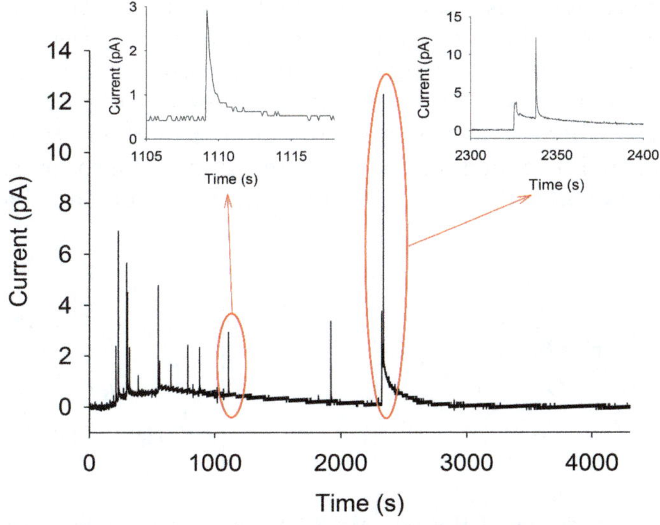

Figure 6.6 Representative amperometric response from a single Int-γ/LPS-stimulated macrophage (450 mV *vs.* SSCE after 19.5 h stimulation). Two amperometric spikes are displayed as insets. Adapted from Amatore *et al.*[22]

under conditions mimicking an *in vivo* inflammatory situation, the presence of ONOO⁻, but not H_2O_2, was again detected amongst the complex cocktail of ROS and RNS produced. The lack of H_2O_2 is probably due to the absence of NOX stimulation by Int-γ/LPS biochemical means that led consequently to very low fluxes of $O_2^{\bullet-}$ (which later gives H_2O_2 by disproportionation) and thus non-discernible fluxes of H_2O_2 even with the low S/N ratio reached here (0.3 pA). Finally, the presence of 1400W, a selective iNOS inhibitor, or peroxynitrite scavengers led to almost complete attenuation or at least diminution of half of the amperometric response of activated RAW 264.7 cells. The majority of the reactive species released by a macrophage are thus likely to be derived from NO and $O_2^{\bullet-}$ that have been co-produced by iNOS. This is likely to be the first direct evidence that ONOO⁻ is released by immunologically activated macrophages.[22] Next, the effects of SOD mimics, either Mn^{II} pentaazamacrocyclic or 1,2-diaminoethane based Mn(II) complexes, on macrophages' production of ROS and RNS were studied. These SOD mimics are metalloenzymes involved in the control of $O_2^{\bullet-}$ concentration and, as a consequence, the limitation of ONOO⁻ production. Efficient SOD mimics need to combine important properties such as choice of the metal, ligand, redox potential and chemical activity towards $O_2^{\bullet-}$, *etc.* In the studies reported here, estimation of the efficiency of Mn-SOD mimics (Scheme 6.1) in real conditions towards an oxidative stress situation at the single cell level (macrophages) was proposed using amperometry.

The first complex tested (Mn^{II} pentaazamacrocyclic SOD mimics) had a rigid pentadentate ligand, relatively good SOD activity of 10^7 L mol⁻¹ s⁻¹ and, interestingly, NO dismutase activity (of 10^3 L mol⁻¹ s⁻¹). Pre-incubation with RAW 264.7 macrophages under the biochemical stimulation cited above (Int-γ/LPS) showed a total disappearance of ONOO⁻ and NO, consistent with its chemical properties. This is probably the first time that a chemical compound with promising pharmacological properties (protective effects for the treatment of inflammation and inflammation-related diseases) has been reported to efficiently remove both $O_2^{\bullet-}$ and NO, and thus ONOO⁻, by transforming them into more benign species and without altering the normal enzymatic pathways for their production. The direct reaction of the complex itself with ONOO⁻ was also proved by CV.[35] The other 1,2-diaminoethane based Mn(II)

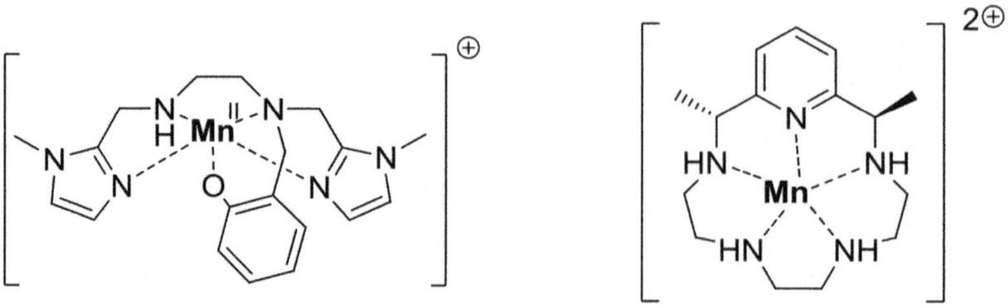

Scheme 6.1 Structures of the two Mn-SOD mimics whose activities have been investigated in real biological conditions by using the "artificial synapse" configuration.[35,36]

complex (SOD activity of 7×10^6 L mol^{-1} s^{-1}) was tested in the same conditions of incubation with Int-γ/LPS for iNOS and PMA for NADPH activation. This SOD mimic was shown to reduce the flux of ONOO$^-$ by a factor of five and, surprisingly, with a smaller increase in H_2O_2 than would have been expected in the case of pure SOD activity. This suggests again a direct reaction of the complex with ONOO$^-$ and H_2O_2 (catalase activity).[36]

All the examples presented here, either with physical or biochemical stimulation of single cells, established the power and versatility of direct amperometric measurements by "artificial synapses" coupled to the modification of carbon UMEs by the black Pt dendritic deposit for the detection of ROS/ RNS such as ONOO$^-$.

6.3 Analytical Evolution of Electrochemical Detection

6.3.1 Adapted Protocols from the "Artificial Synapse" Configuration

As described in the previous sections, many benefits can be taken from the electrochemical detection of peroxynitrite during an oxidative stress situation by using C-Pt fiber UMEs in the "artificial synapse" configuration: real time measurements, quantification of fluxes and amounts of matter, and detection of peroxynitrite with other important reactive species. Unfortunately, two main drawbacks remain and possibly limit the use of such methodology in certain cases. The first drawback is related to the analysis at the single cell level by chronoamperometry. In order to overcome the cell variability, between 30 and 50 measurements are needed for each potential value in order to provide statistically significant information. Because four species are considered, this means that examining the effect of a single experimental condition or treatment requires from 120 to 200 measurements at the single cell level, which is time consuming, especially in the case of biochemical stimulation (length of time of release being ~1 h for one cell). The second drawback is related to the experimental configuration. The electrochemical sensor is located in the extracellular medium and is thus unable to detect intracellular ROS/RNS production, which can be a matter of interest especially in the case of phagocytosis.

One way to reduce the number of experiments with single cells is to modify the electrochemical technique involved in the methodology from single applied potential chronoamperometry to triple-potential-step chronoamperometry.[19] In this way, a repetitive sequence of three potentials (+300, +450 and +650 mV *vs.* SSCE) can be applied to the C-Pt fiber UME in order to achieve, on the same single cell, measurements usually performed on three different cells. The analytical feasibility of this strategy has been demonstrated on Int-γ/LPS/PMA-stimulated macrophages and consistent amounts of ROS/ RNS (H_2O_2, ONOO$^-$ and NO) were found after appropriate data treatment.

However, each change of working potential during the recording induces a significant and undesired charging capacitive current that combines with the faradaic information. As a consequence, the relatively large time constants for C-Pt fiber UMEs impose a "blank-time" period in which the capacitive current cannot be neglected and thus limits the temporal sampling resolution (optimized here to one point per 60 s). This rather weak sampling frequency thus limits the scope of the triple step method to slow kinetics of release, but remains suitable for immune-stimulated oxidative stress situations. Furthermore, the fourth potential value (+850 mV *vs.* SSCE) was not included into the sequence due to great dispersion from the baseline current. Therefore, the potential-step method did not measure NO_2^- in this case.

A second and complementary strategy to prevent long and tedious measurements is to extend the protocol to a cell population without losing the analytical benefits of the "artificial synapse" configuration.[37] It can be done by using a microfluidic device made with a detection chamber containing a three electrode system (including a Pt band working electrode) in which macrophages can be cultured before appropriate stimulation. ROS/RNS production was thus measured by amperometry at the surface of the platinized microelectrode at the cell population level with good reproducibility.

As stated above, depending on the stimulation, ROS/RNS can be partially produced in the intracellular medium, while the "artificial synapse" configuration involves an extracellular sensor. This is why inserting a nanoelectrode within a single cell has recently been considered and has been demonstrated as a noninvasive technique.[38] However, electrochemically detecting the four species mentioned above (H_2O_2, $ONOO^-$, NO and NO_2^-) still necessitates a black Pt surface. Such modification was performed through the etching of the nanoelectrode glass tip and subsequent black Pt electrodeposition within the recessed cavity.[39] The ability of corresponding black Pt nanoelectrodes to electrochemically detect H_2O_2, $ONOO^-$, NO and NO_2^- has been demonstrated with good reproducibility. Finally, the manual insertion of such a nanoelectrode within a single living macrophage has also been performed and allowed a first comparison between intracellular and extracellular ROS/RNS production.

6.3.2 Other Examples

While they will be mentioned in more details in other chapters of this book, complementary approaches have been considered by other groups. As an example, Bedioui and colleagues reported recent studies related to the use of microelectrode arrays for which NO and $ONOO^-$ are the two species of interest.[40] Arrays (from 7 to 110 UMEs) were designed to gather two networks, each one devoted to the detection of only one species, NO_2^- or peroxynitrite. The whole strategy consisted of associating two selective electrochemical methods of detection at two different parts of the microdevice. On the one hand, selective oxidation of NO was allowed through a chemical modification of the gold (Au) surface electrodes [thin layers of poly(eugenol) and poly(phenol)] at +0.8 V *vs.* Ag/AgCl, while on the other hand, electrochemical detection of peroxynitrite was ingeniously performed on uncoated Au electrodes without any further chemical modification of the surface through reduction of ONOOH in PBS at a moderate potential value (-0.1 V *vs.* Ag/AgCl).[41] The simultaneous and selective detection of NO and $ONOO^-$ has thus been demonstrated with standard solutions or precursors in the presence of interferents (H_2O_2, dopamine, ascorbic acid, NO_2^-). It is important to note that the electrochemical detection of oxidative stress strongly depends on the purpose of the study and that H_2O_2 and NO_2^- are viewed here as interferents while they are considered as species of interest in other investigations.

This type of microelectrode array has been extended to the detection of NO/$ONOO^-$ at the cell population level with arrays containing over 2400 Au UMEs.[42] Suspensions of PMA-activated HL60 cells (3 million) producing $O_2^{\cdot-}$ were tested in the presence of an exogenous NO donor. The feasibility of using such devices for biological systems was shown through the successful simultaneous recording of NO/$ONOO^-$ in that study. It is worth noting that the inter-electrode distance is significantly larger (at least ten times) than the UME dimensions ($r = 20$ μm) to avoid coupling of the diffusion layers and to ensure that the obtained current will only track the concentration of the species released and no cross-electrochemical feedback. Other recent works have examined peroxynitrite coupled detection with electrophoresis and electrochemistry,[43] electrochemical detection of NO, superoxide and peroxynitrite,[44] and electrochemical monitoring in microchannels with highly sensitive Pt-black coated Pt electrodes for NO_2^-, NO, H_2O_2 and peroxynitrite analytical detection.[45,46] These studies suggest that many more works dealing with detection of peroxynitrite from cells in microsystems will blossom in the near future.

6.4   Conclusion

In this chapter we have tried to show how peroxynitrite emerged as an important contributing agent in oxidative stress thanks to the development of electrochemical measurements with "artificial synapses" at platinized-carbon UMEs. This methodology has indeed provided many examples of the unsuspected biological involvement of peroxynitrite, whose role in oxidative stress is now considered to be as important, if not more, than that of H_2O_2. However, as discussed in this chapter, the "artificial synapse" methodology is restricted to single cell analysis. Single cell analysis provides information on cellular variability, but this is not really a drastic piece of information at the present level of research on oxidative stress. Conversely, with single cell measurements a huge amount of experimental data is required to produce statistically significant information, precisely to eliminate the effect of cellular variability. This has stimulated the development, in our group and among others, of microchips in which a batch of a few tens of cells can be analyzed in order to produce information insensitive to cellular variability and relative to the action of antioxidants or drugs effecting oxidative stress. These microchips are opening a new era and we have no doubt that their tandem use with single cell analysis will open new perspectives of ONOO⁻ and its role in health.

References

1. C. Amatore, S. Arbault, M. Guille and F. Lemaitre, *Chem. Rev.*, 2008, **108**, 2585.
2. S. F. Peteu, R. Boukherroub and S. Szunerits, *Biosens. Bioelectron.*, 2014, **58**, 359.
3. W. H. Koppenol, J. J. Moreno, W. A. Pryor, H. Ischiropoulos and J. S. Beckman, *Chem. Res. Toxicol.*, 1992, **5**, 834.
4. G. Ferrer-Sueta and R. Radi, *ACS Chem. Biol.*, 2009, **4**, 161.
5. W. Wang, S.-H. Zhang, L.-M. Li, Z.-L. Wang, J.-K. Cheng and W.-H. Huang, *Anal. Bioanal. Chem.*, 2009, **394**, 17.
6. L. Mellander, A.-S. Cans and A. G. Ewing, *ChemPhysChem*, 2010, **11**, 2756.
7. A.-S. Cans and A. G. Ewing, *J. Solid State Electrochem.*, 2011, **15**, 1437.
8. S. Arbault, P. Pantano, J. A. Jankowski, M. Vuillaume and C. Amatore, *Anal. Chem.*, 1995, **67**, 3382.
9. S. Arbault, M. Edeas, S. Legrand-Poels, N. Sojic, C. Amatore, J. Piette, M. Best-Belpomme, A. Lindenbaum and M. Vuillaume, *Biomed. Pharmacother.*, 1997, **51**, 430.
10. S. Arbault, P. Pantano, N. Sojic, C. Amatore, M. BestBelpomme, A. Sarasin and M. Vuillaume, *Carcinogenesis*, 1997, **18**, 569.
11. A. Lachgar, N. Sojic, S. Arbault, D. Bruce, A. Sarasin, C. Amatore, B. Bizzini, D. Zagury and M. Vuillaume, *J. Virol.*, 1999, **73**, 1447.
12. W. A. Pryor, R. Cueto, X. Jin, W. H. Koppenol, M. Nguschwemlein, G. L. Squadrito, P. L. Uppu and R. M. Uppu, *Free Radical Biol. Med.*, 1995, **18**, 75.
13. C. Amatore, S. Arbault, D. Bruce, P. de Oliveira, M. Erard and M. Vuillaume, *Chem.–Eur. J.*, 2001, **7**, 4171.
14. C. Amatore, S. Arbault and M. Erard, *Anal. Chem.*, 2008, **80**, 9635.
15. C. Amatore, S. Arbault, D. Bruce, P. de Oliveira, M. Erard, N. Sojic and M. Vuillaume, *Analusis*, 2000, **28**, 506.
16. C. Amatore, S. Arbault, D. Bruce, P. de Oliveira, M. Erard and M. Vuillaume, *Faraday Discuss.*, 2000, **116**, 319.
17. J. S. Cortes, S. G. Granados, A. A. Ordaz, J. A. L. Jimenez, S. Griveau and F. Bedioui, *Electroanalysis*, 2007, **19**, 61.
18. C. Amatore, S. Arbault, M. Guille and F. Lemaitre, *Actual. Chim.*, 2011, 25.
19. C. Amatore, S. Arbault and A. C. W. Koh, *Anal. Chem.*, 2010, **82**, 1411.
20. S. Arbault, N. Sojic, D. Bruce, C. Amatore, A. Sarasin and M. Vuillaume, *Carcinogenesis*, 2004, **25**, 509.
21. C. Amatore, S. Arbault, C. Bouton, K. Coffi, J. C. Drapier, H. Ghandour and Y. H. Tong, *ChemBioChem*, 2006, 7, 653.

22. C. Amatore, S. Arbault, C. Bouton, J.-C. Drapier, H. Ghandour and A. C. W. Koh, *ChemBioChem*, 2008, **9**, 1472.
23. Y. Verchier, B. Lardy, M. V. C. Nguyen, F. Morel, S. Arbault and C. Amatore, *Biochem. Biophys. Res. Commun.*, 2007, **361**, 493.
24. C. Amatore, S. Arbault, D. C. M. Ferreira, I. Tapsoba and Y. Verchier, *J. Electroanal. Chem.*, 2008, **615**, 34.
25. D. C. M. Ferreira, I. Tapsoba, S. Arbault, Y. Bouret, M. S. Alexandre Moreira, A. V. Pinto, M. O. F. Goulart and C. Amatore, *ChemBioChem*, 2009, **10**, 528.
26. C. Amatore, S. Arbault, G. Jaouen, A. C. W. Koh, W. K. Leong, S. Top, M.-A. Valleron and C. H. Woo, *ChemMedChem*, 2010, **5**, 296.
27. R. Hu, M. Guille, S. Arbault, C. J. Lin and C. Amatore, *Phys. Chem. Chem. Phys.*, 2010, **12**, 10048.
28. C. Lu, J. M. Heldt, M. Guille-Collignon, F. Lemaitre, G. Jaouen, A. Vessieres and C. Amatore, *ChemMedChem*, 2014, **9**, 1286.
29. J. MacMicking, Q. W. Xie and C. Nathan, *Annu. Rev. Immunol.*, 1997, **15**, 323.
30. S. S. Marla, J. Lee and J. T. Groves, *Proc. Natl. Acad. Sci. U. S. A.*, 1997, **94**, 14243.
31. D. M. Porterfield, J. D. Laskin, S. K. Jung, R. P. Malchow, B. Billack, P. J. S. Smith and D. E. Heck, *Am. J. Physiol.*, 2001, **281**, L904.
32. P. Pacher, J. S. Beckman and L. Liaudet, *Physiol. Rev.*, 2007, **87**, 315.
33. A. Di, B. Krupa, V. P. Bindokas, Y. M. Chen, M. E. Brown, H. C. Palfrey, A. P. Naren, K. L. Kirk and D. J. Nelson, *Nat. Cell Biol.*, 2002, **4**, 279.
34. V. Braun and F. Niedergang, *Biol. Cell*, 2006, **98**, 195.
35. M. R. Filipovic, A. C. W. Koh, S. Arbault, V. Niketic, A. Debus, U. Schleicher, C. Bogdan, M. Guille, F. Lemaitre, C. Amatore and I. Ivanovic-Burmazovic, *Angew. Chem., Int. Ed.*, 2010, **49**, 4228.
36. A.-S. Bernard, C. Giroud, H. Y. V. Ching, A. Meunier, V. Ambike, C. Amatore, M. Guille Collignon, F. Lemaitre and C. Policar, *Dalton Trans.*, 2012, **41**, 6399.
37. C. Amatore, S. Arbault, Y. Chen, C. Crozatier and I. Tapsoba, *Lab Chip*, 2007, **7**, 233.
38. P. Sun, F. O. Laforge, T. P. Abeyweera, S. A. Rotenberg, J. Carpino and M. V. Mirkin, *Proc. Natl. Acad. Sci. U. S. A.*, 2008, **105**, 443.
39. Y. Wang, J.-M. Noel, J. Velmurugan, W. Nogala, M. V. Mirkin, C. Lu, M. Guille Collignon, F. Lemaitre and C. Amatore, *Proc. Natl. Acad. Sci. U. S. A.*, 2012, **109**, 11534.
40. D. Quinton, A. Girard, L. T. T. Kim, V. Raimbault, L. Griscom, F. Razan, S. Griveau and F. Bedioui, *Lab Chip*, 2011, **11**, 1342.
41. D. Quinton, S. Griveau and F. Bedioui, *Electrochem. Commun.*, 2010, **12**, 1446.
42. L. T. O. Thi Kim, V. Escriou, S. Griveau, A. Girard, L. Griscom, F. Razan and F. Bedioui, *Electrochim. Acta*, 2014, **140**, 33.
43. M. K. Hulvey, C. N. Frankenfeld and S. M. Lunte, *Anal. Chem.*, 2010, **82**, 1608.

44. F. Bedioui, D. Quinton, S. Griveau and T. Nyokong, *Phys. Chem. Chem. Phys.*, 2010, **12**, 9976.
45. Y. Li, C. Sella, F. Lemaitre, M. Guille Collignon, L. Thouin and C. Amatore, *Electroanalysis*, 2013, **25**, 895.
46. Y. Li, C. Sella, F. Lemaitre, M. Guille Collignon, L. Thouin and C. Amatore, *Electrochim. Acta*, 2014, **144**, 111.

Electrophoretic Methods for Separation of Peroxynitrite and Related Compounds

JOSEPH M. SIEGEL[a,b], RICHARD P. S. DE CAMPOS[a,b,c,d], DULAN B. GUNASEKARA[a,b,e], JOSÉ A. F. DA SILVA[c,d], AND SUSAN M. LUNTE*[a,b]

[a]Ralph N. Adams Institute for Bioanalytical Chemistry, University of Kansas, Lawrence, KS, USA; [b]Department of Chemistry, University of Kansas, Lawrence, KS, USA; [c]Chemistry Institute, State University of Campinas, SP, Brazil; [d]National Institute of Science and Technology in Bioanalytics, INCTBio, Brazil; [e]Department of Chemistry, University of North Carolina, Chapel Hill, NC, USA
*E-mail: slunte@ku.edu

7.1   Introduction

Peroxynitrite ($ONOO^-$) is a powerful oxidant produced from the reaction of nitric oxide (NO) and superoxide ($O_2{}^{\bullet-}$). In healthy cells, a balance exists between endogenous antioxidants and pro-oxidants to prevent oxidation, nitrosation, and nitration reactions with cellular macromolecules, as well as reactions of peroxynitrite with small molecules such as carbon dioxide. However, when this system is not in homeostasis, $ONOO^-$ can cause cytotoxicity by reacting with important biomolecules such as proteins, nucleic acids, and

RSC Detection Science Series No. 7
Peroxynitrite Detection in Biological Media: Challenges and Advances
Edited by Serban F. Peteu, Sabine Szunerits, and Mekki Bayachou
© The Royal Society of Chemistry 2016
Published by the Royal Society of Chemistry, www.rsc.org

lipids, resulting in inhibited or altered function.[1,2] For this reason, peroxynitrite has been linked to a number of disease states including cancer, stroke, myocardial infarction, atherosclerosis, hypertension, chronic heart failure, Alzheimer's disease, Parkinson's disease, multiple sclerosis, and amyotrophic lateral sclerosis.[1,3,4]

Peroxynitrite is known to react with numerous intracellular molecules or degrade to produce other reactive nitrogen and oxygen species (RNOS), including nitrite (NO_2^-), nitrate (NO_3^-), nitrogen dioxide, NO, carbonate radical, and hydroxyl radical (Figure 7.1).[1,2,5–8] Many of these species interfere with fluorescence and electrochemical detection methods for peroxynitrite.[9] Therefore, to better understand the role of $ONOO^-$ in disease, a separation method to isolate peroxynitrite from potential interference is necessary. Peroxynitrite naturally degrades into NO_2^- and NO_3^-.[5] Several capillary electrophoresis (CE)- and microchip electrophoresis (ME)-based methods have been reported to measure these two degradation products in biological matrices.[10–15] Less work has been done on the separation and direct detection of $ONOO^-$.

Peroxynitrite is difficult to detect directly in biological samples due to its very short half-life (<10 ms), which is due to its many reactions with intracellular macromolecules and small molecules. However, under alkaline conditions, the half-life is greatly increased because the compound is kept in its more stable anionic form.[16] At pH values lower than 10, a significant amount of the protonated form of peroxynitrite [peroxynitrous acid (ONOOH)] exists that readily degrades into NO_2^- and isomerizes into NO_3^-.[17] CE and ME methods are particularly well suited for separations at a high pH because, in contrast to most chromatographic methods, there is no silica stationary phase that can be dissolved. Additionally, electrophoresis typically provides faster and more efficient separation for small ions than chromatography. Therefore, the use of CE and ME is advantageous for peroxynitrite separations because the sample can be analyzed quickly in a high pH run buffer.

Currently, only ultraviolet (UV) absorbance and electrochemical detection (EC) have been used to directly detect peroxynitrite following CE and ME.[18–20]

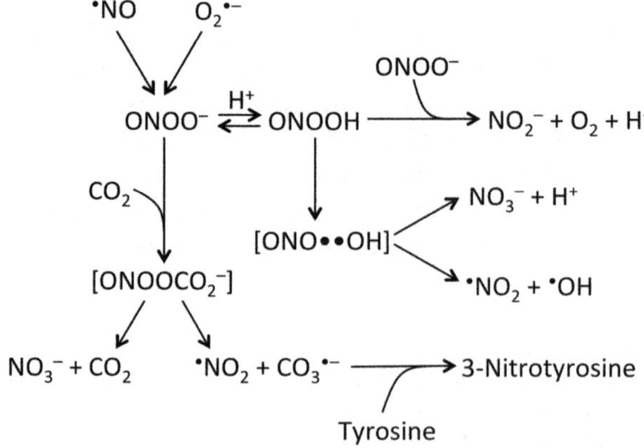

Figure 7.1 Diagram of $ONOO^-$ reaction pathways. Adapted from ref. 8 with permission from Elsevier.

These two techniques are easily coupled to conventional CE, while electrochemical detection is more compatible with ME. This is because the planar format of ME makes it more challenging to incorporate absorbance detection without the use of fiber optics.

This chapter reviews the electrophoretic methods that have been developed for the separation and subsequent detection of peroxynitrite as well as its metabolites and degradation products.

7.2   Electrophoresis

Electrophoresis describes a family of techniques that separates molecules based on the ratio of their charge-to-hydrodynamic radius. In this method, the capillary or channel is first filled with a buffered solution, followed by the injection of sample and subsequent application of a high voltage to induce the separation. In conventional slab gel electrophoresis, the gel is incorporated into the system to cause molecular sieving of large molecules as well as to reduce Joule heating. CE and ME can be used to separate both large and small molecules.[21–28] Once the capillary or channel is filled with run buffer, the sample is injected at one end and then a voltage is applied across the separation region. Following the application of an electric field, anions migrate toward the positive electrode (anode) and cations migrate toward the negative electrode (cathode). Neutral molecules will not move because they are unaffected by the electric field (Figure 7.2A). The velocity at which the ion will migrate toward its respective electrode (v) is defined as:

$$v = \mu E \tag{7.1}$$

where E is the strength of the electric field and μ is the ion's electrophoretic mobility.[21,22] Since each ion has a unique electrophoretic mobility, it is possible to separate multiple ions simultaneously in an electric field. The electrophoretic mobility is defined as:

$$\mu = \frac{q}{6\pi\eta r} \tag{7.2}$$

where q, η, and r are the net charge of the molecule, viscosity of the run buffer, and solvated ionic radius of the molecule, respectively.[22]

7.2.1   CE

CE is a commonly used separation technique, and, therefore, its theory and applications have been heavily reviewed.[21–25]

In CE, separations are performed in a fused silica capillary that is typically 20–100 cm long with an internal diameter of 10–200 μm (Figure 7.2B). A bulk flow called the electroosmotic flow (EOF) is generated due to the presence of an electric double layer (EDL) at the wall of the fused silica capillary. The wall is negatively charged above pH 3 due to the presence of ionized silanol groups. As the pH increases, more silanol groups are deprotonated, resulting in a more

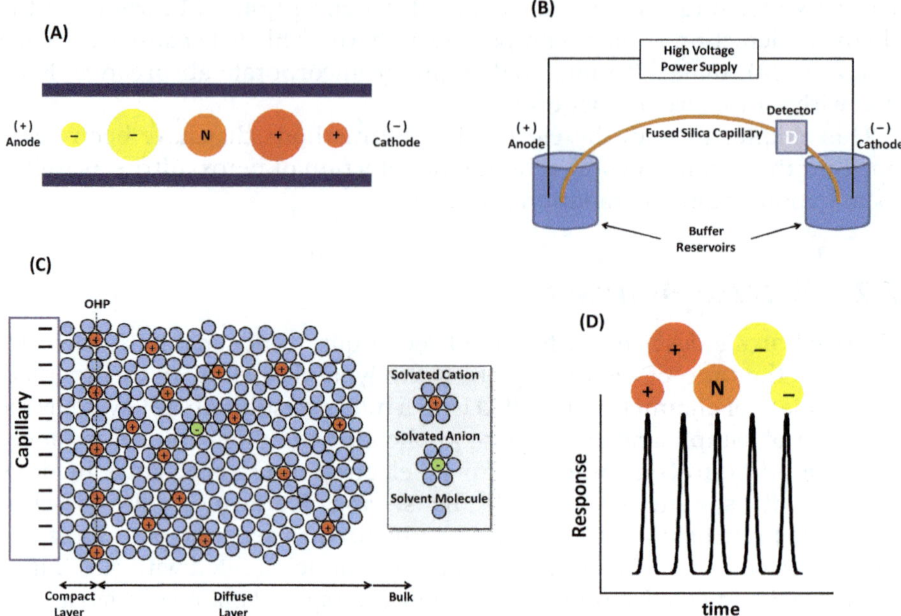

Figure 7.2 (A) Diagram of the separation of molecules in an electrophoretic system based on their charge-to-hydrodynamic radius ratio. (B) Schematic of a typical CE setup. (C) Schematic of the electrical double layer on a fused silica capillary. (D) Sample electropherogram obtained under normal polarity separation conditions. OHP: outer Helmholtz plane.

negatively charged wall and thicker EDL. It is the attraction of the solvated cations present in the background electrolyte (BGE) to the capillary wall that forms the EDL. The EDL consists two parts: (1) a compact layer of solvated cations that are strongly attracted to the negatively charged wall; and (2) a diffuse layer of solvated cations that have a weaker attraction to the wall (Figure 7.2C). When a voltage is applied across the capillary, the solvated cations in the diffuse layer migrate toward the cathode and drag the bulk solution with them, causing the EOF. In the presence of this EOF, the apparent (experimental) mobility of a molecule (μ_{app}) in the capillary will be the vectorial sum of the molecule's actual electrophoretic mobility (μ_{ep}) and the mobility of the EOF (μ_{eof}).

$$\mu_{\text{app}} = \mu_{\text{ep}} + \mu_{\text{eof}} \tag{7.3}$$

In "normal polarity" CE, a positive voltage is applied across the capillary and detection occurs at the cathode. In this case, cations are moving in the same direction as the EOF and, thus, will be detected first. Meanwhile, anions will be attracted to the anode but will still be dragged to the cathode by the EOF as long as the μ_{eof} is larger than the μ_{ep} of the compound of interest. The EOF also causes the movement of neutral species down the capillary at the same velocity as the bulk flow (Figure 7.2D). The major advantage of EOF for CE separations is the ability to detect most of the molecules present in a system at one end of the capillary in a single run regardless of their net charge.

The magnitude of the EOF is dependent on a number of factors. In particular, the ionic strength of the run buffer used for the separation will change the magnitude of the EOF because it directly influences the zeta potential of the EDL. The zeta potential is defined as the potential at the shear plane between the adsorbed and mobile layers of the EDL. Therefore, it describes the strength of electrical attraction between the buffer ions and the charged wall. A decrease in the ionic strength will cause an increase in the zeta potential, thus increasing the magnitude of the EOF. This faster EOF will lead to shorter migration times; however, it could also lead to decreased resolution. Additionally, high ionic strengths can lead to Joule heating within the capillary, causing convection and a reduction in the separation efficiency. In extreme cases, the temperature increase can cause the buffer to outgas and form bubbles, which will interrupt the electrical current through the capillary. Therefore, the run buffer must be optimized to produce the ideal EOF strength and provide the best separation possible.

Surfactants can be added to the CE run buffer to dynamically modify the EOF within the capillary. A cationic surfactant, consisting of a positively charged head group and a nonpolar tail, can be added to the run buffer to reverse the EOF. These surfactants adsorb to the walls of the capillary and form a bilayer that creates a positively charged wall. When using reverse polarity (applying a negative electric field across the capillary), the EOF will flow toward the anode, and the order in which the molecules migrate will be reversed. This approach is ideal for the separation of small, negatively charged molecules such as $ONOO^-$, as well as its anionic metabolites and degradation products.

7.2.1.1   *Sample Injection in CE*

Samples are introduced into the capillary for CE by either hydrodynamic or electrokinetic injection.[22] Hydrodynamic injections are based on a pressure difference between the sample and the capillary. The amount of sample that enters the capillary is dependent on the injection time, pressure, and run buffer viscosity. On the other hand, electrokinetic injections use an applied voltage to introduce ions into the capillary. This form of injection is size and charge biased because each sample component is injected based on its electrophoretic mobility, which leads to higher amounts of small ions of the same charge being injected compared with larger ions.[22] Therefore, this approach can be advantageous for some analyses because the injection process can be used to selectively inject small anions while minimizing the amount of large anions, neutral molecules, and cations that are injected into the capillary.

7.2.2   **ME**

ME separations are based on the same principles as those in CE.[23] Analytes are separated based on their electrophoretic mobility, while also taking the EOF into consideration. However, separations on microchips are accomplished in a separation channel that has been fabricated in a planar substrate instead of a capillary.[26–28] Therefore, ME has some distinct advantages and disadvantages compared with CE.

The separation channels used in ME are normally much shorter than capillaries used in CE. Typical channel lengths are 5–15 cm compared with the 20–100 cm capillaries needed for CE. However, the electric field applied across the separation channel is still comparable to that of a CE system, making it possible to perform sub-minute separations on these devices. This makes it an ideal approach for the separation of unstable or highly reactive species, such as $ONOO^-$. A simple-t microchip design, as shown in Figure 7.3, is commonly used for ME. Typically, the separation and sampling channels and reservoirs are filled with run buffer and sample, respectively, and a high voltage is applied across the sample and separation channels, using a high-voltage power supply, by placing platinum (Pt) electrodes in reservoirs at both ends of the channels.[29]

Microchips can be fabricated from glass, silicon, low-temperature co-fired ceramics (LTCC), and polymer materials, such as poly(dimethylsiloxane) (PDMS), poly(methyl methacrylate) (PMMA), cyclic olefin copolymers (COC), and polyesters, among others.[27,30] Conventional microfluidic devices made of glass or polymers are commercially available and can be fabricated in-house using photolithography or, for some polymeric substrates, hot embossing techniques.

The fabrication of microchips for electrophoresis consists of transferring micrometric structures, such as reaction microchambers and microchannels, from a photolithography mask to a substrate. For example, glass microchips are produced by exposing a glass substrate previously coated with a positive photoresist layer and a chromium layer to UV radiation. Next, the patterned structure on the substrate is removed using a photoresist developer and the

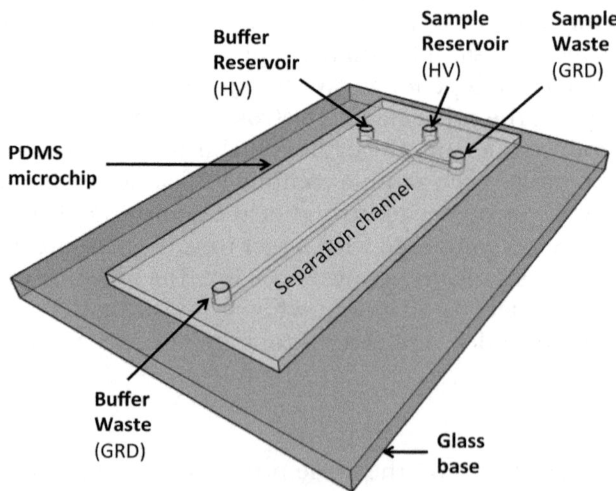

Figure 7.3   Schematic of a ME system that utilizes a simple-t microchip design with a gated injection system. GRD: ground; HV: high voltage.

exposed chrome surface is etched with an etchant. The exposed glass is then etched with buffered oxide etchant to provide the channels. To complete the microdevice, the substrate containing the channels is sealed against a second piece of glass. This process usually requires the exposure of both substrates to high temperatures inside a vacuum oven, which creates an irreversible bond between the two glass substrates.[31] Figure 7.4A summarizes the fabrication of glass microdevices.

The major advantage of glass-based microchips is that the surface chemistry of the separation channel is similar to that of a fused silica capillary. This makes transferring methods from CE to ME easier, and also, the presence of the free silanols leads to a generally reproducible EOF. However, the fabrication process for these microchips is time-consuming and costly. Additionally,

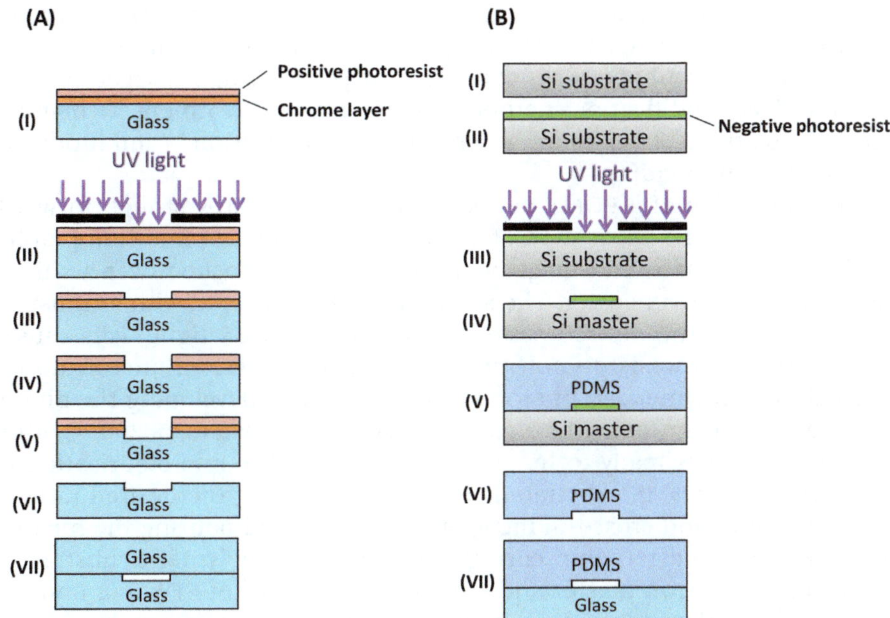

Figure 7.4   (A) Fabrication of glass microfluidic devices. (I) Soda-lime glass is coated with chromium and positive photoresist. (II) The substrate is covered with a mask containing the microfeatures and exposed to UV. (III) The photoresist is developed to remove UV exposed photoresist. (IV) The unprotected chrome layer is etched. (V) Glass is etched. (VI) Photoresist and chromium layers are removed to reveal the microfeatures. (VII) The substrate is sealed against a flat glass substrate to form the final microdevice. (B) Fabrication of PDMS microdevices. (I) A silicon substrate is used as the master substrate. (II) The negative photoresist layer is deposited. (III) The substrate is covered with a mask containing the microfeatures and exposed to UV to polymerize the photoresist. (IV) Development of photoresist removes all non-exposed photoresist, leaving only the microfeatures. (V) PDMS is poured over the silicon master. (VI) PDMS containing the microfeatures is peeled off the silicon master and (VII) sealed against another substrate (*e.g.*, glass or PDMS) to form the enclosed microchip.

depending on the nature of the analytes, the microchannels can become clogged during experiments; this blockage can be hard to remove on irreversibly bonded devices.

For these reasons, polymeric substrates are usually chosen for method development. Of all the polymeric substrates available, PDMS is most commonly used due to its elastomeric properties, optical transparency, biocompatibility, low cost, and ease of fabrication.[32,33] These properties make PDMS ideal for device prototyping and analytical method development. The major disadvantage of PDMS for ME systems is that the polymer surface is hydrophobic and contains relatively few silanol groups. The absence of a uniformly charged surface causes the EOF to be unstable and leads to changes in the migration times of analytes over the lifetime of the chip.[34,35] Thus, to improve the stability of the EOF on a PDMS device, ionic surfactants, such as sodium dodecyl sulfate (SDS), are often added to the running buffer as a dynamic surface modification agent.[36,37] The hydrophobic tail of the surfactant interacts with the hydrophobic walls of the channel, generating a charged surface. Other surface modification methods include plasma treatment,[38] chemical vapor deposition,[39] layer-by-layer deposition,[40] silanization,[41] and modification by covalent bonding.[32]

The fabrication of a PDMS device (Figure 7.4B) is fairly straightforward. The first step is the fabrication of a silicon master using photolithography. To fabricate the master, a negative photoresist is deposited over a silicon substrate with a spin coater. The substrate is then covered with a transparency mask with the desired pattern and exposed to UV light. This causes the exposed photoresist to polymerize, forming the desired microfeatures. Then all non-polymerized photoresist is removed by developing the photoresist, and the features are hardened by baking the master at a 200 °C. After this, the master is ready to be used as a mold for the microfabrication of the device. The PDMS substrate is prepared by mixing the desired ratio of its pre-polymer and cross-linking agent solutions, and pouring the mixture over the silicon master. After curing the mixture, the polymeric substrate is peeled off the silicon master and sealed to a flat piece of PDMS or another substrate (*e.g.*, glass containing working electrodes for amperometric detection) to form the final microdevice in which the electrophoresis separation will be performed.

7.2.2.1   *Sample Injection in ME*

ME systems most commonly use either a gated or pinched injection for sample introduction. The gated injection is considered an electrokinetic injection. To perform this type of injection, a high voltage is applied to both the buffer and sample reservoirs, which establishes a "gate" at the intersection of the two voltages. The high voltage in the buffer reservoir is then floated for a short time to open the gate and permit a small amount of sample to enter the separation channel (Figure 7.5A). With a pinched injection, a high voltage is first applied to the sample reservoir, and then sample is injected by applying

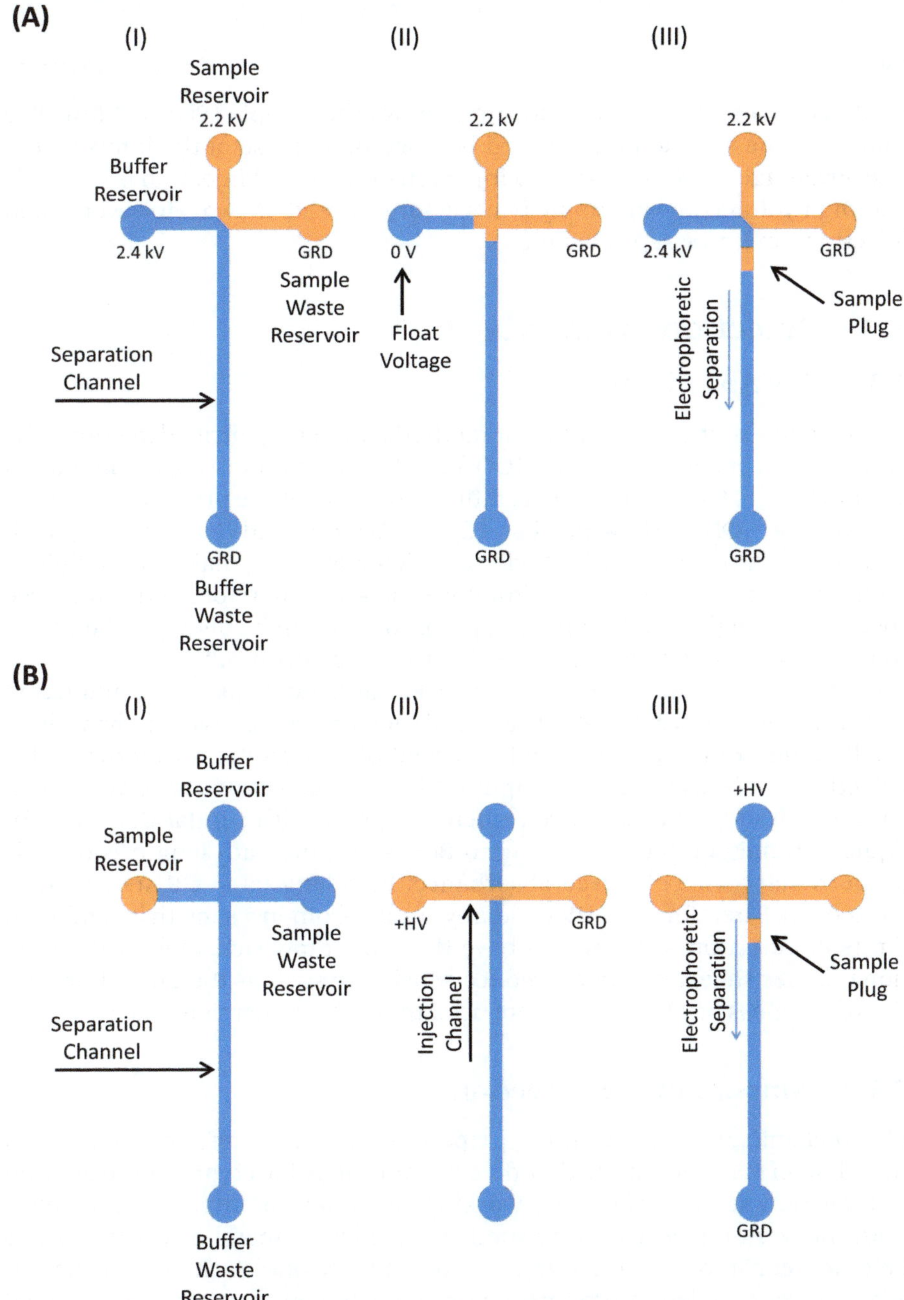

Figure 7.5 (A) ME schematic for the gated injection process. (I) A gate is established upon application of an external voltage to both the buffer and sample reservoirs. (II) The voltage at the buffer reservoir is floated to allow the sample to enter the separation channel. (III) The gate is reestablished, thereby injecting the sample plug and starting the separation. (B) ME layout for pinched injection. (I) Sample is loaded into sample reservoir. (II) The injection channel is filled upon application of an electric field between the sample and sample waste reservoirs. (III) The injection occurs after switching the applied electric field to between the buffer and buffer waste reservoirs. GRD: ground; HV: high voltage.

a voltage in a perpendicular direction across the sample (Figure 7.5B). The volume of the sample plug injected is based on the size of the intersection. The advantage of using a pinched injection over a gated injection is that only one high voltage power supply is necessary instead of two. However, gated injections are more reproducible.

7.3   Detection Strategies

7.3.1   UV Absorbance

UV detection is one of the simplest methods for peroxynitrite detection. The anionic form of peroxynitrite ($ONOO^-$) exhibits a maximum absorbance at λ = 302 nm (ε = 1670 M^{-1} cm^{-1}).[42] This differs from that of the protonated form of peroxynitrite (ONOOH), which has a λ_{max} of 240 nm, and NO_2^- and NO_3^- with a λ_{max} of 214 nm. The use of a diode array detector (DAD) makes it possible to monitor multiple wavelengths simultaneously and therefore detect all three species in a single run. In this case, peroxynitrite can be distinguished from NO_2^- and NO_3^- based on its unique absorbance at 302 nm.[18]

UV absorbance can be coupled with CE, but is rarely used in conjunction with ME. Since most fused silica capillaries are coated with polyimide, a small section of the coating can be burned off to produce a window for UV detection. A disadvantage of coupling CE to UV detection is the very small path length of the detection cell when compared with similar detectors for liquid chromatography. According to Beer's law, the path length is directly proportional to the observed absorbance. Path lengths in CE are generally 25–100 µm. Therefore, CE-UV generally suffers from low sensitivity and high limits of detection. ME systems have the same drawback, along with additional issues that can severely reduce sensitivity such as the optical properties of the glass or plastic substrate used to produce the microchip.

7.3.2   Amperometric Detection

Electrochemical detection in the amperometric mode (referred to as EC in this chapter) is a technique that detects molecules based on their oxidation or reduction at the surface of a working electrode. In amperometric detection, the working electrode is usually composed of metal or carbon and is held at a constant potential *vs.* a reference electrode. A change in the faradaic current, which is proportional to analyte concentration, is observed when there is an oxidation or reduction at the electrode surface.[43,44] Only electrochemically active species can be monitored with this technique, and most biological samples contain many interfering electrochemically active species.[45] Since EC does not have a separation aspect, it can be coupled with either CE or ME to increase the selectivity by separating the analytes of interest from interferents.

Microelectrodes are the most commonly employed working electrodes for ME and CE.[46,47] They provide high signal-to-noise ratios (S/N) and are compatible with the dimensions of the capillary or separation channel.[43,44,48] Microelectrodes can be incorporated into ME more easily than CE due to the planar format of ME devices. The most common microelectrode material used for EC is carbon; however, metal microelectrodes [*e.g.*, Pt or palladium (Pd)] facilitate better oxidation of $ONOO^-$ and related compounds. Metal electrodes can be fabricated either on top of or into a trench etched in a glass substrate, and a PDMS microchip containing the separation channel is aligned on top to complete the device.

Metal sputtering is the most commonly employed protocol for the fabrication of metal electrodes on a glass substrate. First the glass substrate is thoroughly cleaned and a layer of titanium (Ti) is deposited on the substrate to increase metal adhesion. This is followed by the deposition of a second layer consisting of the desired metal (Pt or Pd). Positive photoresist is then spin-coated on top of the metal layer. To produce the microfeatures, a mask containing the design of the microelectrodes is then placed over the substrate and exposed to UV light. The non-exposed photoresist is then removed by the developer. Excess metal is removed using aqua regia and Ti etchant. The remaining photoresist is then removed with acetone or isopropyl alcohol, revealing the desired metal electrodes (Figure 7.6).

Even though this fabrication method is commonly used, the microelectrodes produced are very thin (approximately 100–200 nm) and can be easily destroyed. When using these microelectrodes during the reversible sealing and subsequent removal of the PDMS layer from a glass substrate, the microelectrodes can peel off the glass. This limits the number of times the electrode plates can be used. More recently, the Lunte group has developed an alternative fabrication method that consists of controlled etching of a trench in the glass prior to metal deposition. This method provides electrodes that are more rugged and can therefore be reused numerous times with new PDMS substrates.[49]

7.3.3　Conductivity Detection

The most popular way to perform conductivity detection in ME or CE is using capacitively coupled contactless conductivity detection (C^4D). C^4D has been extensively used in CE and ME because it is a universal detector that can provide label-free detection of the analytes of interest. In this mode of detection, the electrodes do not come into contact with the solution inside the capillary or separation channel. The detector response in C^4D is proportional to the conductivity of this solution when passing the electrodes.[50] When an analyte plug passes the electrode region, the conductivity in that region will increase or decrease based on the difference in conductivity of the given analyte and the BGE, thereby generating a signal.[51,52] Since no reaction occurs at the surface of the electrodes, no electrode fouling is observed, and larger electrodes (width up to 1 mm) can be used to simplify fabrication.[32,53] Additionally, C^4D is a non-destructive detection method that can be performed using non-precious metals as electrodes, such as copper, aluminum foil, and adhesive

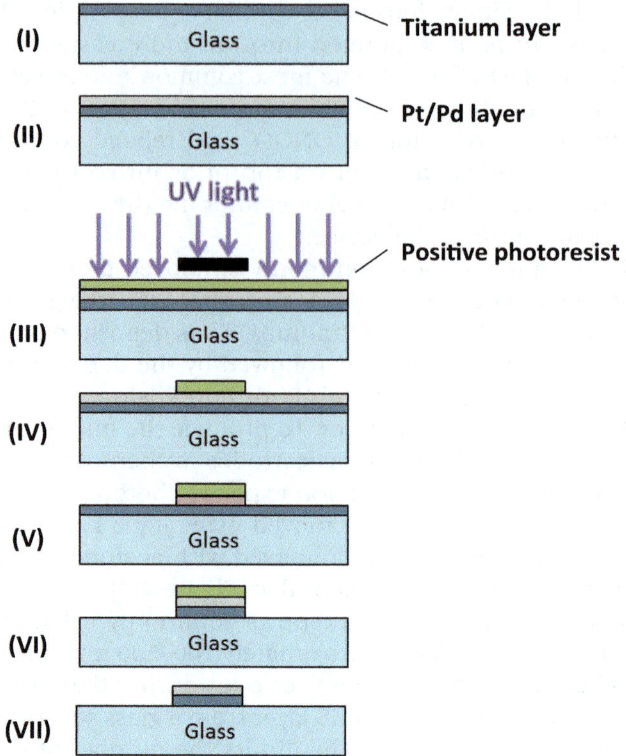

Figure 7.6   Fabrication of metal electrodes on glass substrate. (I) A layer of Ti is deposited on top of the glass substrate. (II) A layer of the desired electrode metal is deposited on top of the glass substrate. (III) The substrate is spin-coated with positive photoresist, covered with a transparency mask containing the microfeatures, and exposed to UV light. (IV) The photoresist is developed to remove all exposed photoresist. (V) Excess electrode metal is removed with aqua regia. (VI) Excess Ti is removed with Ti etchant. (VII) Non-exposed photoresist is removed to uncover the microelectrode.

metal tape.[50] This can provide simple and low cost fabrication methods that can be coupled with other detection methods.[54–56]

In CE, C^{4}D is usually performed by placing two tubular electrodes around the silica capillary, whereas in ME the electrodes are built into the planar geometry and isolated from the microchannel by a dielectric material layer.[57]

For ME-C^{4}D of ONOO$^-$ degradation, Vázquez *et al.* fabricated C^{4}D electrodes using a commercial printed circuit board (PCB) substrate containing a copper layer on its surface (Figure 7.7).[53] The electrode geometry was designed using graphical software and then printed on a sheet of wax paper using a toner cartridge printer. The printed toner layer was then heat-transferred to the PCB surface using a hot embossing machine. Then, the copper not protected by the toner layer was etched with an iron(III) chloride solution.

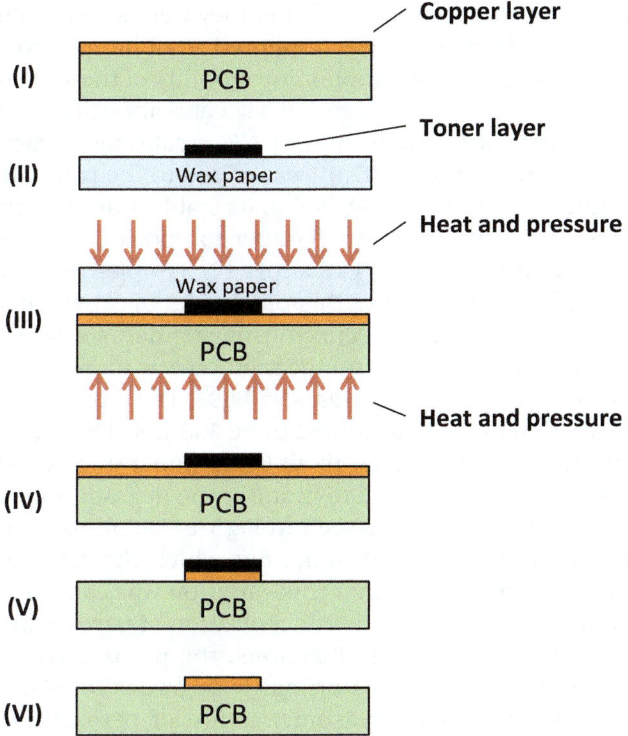

Figure 7.7   Fabrication of copper electrodes for C⁴D detection. (I) A PCB substrate containing a copper layer is used. (II) Electrode geometry is printed in toner on waxed paper. (III) The substrate and waxed paper are put together and submitted to heat and pressure. (IV) The toner layer is transferred to the PCB substrate. (V) The copper layer is etched. (VI) The toner layer is removed to reveal the copper electrode. Adapted from ref. 53 with permission of The Royal Society of Chemistry.

Next, the toner layer was removed using acetonitrile to reveal the copper electrodes. Using this approach, the authors demonstrated a simpler and more cost-effective method than the more common photolithographic approach needed to produce Pt electrodes on a glass substrate.

7.4   Electrophoretic Separation and Direct Detection of Peroxynitrite

7.4.1   CE-UV

Frankenfeld *et al.* were the first to develop a CE-UV method for the detection of peroxynitrite using a Hewlett-Packard[3D] CE system with a DAD.[18] The separation conditions for NO_2^-, NO_3^-, and peroxynitrite were initially optimized using NO_2^- and NO_3^- standards due to the instability of peroxynitrite. Since these are both degradation products of peroxynitrite, they were also present in the per-oxynitrite standard. The negative electrophoretic mobilities of NO_2^-, NO_3^-, and $ONOO^-$ are greater than the electrophoretic mobility of the EOF, so the anions do not migrate out of the cathodic end of the capillary using normal polarity. Therefore, to separate these molecules by CE, a cationic surfactant [dodecyl trimethylammonium bromide (DTAB)] was added to the run buffer to reverse the EOF. In addition, to keep the $ONOO^-$ in its stable anionic form, run buffer at pH 12 was used, and the optimal run buffer composition was determined to be 25 mM K_2HPO_4 and 7.5 mM DTAB at pH 12. Samples were injected using an electrokinetic injection because this provided a bias toward small anions when applying negative polarity. Peroxynitrite standards were run under the optimized conditions at 302 nm, and the method was found to be linear from 0.545 to 35 mM of peroxynitrite with an R^2 = 0.995. The limit of detection (LOD; S/N = 3) was experimentally determined to be 353 µM. The separation of per-oxynitrite from NO_3^- and NO_2^- took about 6 min under these conditions.[18]

The method was then employed to monitor the degradation of a peroxyni-trite standard at pH 12. As can be seen in Figure 7.8, peroxynitrite degrades at a constant rate until about 80 min, after which the rate begins to vary. Frankenfeld *et al.* hypothesized that this variation was caused by the modes of decomposition shifting when the concentration of peroxynitrite decreases to a certain point.[18] At high concentrations, the peroxynitrite sample con-taining both $ONOO^-$ and ONOOH primarily degrades to NO_2^-; however, as the concentration decreases, the isomerization of peroxynitrite into NO_3^- becomes the prevalent path of degradation.[17]

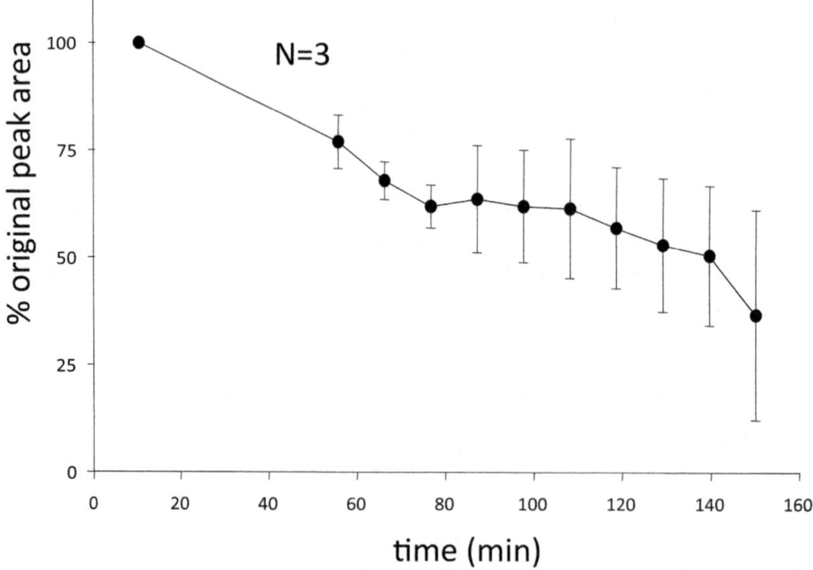

Figure 7.8   Degradation of peroxynitrite in alkaline conditions (pH 12). Reprinted from ref. 18 with permission from Elsevier.

This method was also used to analyze two peroxynitrite samples that were synthesized using different procedures to determine their purity. Figure 7.9A shows an electropherogram obtained from a peroxynitrite sample prepared at the University of Kansas in Dr Schöneich's laboratory. This sample was prepared by bubbling ozone gas through a sodium azide solution at pH 12 and subsequently removing excess ozone from the solution with a potassium iodide trap.[58] The other peroxynitrite sample was a commercially available peroxynitrite standard; its respective electropherogram can be seen in Figure 7.9B. Commercial peroxynitrite samples are usually prepared by reacting isoamyl NO_2^- and hydrogen peroxide under alkaline conditions. Excess hydrogen peroxide is then removed using a MnO_2 column.[59] The concentrations of peroxynitrite in both samples are comparable, and the profiles of the electropherograms obtained at 214 nm differ only by the presence of an unidentified peak and azide peak in the commercially available sample and ozone–azide sample, respectively.[18]

CE with UV detection was shown to be able to successfully separate and detect peroxynitrite and its degradation products and synthesis precursors. However, the LOD was not low enough to detect peroxynitrite in biological samples.[18] Furthermore, even though this method is very selective, the separations take several minutes. Since peroxynitrite is highly reactive and can degrade quickly, a separation method with a better temporal resolution, such as ME, would be highly beneficial.

7.4.2 ME-EC

In an attempt to improve both the temporal resolution (separation time) and LODs for peroxynitrite, Hulvey *et al.* evaluated ME-EC for the direct detection of peroxynitrite.[19] A 3.5 cm simple-t PDMS/glass hybrid microchip with a 50 μm

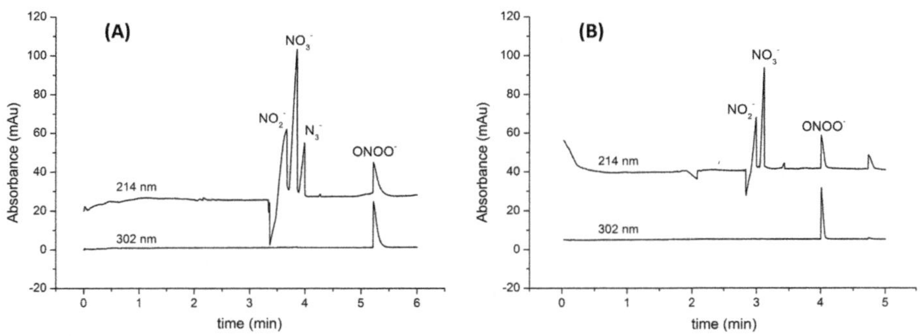

Figure 7.9   Comparison of electropherograms of (A) a peroxynitrite sample synthesized at the University of Kansas using the reaction of ozone and azide, and (B) a commercially available peroxynitrite standard. Parameters: run buffer 25 mM K_2HPO_4 and 7.5 mM DTAB at pH 12, electric field -323 V cm^{-1}, electrokinetic injection -20 kV for 1 s, UV detection at 302 and 214 nm. Reprinted from ref. 18 with permission from Elsevier.

Pd working electrode aligned at the end of the channel was employed. This ME-EC system was optimized for the separation and detection of peroxynitrite and NO_2^- standards. The optimized voltages applied to the buffer and sample reservoirs were −1400 and −1200 V, respectively. These voltages were chosen because they offered the best separation while still maintaining short run times (50 s). Boric acid was used as the run buffer and was tested over a pH range of 10.5–12. The use of a pH 12 buffer resulted in the best peroxynitrite stability, but the high pH caused an increase in the channel currents, which led to Joule heating and an increase in background noise. Therefore, a slightly lower pH of 11 was used for the separation. Since the EOF had to be reversed, 2 mM tetradecyltrimethylammonium bromide (TTAB) was also employed. The final run buffer composition chosen was 10 mM boric acid and 2 mM TTAB at pH 11.[19]

Peroxynitrite standards purchased from Cayman Chemicals were analyzed with this method. The identity of the peroxynitrite peak was confirmed based on three criteria: comparison of relative migration times to previously performed CE experiments, characterization by voltammetric ratio, and the degradation pattern. The voltammetric ratio for each peak was determined by measuring the current response at +900 and +1100 mV *vs.* Ag/AgCl reference electrode during two separate runs. The peroxynitrite peak exhibited a large current ratio, which indicates that it is more easily oxidized. The peak also degraded very quickly, which is characteristic of a highly unstable or reactive species (Figure 7.10). After the identity of the peak was confirmed, a calibration curve was created. This method provided lower LODs than the CE-UV method. The LOD (S/N = 3) for peroxynitrite was experimentally determined

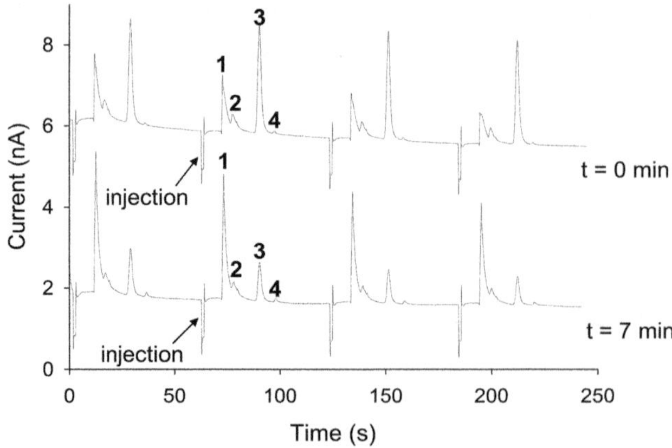

Figure 7.10 Separation and identification of $ONOO^-$ and NO_2^- peaks in a commercially available sample with ME-EC. Peak 1 = unidentified; peak 2 = NO_2^-; peak 3 = $ONOO^-$; peak 4 = unidentified. The $ONOO^-$ peak shows significant degradation after 7 min. Parameters: run buffer 10 mM boric acid and 2 mM TTAB at pH 11, electric field −283 V cm^{-1}, 1 s gated injection, working electrode potential of +1100 mV *vs.* Ag/AgCl. Reprinted with permission from ref. 19. Copyright 2010 American Chemical Society.

to be 2.4 μM. The linear range was found to be 3.12–100 μM, and the R^2 for the calibration curve was 0.9979.[19]

This method was then applied to measure the peroxynitrite produced by 3-morpholinosydnonimine-*N*-ethylcarbamide (SIN-1). SIN-1 is a metabolite of the drug molsidomine, which is a potent vasodilator.[60] Under alkaline conditions, SIN-1 produces NO and superoxide in excess, thereby resulting in the formation of peroxynitrite (Figure 7.11). This reaction provides a means of creating a detectable amount of peroxynitrite over time. SIN-1 has been shown to produce ONOO⁻ at a yield of approximately 1% based on the original concentration of SIN-1.[61] Electropherograms for the SIN-1 samples are shown in Figure 7.12A, where it can be seen that the ONOO⁻ concentration increases over time. Figure 7.12B is a graph of the ONOO⁻ concentration obtained for each run. This plot shows that the ONOO⁻ concentration continuously increases to about 12.5 μM, at which point the curve begins to plateau. The expected final concentration of ONOO⁻ was estimated to be 48 μM based on a 1% reaction yield. Hulvey *et al.* proposed that this low yield was likely due to the reaction between peroxynitrite and ambient carbon dioxide while the sample was sitting in the reservoir of the microchip.[19]

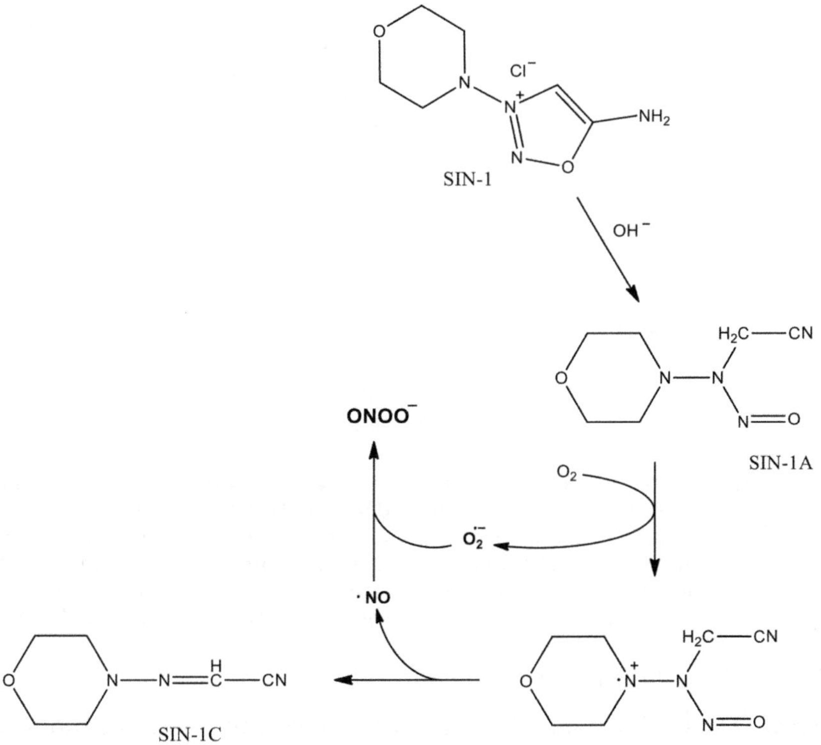

Figure 7.11 Diagram of peroxynitrite production by SIN-1. Reprinted with permission from ref. 19. Copyright 2010 American Chemical Society.

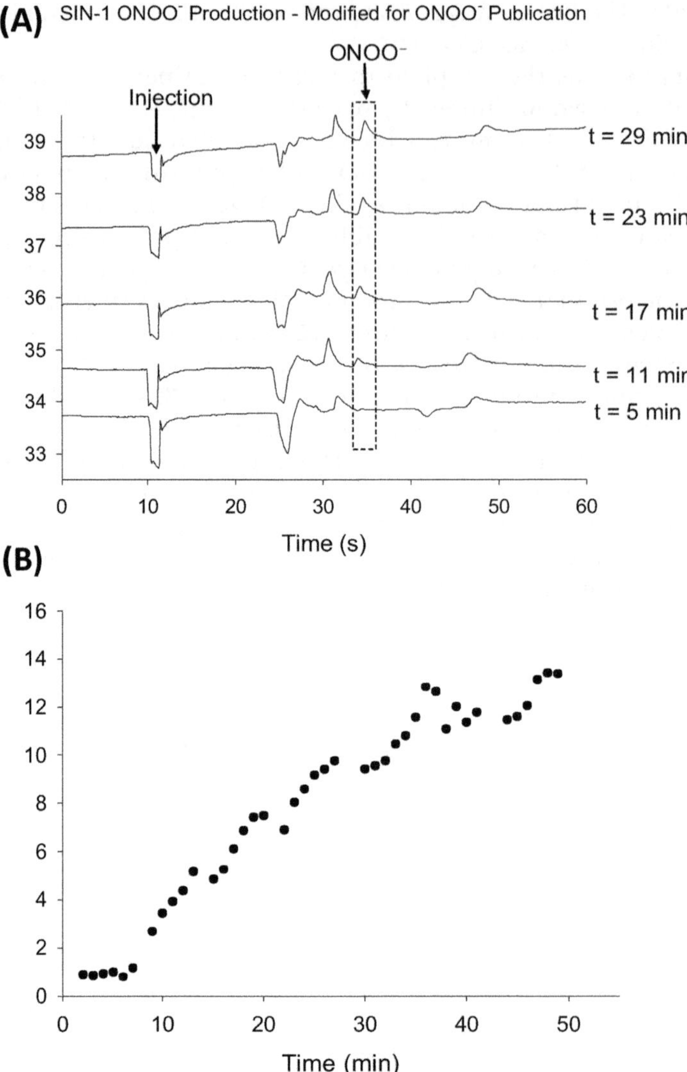

Figure 7.12　(A) Electropherograms depicting ONOO⁻ production by SIN-1 over time. ONOO⁻ peak labeled by dashed box. (B) ONOO⁻ generation by SIN-1 over 50 min. Each point corresponds to the ONOO⁻ peak obtained during that injection. Parameters: run buffer 10 mM boric acid and 2 mM TTAB at pH 11, electric field -283 V cm^{-1}, 1 s gated injection. Reprinted with permission from ref. 19. Copyright 2010 American Chemical Society.

7.4.2.1　In-Channel Electrode Alignment

The ONOO⁻ detection method developed by Hulvey *et al.* (described above) was further optimized by Gunasekara *et al.* by the implementation of in-channel working electrode alignment.[20] In Hulvey's work, the working electrode was aligned end-channel, or after the intersection of the separation channel and

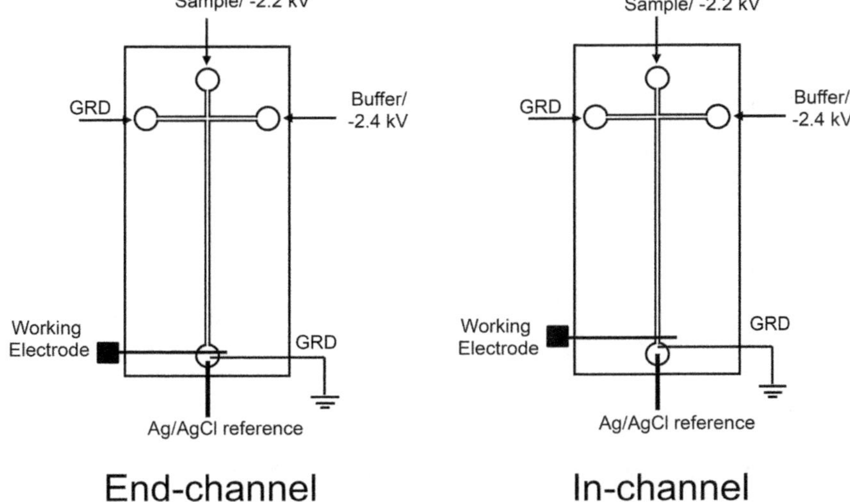

Figure 7.13   Schematics for end-channel and in-channel electrode alignment. GRD: ground. Adapted from ref. 20 with permission. Copyright © 2011 WILEY-VCH Verlag GmbH & Co. KGaA, Weinheim.

the waste reservoir (Figure 7.13).[19] Upon exiting the separation channel, the analytes can diffuse into the waste reservoir, which leads to band broadening and decreased resolution. On the other hand, with the in-channel alignment, the working electrode is placed in the separation channel (Figure 7.13). This alignment allows for the detection of analytes prior to their reaching the waste reservoir, reducing (or eliminating) band broadening while still maintaining the detector sensitivity.[62,63] However, since the working electrode that is placed in the channel is not decoupled from the electric field, an electrically isolated potentiostat must be employed or the electronics will be destroyed.[63]

A standard consisting of NO_2^- and possible cellular interferences was used to illustrate the benefits of using in-channel alignment over end-channel alignment. Gunasekara *et al.* witnessed a 2.4-fold decrease in LOD, 1.7-fold increase in sensitivity, and 1.6-fold increase in the number of theoretical plates for NO_2^- when using in-channel detection with a Pt working electrode. They also utilized the in-channel working electrode alignment to monitor the degradation of peroxynitrite in commercial samples.[20] Figure 7.14 depicts a series of injections obtained with ME-EC. The in-channel alignment was found to provide better resolution and separation efficiencies for a mixture of $ONOO^-$ and NO_2^- than those previously reported by Hulvey *et al.*[19]

In the future, the enhanced separation efficiency of in-channel alignment in conjunction with new electrode configurations, such as dual electrodes, will offer a better means of detecting and identifying peroxynitrite in very complex matrices. Commercially available peroxynitrite samples are often contaminated with hydrogen peroxide, and biological samples contain many electrochemically

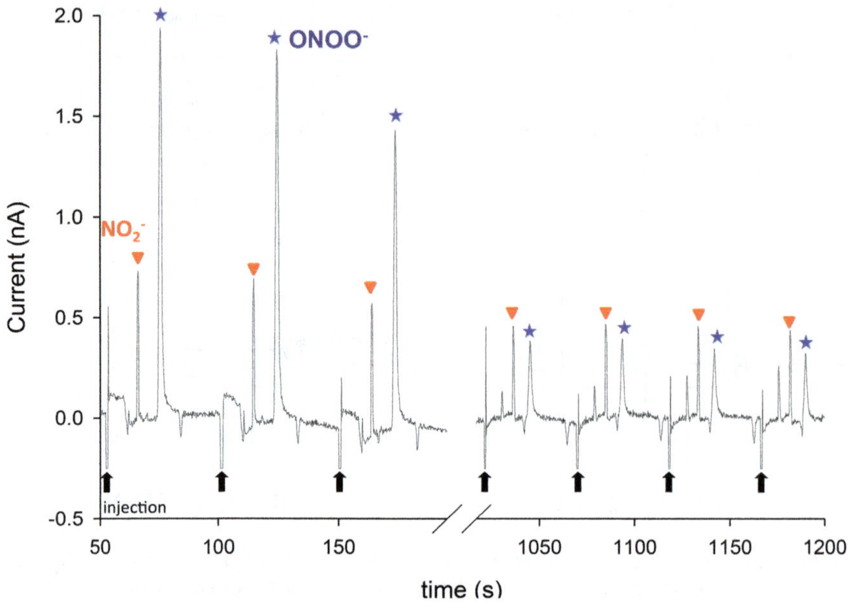

Figure 7.14   Separation of ONOO⁻ (100 μM) from interferents in a commercially available sample of ONOO⁻ using an in-channel working electrode alignment. New injections are indicated by ⬆. Parameters: run buffer 10 mM boric acid and 2 mM TTAB at pH 11, 1 s gated injection. Adapted from ref. 20 with permission. Copyright © 2011 WILEY-VCH Verlag GmbH & Co. KGaA, Weinheim.

active species that could potentially interfere with the detection of peroxynitrite. Although migration time can be used to make a tentative identification of peroxynitrite and related species in an electropherogram, the implementation of dual electrode detection systems makes it possible to obtain voltammetric information for the compounds of interest. Current ratios obtained at two different potentials can provide an additional means of post-separation species identification as well as supply information regarding peak purity.[64,65]

7.5   Indirect Detection of Peroxynitrite by Electrophoretic Methods

Direct detection of peroxynitrite, especially from living, native, or stimulated cells, can be a difficult task due mainly to its high reactivity and instability, but also because it exists in low concentrations.[66] On the other hand, ONOO⁻ metabolites and products from reactions with biomolecules tend to have higher stability and can therefore be used to indirectly monitor peroxynitrite production. Although many papers can be found regarding the detection of NO_3^- and NO_2^- following electrophoretic separation,[10–15] methods to separate and detect them simultaneously as an indicator of peroxynitrite degradation in a cellular environment are still being developed.

7.5.1   ME-EC/C^{4}D

Vázquez *et al.* combined amperometric detection and C^{4}D coupled to ME to monitor the two main peroxynitrite degradation products, NO_3^- and NO_2^-.[53] Since NO_3^- is not electrochemically active, the amperometric detector was only used to monitor NO_2^-. On the other hand, C^{4}D was able to detect both analytes. The device was used to monitor NO_3^- and NO_2^- that were generated through the degradation of ONOOH at pH 3.5.[53]

This was the first report of the integration of both detection techniques on a PDMS/glass hybrid chip. The advantage of this approach is that the detection systems are reusable for several microfluidic devices. To assemble the device, a 40 μm wide Pd working electrode for amperometric detection was fabricated over a 400 μm thick glass substrate using photolithography. The C^{4}D copper electrodes were fabricated over a PCB and isolated from the microchannel using a glass substrate containing the Pd electrodes. To assemble the microdevice, the glass substrate containing the Pd electrode was simply taped to the PCB substrate containing the Cu electrodes and the PDMS microchip was reversibly sealed to the glass substrate. This system utilized an Ag/AgCl reference electrode and a Pt auxiliary electrode. The final layout of the microchip is shown in Figure 7.15.[53]

An important aspect to consider when coupling these two detection methods is their mutual compatibility. In particular, the choice of BGE is critical. BGEs commonly used for amperometric detection, such as borate and phosphate, are considered highly conductive for C^{4}D and can cause high background noise and low sensitivity for high mobility analytes. Moreover, when separating two species with similar electrophoretic mobilities (in this case, NO_3^- and NO_2^-) the pH of the BGE plays a major role, emphasizing the necessity of optimized separation conditions.

In their work, Vázquez *et al.* were able to achieve the separation of NO_3^- and NO_2^- standards in about 60 s using a run buffer consisting of 30 mM lactic acid with 0.75 mM tetradecyltrimethylammonium hydroxide (TTAOH) at pH 3.5.[53] TTAOH was used to reverse the EOF since the analytes of interest have large negative electrophoretic mobilities. The authors reported that lactic acid provided the conductivity and ionic strength necessary for the detection of NO_2^- and NO_3^- with C^{4}D (limits of detection of 308 and 67 μM, respectively). In addition, a run buffer with a pH near the pK_a of NO_2^- was required for the ME separation due to the similarity in electrophoretic mobility of the two analytes. NO_2^- could also be detected by amperometry, which showed a better S/N and a lower LOD of approximately 67 μM. The electropherograms can be seen in Figure 7.16A. The difference in migration times for NO_3^- in the C^{4}D and amperometric electropherograms is due to the 5 mm difference in effective length of the separation channel for each detector.[53] This difference can be avoided by plotting the electropherograms using effective mobility as the *x*-axis, instead of migration time.

The system was evaluated for the actual detection of NO_3^- and NO_2^- generated from the degradation of a 60 mM peroxynitrite sample. Under acidic conditions, peroxynitrite is in its protonated form, ONOOH, which is very

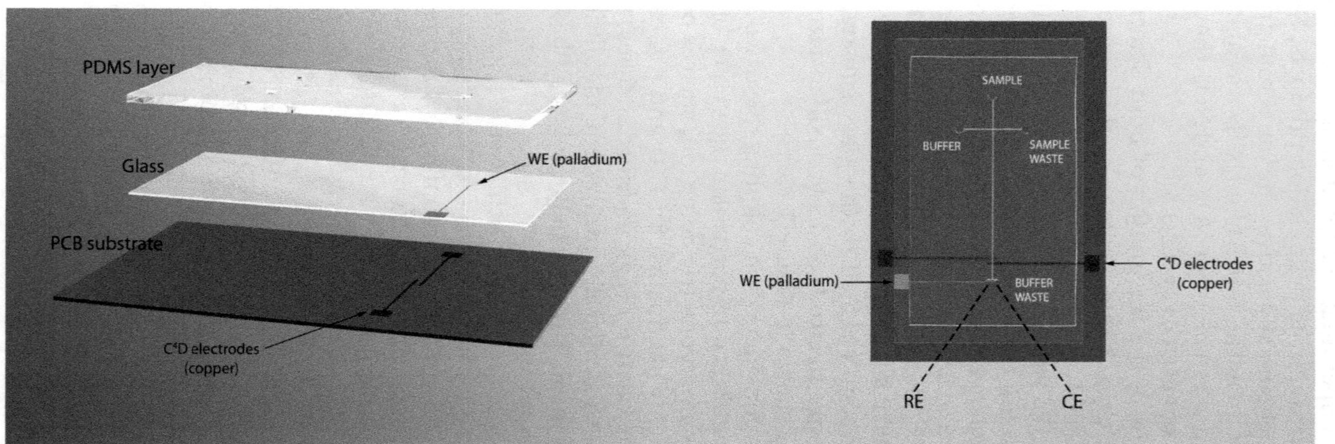

Figure 7.15   Layout of the hybrid PDMS/glass microdevice with C^4D and amperometric dual detection. Channel dimensions: 20 mm long injection channel, 57 mm long separation channel, 30 μm wide and 12 μm deep. CE: counter electrode; RE: reference electrode; WE: working electrode. Adapted from ref. 53 with permission of The Royal Society of Chemistry.

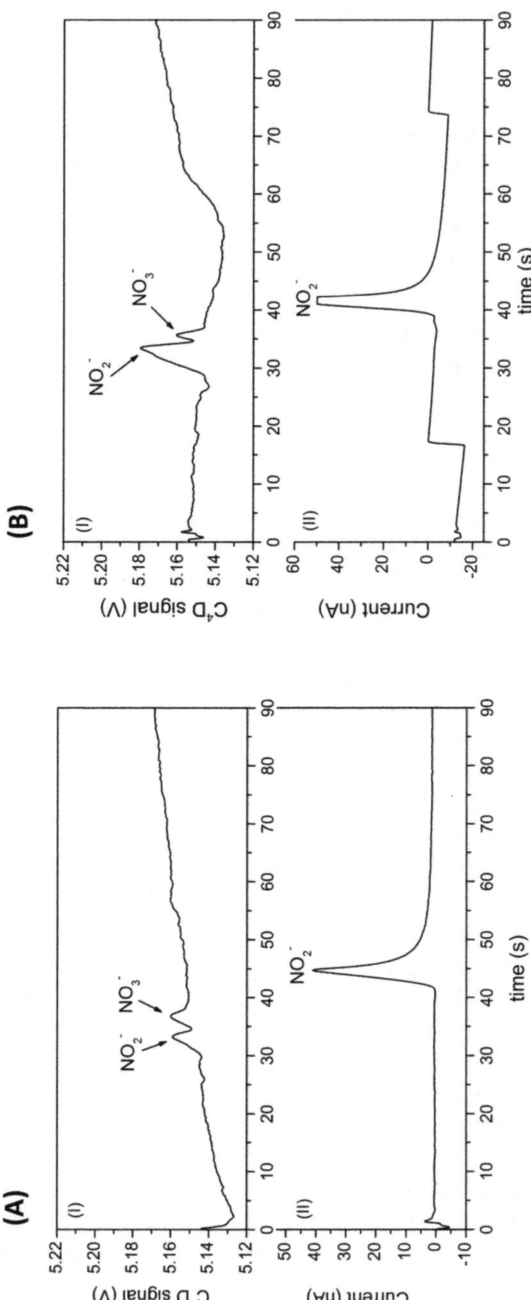

Figure 7.16 Electropherograms for the separation of NO_3^- and NO_2^- using the hybrid PDMS/glass microdevice with dual detection: (A) standard and (B) peroxynitrite degradation products. Parameters: (I) C⁴D detection (frequency 450 Hz, effective length 42 mm); (II) amperometric detection (working electrode applied voltage 850 mV, effective length 47 mm). Parameters: run buffer 30 mM lactic acid and 0.75 mM TTAOH at pH 3.5, sample voltage 1 kV, separation voltage 1.5 kV, 1 s gated injection. Reproduced from ref. 53 with permission of The Royal Society of Chemistry.

unstable and rapidly degrades into NO_3^- and NO_2^-. As with the standards, both degradation products were detected using C^4D, and NO_2^- was simultaneously detected by amperometry (Figure 7.16B). The degradation of peroxynitrite produced more NO_2^- than NO_3^-, with estimated amounts of 22.2 and 1.1 mM, respectively.[53]

Since the main focus of the paper was to show the possibility of creating a dual electrochemical detection system that could be easily used with PDMS microchips, no further investigation of the peroxynitrite degradation was performed. The fabrication of such microdevices was an important first step toward the development of simple and robust methods for label-free indirect detection of peroxynitrite.

7.5.2 Monitoring Peroxynitrite Through its Reaction Products with Biomolecules

In addition to the use of NO_3^- and NO_2^- as indirect indicators of peroxynitrite degradation, other molecules, such as nucleic acids, proteins, and amino acids, can be monitored. Peroxynitrite reacts *in vivo* with these substances, resulting in nitration, nitrosation, or oxidation products that are indicative of nitrosative and oxidative stress.[67] The structure of the biomolecule is altered by the addition of nitro groups, which alters its electrophoretic mobility. For this reason, the products of nitration caused by peroxynitrite can be analyzed using CE.

Althaus *et al.* studied the effects of peroxynitrite on the migration time of plasminogen activator inhibitor-1 monoclonal antibody.[68] They observed that the migration time of the antibody increased as a function of peroxynitrite concentration due to the nitration of different amino acid sites. At low concentrations, the peroxynitrite reaction produced mainly nitrotyrosine residues due to the high affinity between peroxynitrite and the tyrosine reactive site of the antibody. The reaction of peroxynitrite with other amino acids present in the antibody, such as cysteine, lysine, proline, and serine, happened only at higher concentrations of $ONOO^-$ (0.33 mM) and was reported as non-specific. After submitting the antibody to acid hydrolysis, nitrotyrosine was detected using high performance liquid chromatography coupled to UV detection (λ = 280 nm) and dual electrode electrochemical detection (reduction at −2000 mV followed by oxidation at +450 mV) with the latter being 50–100 times more sensitive.

In addition to the above example, several methods for detection of nitrotyrosine as a biomarker of oxidative stress using CE can be found in the literature.[69–72] Detection approaches include UV absorbance[69–71] and laser-induced fluorescence (LIF) using various dyes such as fluorescein isothiocyanate (FITC) and 7-fluoro-4-nitro-2,1,3-benzoxadiazole (NBD-F).[69]

Peroxynitrite can also promote post-translational acetylation of biomolecules in the presence of exogenous diacetyl or endogenous methylglyoxal groups through a mechanism that produces free acetyl radicals in physiological conditions.[73,74] Proteins can be naturally acetylated and deacetylated at the N-terminal of their lysine residue as part of gene regulation, but the presence of peroxynitrite can change this process. Peroxynitrite-induced acetylation alters the stability and functionality of the affected proteins.[73] With that in mind, Alves *et al.* reported a method using CE-electrospray ionization-tandem mass spectrometry (CE-ESI-MS/MS) to separate and identify 20 acetylated amino acids, peptides, and proteins formed in concomitant acetylation caused by transacetylase and the diacetyl/peroxynitrite system.[73] The method was valuable for the identification of biomolecules with structure modifications related to their gain or loss of biological function. CE-ESI-MS/MS was used to distinguish the differences between N-terminal and inner acetylation of L-lysine residues. The authors also stated that the method could be used to identify other free acetylated amino acids or proteins exposed to diacetyl and peroxynitrite *in vitro*.

DNA,[75] lipid,[76] and protein[77] oxidation due to peroxynitrite reactions have also been reported in the literature, but no correlation analyses between peroxynitrite levels and these products have been performed using electrophoretic methods and, therefore, is not within the scope of this chapter.

7.6   Future Directions

Dual detection systems that use combinations of absorbance, fluorescence, electrochemical, conductivity, and electrochemiluminescence detection in conjunction with either CE or ME have been utilized to detect catecholamines, multivitamins, pharmaceuticals, amino acids, organic and inorganic ions, amphetamines, tertiary amines, and enzymes.[78-85] The use of two detection techniques provides enhanced selectivity and characterization capabilities. Specifically, this approach can be extremely advantageous when studying the biochemistry of peroxynitrite in living systems since peroxynitrite reacts with many intracellular molecules to produce a variety of reaction products. One example of a multiple detection system was previously discussed in this chapter, in which Vázquez *et al.* developed a system with both conductivity and amperometric detection. This made it possible to detect both NO_2^- and NO_3^- because NO_3^- is not electrochemically active.[53] In the future, electrochemical detection could be coupled with other detection systems, such as LIF, to detect other non-electrochemically active species that react with fluorescent probes and enzyme activity that have eluded detection thus far in peroxynitrite studies.

Single cell cytometry has been used previously to study the intracellular production of NO in individual cells with LIF.[86] Figure 7.17 depicts the microchip design used for single cell analysis and the cell lysis process within the chip. In those studies, the cells were labeled with both diaminofluorofluorescein diacetate (DAF-FM DA), a fluorescent probe that is selective toward NO, and 6-carboxyfluorescein diacetate (6-CFDA), a fluorescent molecule used

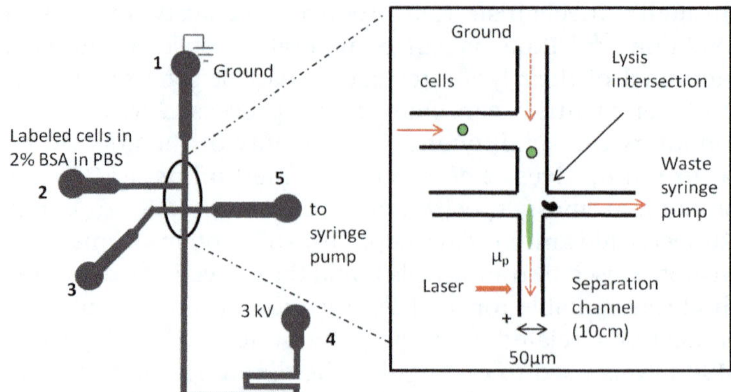

Figure 7.17   Schematic of the microchip design used for single cell analysis. Cell lysis occurs at the lysis intersection (inset). The direction of hydrodynamic flow is represented by solid arrows, while the direction of electrophoretic migration is represented by dashed arrows. BSA: bovine serum albumin. Reprinted with permission from ref. 86. Copyright 2013 American Chemical Society.

to correct for cellular volume.[86] A technique to separate the two molecules was necessary due to their similar fluorescent properties. In the future, this method could be adapted for peroxynitrite detection in single cells through the use of a peroxynitrite-specific fluorescent probe.

Alternatively, the device described above can be modified to integrate electrochemical detection for the direct detection of $ONOO^-$. A particular advantage of single cell cytometry with amperometric detection for the determination of $ONOO^-$ is that the single cell analysis system provides a means of performing on-chip cell manipulation and lysis in a very short time period. Therefore, it should be possible to perform the analysis using a high pH run buffer that would immediately stabilize the intracellular peroxynitrite for ME, which would occur in less than 30 s. Electrochemical detection would also make it possible to employ the method developed by Gunasekara *et al.* to simultaneously detect NO_2^- and intracellular antioxidants (*i.e.*, glutathione) in cell lysates.[87] Due to the speed of the microfluidic-based cell cytometry assay, peroxynitrite and other short-lived species, such as NO, can be detected along with intracellular antioxidants.

In conclusion, electrophoresis is a powerful separation technique that provides a means to better study the chemistry of peroxynitrite and its metabolites, as well as to facilitate its detection in complex biological matrices. There are many modes of detection (*i.e.*, absorbance, EC, and LIF) that can be coupled to electrophoresis in both the capillary and microchip format for the specific detection of peroxynitrite. The use of microchips also offers the ability to perform cell manipulation and lysis on-chip. In the future, these advantages could be used for applications such as the detection of $ONOO^-$ in single cells and coupling an electrophoretic separation with both LIF and EC to study $ONOO^-$ and related metabolites.

Acknowledgements

The authors would like to thank the NSF (CHE-1411993), NIH (R01NS042929 and R21NS061202), FAPESP, CNPq, CAPES, and National Institute of Science and Technology on Bioanalysis (INCTBio), Brazil for financial support. J. M. S. was supported by the Madison and Lila Self Graduate Fellowship. R. P. S. C. received support from the CNPq scholarship and Science Without Borders program. The authors would also like to thank Nancy Harmony for editorial support.

References

1. P. Pacher, J. S. Beckman and L. Liaudet, *Physiol. Rev.*, 2007, **87**, 315.
2. G. Ferrer-Sueta and R. Radi, *ACS Chem. Biol.*, 2009, **4**, 161.
3. F. Torreilles, S. Salman-Tabcheh, M.-C. Guérin and J. Torreilles, *Brain Res. Rev.*, 1999, **30**, 153.
4. C. Szabó, H. Ischiropoulos and R. Radi, *Nat. Rev. Drug Discovery*, 2007, **6**, 662.
5. S. Goldstein, J. Lind and G. Merényi, *Chem. Rev.*, 2005, **105**, 2457.
6. R. Meli, T. Nauser, P. Latal and W. H. Koppenol, *J. Biol. Inorg. Chem.*, 2002, **7**, 31.
7. R. Radi, *J. Biol. Chem.*, 2013, **288**, 26464.
8. O. Augusto, M. G. Bonini, A. M. Amanso, E. Linares, C. C. Santos and S. L. De Menezes, *Free Radical Biol. Med.*, 2002, **32**, 841.
9. R. Radi, G. Peluffo, M. N. Alvarez, M. Naviliat and A. Cayota, *Free Radical Biol. Med.*, 2001, **30**, 463.
10. T. Miyado, Y. Tanaka, H. Nagai, S. Takeda, K. Saito, K. Fukushi, Y. Yoshida, S.-i. Wakida and E. Niki, *J. Chromatogr. A*, 2004, **1051**, 185.
11. É. Szökő, T. Tábi, A. S. Halász, M. Pálfi and K. Magyar, *J. Chromatogr. A*, 2004, **1051**, 177.
12. L. Gao, J. Barber-Singh, S. Kottegoda, D. Wirtshafter and S. A. Shippy, *Electrophoresis*, 2004, **25**, 1264.
13. T. Miyado, H. Nagai, S. Takeda, K. Saito, K. Fukushi, Y. Yoshida, S.-I. Wakida and E. Niki, *J. Chromatogr. A*, 2003, **1014**, 197.
14. R. Kikura-Hanajiri, R. S. Martin and S. M. Lunte, *Anal. Chem.*, 2002, **74**, 6370.
15. P. Troška, R. Chudoba, L. Danč, R. Bodor, M. Horčičiak, E. Tesařová and M. Masár, *J. Chromatogr. B*, 2013, **930**, 41.
16. J. S. Beckman, T. W. Beckman, J. Chen, P. A. Marshall and B. A. Freeman, *Proc. Natl. Acad. Sci. U. S. A.*, 1990, **87**, 1620.
17. R. Kissner and W. H. Koppenol, *J. Am. Chem. Soc.*, 2002, **124**, 234.
18. C. N. Frankenfeld, M. R. Rosenbaugh, B. A. Fogarty and S. M. Lunte, *J. Chromatogr. A*, 2006, **1111**, 147.
19. M. K. Hulvey, C. N. Frankenfeld and S. M. Lunte, *Anal. Chem.*, 2010, **82**, 1608.
20. D. B. Gunasekara, M. K. Hulvey and S. M. Lunte, *Electrophoresis*, 2011, **32**, 832.

21. J. W. Jorgenson and K. D. Lukacs, *Anal. Chem.*, 1981, **53**, 1298.
22. R. Weinberger, *Practical Capillary Electrophoresis*, 2nd edn, Academic Press, San Diego, CA, 2000.
23. J. P. Landers, *Capillary and Microchip Electrophoresis and Associated Micro-techniques*, 3rd edn, CRC Press, Boca Raton, FL, 2008.
24. R. A. Frazier, J. M. Ames and H. E. Nursten, in *Capillary Electrophoresis for Food Analysis: Method Development*, ed. R. A. Frazier, J. M. Ames and H. E. Nursten, The Royal Society of Chemistry, Cambridge, UK, 2000, pp. 1–7.
25. C. D. García, K. Y. Chumbimuni-Torres and E. Carrilho, *Capillary Electrophoresis and Microchip Capillary Electrophoresis: Principles, Applications, and Limitations*, John Wiley & Sons, Inc., Hoboken, NJ, 2013.
26. S. C. Jacobson, R. Hergenroder, L. B. Koutny and J. M. Ramsey, *Anal. Chem.*, 1994, **66**, 1114.
27. C. T. Culbertson, T. G. Mickleburgh, S. A. Stewart-James, K. A. Sellens and M. Pressnall, *Anal. Chem.*, 2014, **86**, 95.
28. A. Manz, N. Graber and H. á. Widmer, *Sens. Actuators, B*, 1990, **1**, 244.
29. N. A. Lacher, K. E. Garrison, R. S. Martin and S. M. Lunte, *Electrophoresis*, 2001, **22**, 2526.
30. A. W. Martinez, S. T. Phillips, G. M. Whitesides and E. Carrilho, *Anal. Chem.*, 2009, **82**, 3.
31. Z. H. Fan and D. J. Harrison, *Anal. Chem.*, 1994, **66**, 177.
32. R. P. S. de Campos, I. V. P. Yoshida and J. A. F. da Silva, *Electrophoresis*, 2014, **35**, 2346.
33. J. R. Anderson, D. T. Chiu, H. Wu, O. J. Schueller and G. M. Whitesides, *Electrophoresis*, 2000, **21**, 27.
34. N. A. Lacher, N. F. de Rooij, E. Verpoorte and S. M. Lunte, *J. Chromatogr. A*, 2003, **1004**, 225.
35. W. K. T. Coltro, S. M. Lunte and E. Carrilho, *Electrophoresis*, 2008, **29**, 4928.
36. H. Makamba, J. H. Kim, K. Lim, N. Park and J. H. Hahn, *Electrophoresis*, 2003, **24**, 3607.
37. G. T. Roman, K. McDaniel and C. T. Culbertson, *Analyst*, 2006, **131**, 194.
38. H. T. Kim and O. C. Jeong, *Microelectron. Eng.*, 2011, **88**, 2281.
39. J. Xu and K. K. Gleason, *Chem. Mater.*, 2010, **22**, 1732.
40. H. Schmolke, S. Demming, A. Edlich, V. Magdanz, S. Büttgenbach, E. Franco-Lara, R. Krull and C.-P. Klages, *Biomicrofluidics*, 2010, **4**, 044113.
41. G. Sui, J. Wang, C.-C. Lee, W. Lu, S. P. Lee, J. V. Leyton, A. M. Wu and H.-R. Tseng, *Anal. Chem.*, 2006, **78**, 5543.
42. G. Yagil and M. Anbar, *J. Inorg. Nucl. Chem.*, 1964, **26**, 453.
43. A. J. Bard and L. R. Faulkner, *Electrochemical Methods: Fundamentals and Applications*, 2nd edn, John Wiley & Sons, Inc., Hoboken, NJ, 2001.
44. P. T. Kissinger and W. R. Heineman, *Laboratory Techniques in Electroanalytical Chemistry*, 2nd edn, Marcel Dekker, Inc., New York, NY, 1996.
45. C. Amatore and S. Arbault, in *Electrochemical Methods for Neuroscience*, ed. A. C. Michael and L. M. Borland, CRC Press, Boca Raton, FL, 2007, pp. 261–284.

46. W. R. Vandaveer, S. A. Pasas-Farmer, D. J. Fischer, C. N. Frankenfeld and S. M. Lunte, *Electrophoresis*, 2004, **25**, 3528.
47. D. B. Gunasekara, M. B. Wijesinghe, R. A. Saylor and S. M. Lunte, in *Electrochemical Strategies in Detection Science*, ed. D. W. M. Arrigan, The Royal Society of Chemistry, Cambridge, UK, 2014.
48. R. M. Wightman, *Science*, 2006, **311**, 1570.
49. D. E. Scott, R. J. Grigsby and S. M. Lunte, *ChemPhysChem*, 2013, **14**, 2288.
50. W. K. T. Coltro, R. S. Lima, T. P. Segato, E. Carrilho, D. P. de Jesus, C. L. do Lago and J. A. F. da Silva, *Anal. Methods*, 2012, **4**, 25.
51. J. G. A. Brito-Neto, J. A. Fracassi da Silva, L. Blanes and C. L. do Lago, *Electroanalysis*, 2005, **17**, 1198.
52. J. G. A. Brito-Neto, J. A. Fracassi da Silva, L. Blanes and C. L. do Lago, *Electroanalysis*, 2005, **17**, 1207.
53. M. Vázquez, C. Frankenfeld, W. K. T. Coltro, E. Carrilho, D. Diamond and S. M. Lunte, *Analyst*, 2010, **135**, 96.
54. W. K. T. Coltro, J. A. F. da Silva and E. Carrilho, *Electrophoresis*, 2008, **29**, 2260.
55. L.-M. Fu, C.-Y. Lee, M.-H. Liao and C.-H. Lin, *Biomed. Microdevices*, 2008, **10**, 73.
56. R. M. Guijt, J. P. Armstrong, E. Candish, V. Lefleur, W. J. Percey, S. Shabala, P. C. Hauser and M. C. Breadmore, *Sens. Actuators, B*, 2011, **159**, 307.
57. J. Lichtenberg, N. F. de Rooij and E. Verpoorte, *Electrophoresis*, 2002, **23**, 3769.
58. R. M. Uppu, G. L. Squadrito, R. Cueto and W. A. Pryor, *Methods Enzymol.*, 1996, **269**, 311.
59. R. M. Uppu and W. A. Pryor, *Anal. Biochem.*, 1996, **236**, 242.
60. W. Schulz, G. Kober, R. Bernauer and M. Kaltenbach, *Am. Heart J.*, 1985, **109**, 694.
61. F. J. Martin-Romero, Y. Gutiérrez-Martin, F. Henao and C. Gutiérrez-Merino, *J. Fluoresc.*, 2004, **14**, 17.
62. D. J. Fischer, M. K. Hulvey, A. R. Regel and S. M. Lunte, *Electrophoresis*, 2009, **30**, 3324.
63. R. S. Martin, K. L. Ratzlaff, B. H. Huynh and S. M. Lunte, *Anal. Chem.*, 2002, **74**, 1136.
64. S. M. Lunte, D. B. Gunasekara, P. Pichetsurnthorn and D. M. dos Santos, *Proceedings of μTAS*, San Antonio, TX, 2014.
65. D. B. Gunasekara, PhD Thesis, University of Kansas, 2014.
66. C. Amatore, S. Arbault, D. Bruce, P. de Oliveira, M. Erard and M. Vuillaume, *Chem.–Eur. J.*, 2001, **7**, 4171.
67. M. B. Feeney and C. Schöneich, *Antioxid. Redox Signaling*, 2013, **19**, 1247.
68. J. Althaus, G. Fici, S. Plaisted, F. Kezdy, C. Campbell, J. Hoogerheide and P. Von Voigtlander, *Microchem. J.*, 1997, **56**, 155.
69. C. Massip, P. Riollet, E. Quemener, C. Bayle, R. Salvayre, F. Couderc and E. Causse, *J. Chromatogr. A*, 2002, **979**, 209.
70. N. Maeso, A. Cifuentes and C. Barbas, *J. Chromatogr. B*, 2004, **809**, 147.

71. H. Ren, L. Wang, X. Wang, X. Liu and S. Jiang, *J. Pharm. Biomed. Anal.*, 2013, **77**, 83.
72. V. V. Shumyantseva, E. V. Suprun, T. V. Bulko and A. I. Archakov, *Biosens. Bioelectron.*, 2014, **61**, 131.
73. A. N. Alves, L. D. Jedlicka, J. Massari, M. A. Juliano, E. J. Bechara and N. A. Assunção, *J. Braz. Chem. Soc.*, 2013, **24**, 1983.
74. J. Massari, R. Tokikawa, L. Zanolli, M. F. M. Tavares, N. A. N. Assunção and E. J. H. Bechara, *Chem. Res. Toxicol.*, 2010, **23**, 1762.
75. S. W. Ballinger, C. Patterson, C.-N. Yan, R. Doan, D. L. Burow, C. G. Young, F. M. Yakes, B. Van Houten, C. A. Ballinger and B. A. Freeman, *Circ. Res.*, 2000, **86**, 960.
76. E. D. Hall, M. R. Detloff, K. Johnson and N. C. Kupina, *J. Neurotrauma*, 2004, **21**, 9.
77. H. Ischiropoulos and A. B. Al-Mehdi, *FEBS Lett.*, 1995, **364**, 279.
78. J. A. Lapos, D. P. Manica and A. G. Ewing, *Anal. Chem.*, 2002, **74**, 3348.
79. M. Novotný, F. Opekar, I. Jelínek and K. Štulík, *Anal. Chim. Acta*, 2004, **525**, 17.
80. H. Qiu, X. B. Yin, J. Yan, X. Zhao, X. Yang and E. Wang, *Electrophoresis*, 2005, **26**, 687.
81. C. Liu, Y.-y. Mo, Z.-g. Chen, X. Li, O.-l. Li and X. Zhou, *Anal. Chim. Acta*, 2008, **621**, 171.
82. J. Sun, X. Xu, C. Wang and T. You, *Electrophoresis*, 2008, **29**, 3999.
83. B. Yuan, J. Huang, J. Sun and T. You, *Electrophoresis*, 2009, **30**, 479.
84. F. Shen, Y. Yu, M. Yang and Q. Kang, *Microsyst. Technol.*, 2009, **15**, 881.
85. O. Ordeig, P. Ortiz, X. Muñoz-Berbel, S. Demming, S. Büttgenbach, C. S. Fernaández-Saánchez and A. Llobera, *Anal. Chem.*, 2012, **84**, 3546.
86. E. C. Metto, K. Evans, P. Barney, A. H. Culbertson, D. B. Gunasekara, G. Caruso, M. K. Hulvey, J. A. Fracassi da Silva, S. M. Lunte and C. T. Culbertson, *Anal. Chem.*, 2013, **85**, 10188.
87. D. B. Gunasekara, J. M. Siegel, G. Caruso, M. K. Hulvey and S. M. Lunte, *Analyst*, 2014, **139**, 3265.

Investigation of Peroxynitrite–Biomembrane Interactions Using Biomimetic Interfaces

YING LIU[a], SERBAN F. PETEU[a], AND R. MARK WORDEN*[a]

[a]Chemical Engineering and Materials Science, Michigan State University, East Lansing, MI 48824, USA
*E-mail: worden@msu.edu

8.1 Introduction

Peroxynitrite ($ONOO^-$ or PON for short), and the reactive nitrogen oxide species (RNOS) and reactive oxygen species (ROS) that reversibly form it are clinically relevant oxidants that play an active role in several devastating disease processes, including cancer, diabetes, inflammatory processes, ischemia, and stroke.[1-6] Disease-relevant molecular transformations induced by PON involve cytoplasmic components, such as soluble proteins, as well as biomembrane components, such as membrane proteins and phospholipids. Mechanistic studies that elucidate the role of PON in oxidation and nitration of soluble proteins have been reviewed elsewhere[2,7-17] and will not be discussed here. The focus of this chapter will be on PON activity against biomembrane lipids and the resulting changes in biomembrane properties.

Experimental characterization of biomembrane phenomena is challenging for three reasons. First, the heterogeneous composition of biomembranes, which contain phospholipids, membrane proteins, glycolipids, sterols, *etc.*,

RSC Detection Science Series No. 7
Peroxynitrite Detection in Biological Media: Challenges and Advances
Edited by Serban F. Peteu, Sabine Szunerits, and Mekki Bayachou
© The Royal Society of Chemistry 2016
Published by the Royal Society of Chemistry, www.rsc.org

makes it difficult to distinguish effects caused by different membrane constituents. Second, the composition of cell membranes varies with the type of cell, type of membrane, experimental conditions, *etc.* in a manner that is complex and poorly understood. Third, robust methods are not available to systematically alter the composition of *in vivo* cell membranes to test hypotheses about the effect of a membrane's composition on its properties.

To address these challenges, well-characterized biomimetic interfaces consisting of artificial bilayer lipid membranes (BLM) with a specified composition have been developed and used to mimic biomembrane behavior and facilitate biomembrane research. These biomimetic interfaces are well suited to systematically study the effect of PON on various biomembrane properties. The goals of this chapter are to summarize mechanisms by which PON damages membrane lipids, to describe electrochemical and optical methods to study biomimetic interfaces, and to test hypotheses about PON-associated biomembrane damage. To illustrate the approach, an example application of a biomimetic interface is presented: use of electrochemical impedance spectroscopy (EIS) to measure changes in the electrical resistance (R_m) of a tethered BLM following PON exposure.

8.2 Molecular Mechanisms of the Interaction of PON with Biomembrane Lipids

PON and associated RNOS compose a complex and rapidly evolving web of highly reactive species, many of which have a half-life in the order of milliseconds.[18,19] Biomembrane degradation can be mediated either by the anion ONOO⁻ or its conjugate peroxynitrous acid ONOOH through a variety of lipid peroxidation and nitration mechanisms.[20,21] A simplified conceptual model of lipid peroxidation within a BLM that accounts for a subset of known RNOS reactions is illustrated in Figure 8.1. Numerous *in vivo* and *in vitro* studies of the effects of oxidative processes have demonstrated that PON and associated oxidative species, including hydroxyl and peroxyl radicals, induce reactions involving biomembranes.[23]

A variety of metal ions and metal-containing porphyrins and membrane-associated proteins can also participate in biomembrane oxidation reactions, as well as interconversion of RNOS and ROS molecular species within biomembranes.[24–26] To monitor these effects, amphiphilic iron and manganese porphyrin that have high affinity for biomembranes have been developed and used to elucidate PON decomposition rates at the cytoplasm–biomembrane interface.[27]

Lipid peroxidation or nitration reactions in biomembranes can be initiated *via* abstraction of a hydrogen atom from polyunsaturated fatty acids.[20,21,28] Polyunsaturated fatty acids containing bis-allylic methylene hydrogens are more susceptible to hydrogen abstraction upon exposure to oxidizing radicals than the methylene hydrogens found in fully saturated lipids.[29] As a result, polyunsaturated fatty acids are more likely to be oxidized, resulting in the formation of aldehydes, alcohols, and hydroperoxides.[1]

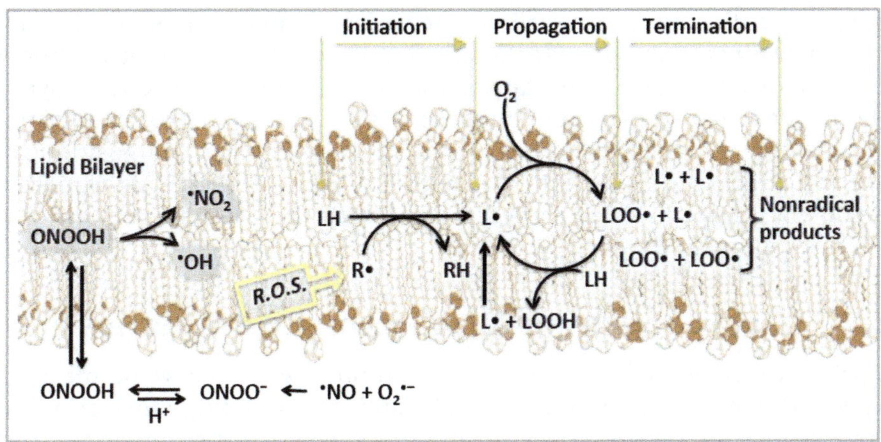

Figure 8.1   Conceptual model of lipid peroxidation in a BLM that takes into account a subset of known RNOS reactions. LH: unsaturated lipid; LOO˙: lipid peroxyl radical; LOOH: lipid hydroxyl peroxide. Reprinted with permission from A. Vasilescu, A. Vezeanu, Y. Liu, L. Hosu, R. M. Worden and S. F. Peteu, *ACS Symposium Series: Instrumental Methods for the Analysis and Identification of Bioactive Molecules*, 2014, 303. Copyright 2014 American Chemical Society.

PON-induced damage has been shown to stimulate state 4 respiration in brain mitochondria, which is associated with increased proton permeability of the mitochondrial inner membrane and loss of proton-motive force.[30] This effect could be attributed to a phospholipid peroxidation mechanism, because lipid-soluble antioxidants prevented the stimulation, and mitochondrial phospholipids are known to have a high content of polyunsaturated fatty acids, predisposing them to peroxidation.

Biomimetic interfaces have also been used to characterize chemical products of ˙NO₂-induced nitration of unsaturated oleyl fatty acids of phospholipids and to elucidate the underlying reaction mechanisms.[31] The rate at which PON diffuses into the hydrophobic core of a biomembrane and then reacts with the double bonds of unsaturated phospholipids strongly influences the lipid degradation rate. Investigations of enthalpy variations among phospholipid head groups suggests that negatively charged head groups can form an electrical barrier that slows penetration of negatively charged oxidative species such as PON.[1] The tendency of certain hydrophobic antioxidants (*e.g.*, quercetin 3-*O*-β-D-glucuronide) to partition into biomembranes is responsible for their potency at preventing intramembrane lipid peroxidation.[32]

The degree to which PON-mediated tyrosine nitration occurs in a biomembrane is difficult to measure and has not been studied extensively. However, the use of a hydrophobic tyrosyl probe (*e.g.*, N-*t*-BOC L-tyrosine *tert*-butyl ester, or BTBE) that partitions into the biomembrane's core, has enabled the relative nitration rates inside and outside the biomembrane to be compared.[33] PON-induced tyrosine oxidation was found to predominantly form a dityrosyl product outside the membrane, whereas nitration was predominant inside. This difference was attributed to variations in translational diffusion coefficients inside and outside the membrane, which affected the likelihood of a biomolecular reaction producing the dityrosyl product. The same hydrophobic probe has been used to investigate mechanisms of tyrosine nitration within biomembranes and to explore the effect of fatty acid composition and free radical scavengers on nitration yields.[25] Additional details of the method are described later in this chapter in the section that discusses optical characterization methods for PON.

8.3   Effects of PON on Biomembrane Properties

A primary function of biomembranes is to provide a selective permeability barrier. Hydroxyl and peroxyl radicals resulting from lipid peroxidation can cause significant increases in permeability, a decrease in membrane electrical resistance, and enhanced lipid exchange between the upper and lower leaflet of the BLM.[29,34–36] RNOS-associated chemical transformations can also affect biomembrane viscosity and fluidity.[32,38] In erythrocyte membranes, nitric oxide (NO) increased membrane fluidity and lipid peroxidation,[37] and PON enhanced passive permeability to potassium and sodium cations while inhibiting anion flux.[38] Furthermore, ion imbalance and membrane structural reorganization induced by PON caused crenation (contraction) of erythrocytes.

Quantitative kinetic models of biomembranes that take into account lipid–lipid and lipid–protein interactions have recently been reviewed.[39] A model based on an a mechanistic understanding of biomembrane dynamics and PON reaction mechanisms was used to predict PON transport rates and phospholipid oxidation rates within the membranes.[40] The model was validated using experimentally measured $Fe(CN)_6^{4-}$ oxidation rates following PON diffusion through the BLM shell of liposomes that have a known phospholipid composition. The model, whose constants can be adjusted to explore hypotheses about the effects of key independent variables, predicted that neutral PON species would rapidly permeate the bilayer, while anionic PON species would be membrane impermeable.

PON transportation rates through biomimetic interfaces have been measured using manganese porphyrins as reporter molecules.[41] Differences in the absorbance spectra of the oxidized and reduced forms of the reporters allowed the dynamics of PON transport to be studied. The calculated permeability coefficient through the biomimetic interface was similar to that of water and about 400 times greater than that of superoxide, indicating that biomembranes are highly permeable to PON.

The ability of hydrophobic antioxidants to partition into biomembranes provides a mechanism to reduce PON transport across biomembranes. For example, the scavenger (−)-epicatechin has an octanol/buffer partition coefficient of 1.5, allowing it to be soluble in both aqueous solutions and the lipophilic biomembrane core.[42] This compound was shown to be stably taken up into cell membranes of murine aortic endothelial cells and to effectively protect them against PON penetration and the resulting nitration of protein tyrosyl residues and oxidation of intracellular dichlorodihydrofluorescein.

The lipid-soluble antioxidant α-tocopherol has also been shown to be an effective PON scavenger in biomembranes.[43,44]

Other common biomembrane constituents, such as cholesterol and cholesterol esters, could also be added to a biomimetic interface to further modulate its properties and allow it to better mimic a true biomembrane. Such constituents are also susceptible to PON-induced reactions. For example, reactions between cholesterol esters and PON have been characterized.[45] While cholesterol does not contain bis-allylic hydrogens and is thus less reactive toward free radicals than polyunsaturated lipids, cholesterol esters derived from unsaturated fatty acids (*e.g.*, 5-cholesten-3β-ol 3-linoleate) were found to react rapidly with PON.

8.4 Electrochemical Methods to Measure the Effects of PON on Biomembrane Lipids

Electrochemical methods appear promising for relatively simple, direct, real-time and label-free PON measurement.[28] In addition, sensor miniaturization to the micro- to nanoscale allows faster response times and measurement in non-mixed, soft solid media, to monitor events in bacteria populations, cells or tissues.[46–48] Two major biomimetic interface platforms have been widely used for electrochemical characterization of biomembranes: (1) the planar BLM (pBLM), in which an unsupported BLM is formed across a small orifice; and (2) the tethered BLM (tBLM), in which some phospholipids in the lower leaflet are chemically linked to an underlying surface (*e.g.*, an electrode). Potential applications of these platforms for PON-related research are described below.

8.4.1 Planar Bilayer Methods

pBLMs have been used extensively in the field of electrophysiology to study transmembrane ion migration mediated by channel proteins,[49] transient lipid pores,[50] and membrane-active toxins.[51–53] The approach consists of using Ag/AgCl electrodes to establish a fixed electrochemical potential across the BLM, and then measuring the resulting ion current through the pBLM. The method's high sensitivity (picoamperes) and high temporal resolution (milliseconds) enable it to capture the dynamics of single transient pore opening and closing events.[54,55] The method also has the potential to be adapted to high throughput operations in a microfluidic setting.[56]

A high-quality pBLM formed with fresh phospholipids typically exhibits an extremely high electrochemical resistivity ($\sim 10^{12}$ Ω cm) and a low, stable baseline current. However, as the phospholipids age or are exposed to oxidants, phospholipid oxidation results in transient instabilities (pores) in the tBLM, as evidenced by current spikes, extended periods of integral conductance, and eventually, BLM rupture. We previously reported that oxidative degradation of phospholipids due to air exposure for 24 h degraded a pBLM's electrical sealing properties, as evidenced by a three-fold increase in the baseline current value, as well as current spikes (as evidenced by sudden increases in current that quickly return to the original current level), and integral conductance (as evidenced by sustained increases in current that do not quickly return to the original level).[22] Additional examples of distinctive, information-rich electronic signatures provided by pBLM systems in response to environmental factors are shown in Figure 8.2.

Because the pBLM method can sensitively measure ion flow arising from oxidation, it is well suited to monitor the effects of PON on biomembrane

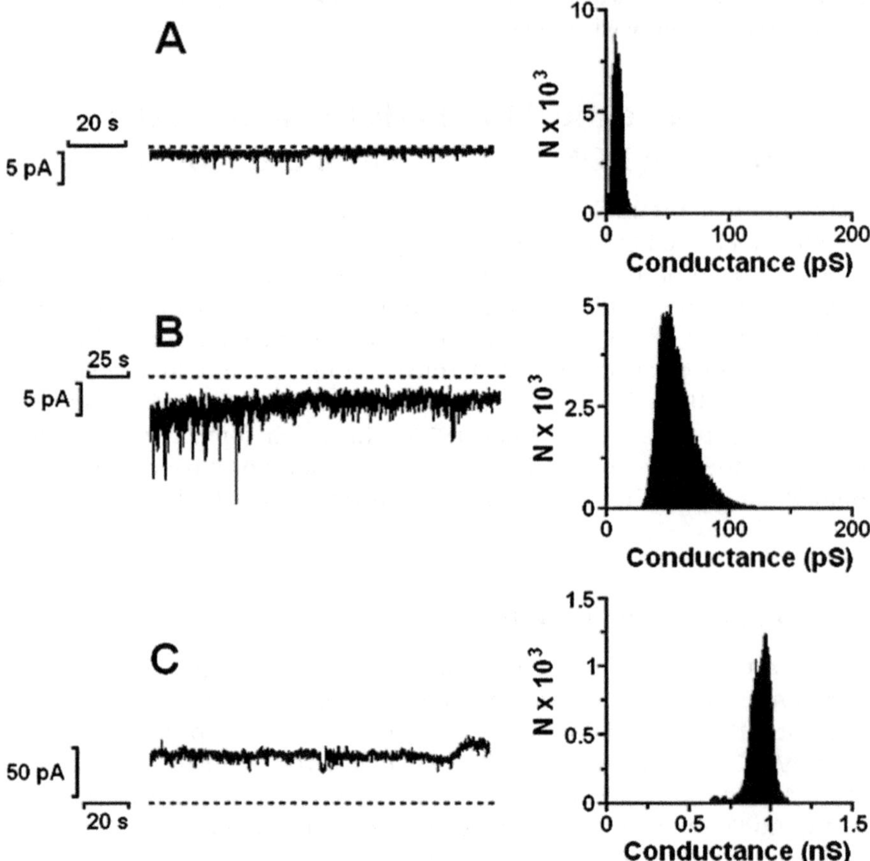

Figure 8.2 Plots of ionic current *vs.* time for pBLMs exposed to nanoparticles. Dotted lines show zero current. (A) Baseline current before nanoparticle addition. (B) Current trace following addition of functionalized silica core nanoparticles. (C) Current trace following addition of functionalized carbon nanotubes. Additional details are provided elsewhere.[57] Reprinted from *J. Biomed. Nanosci. Nanotechnol.*, 2013, 3, 52, A. Negoda, Y. Liu, R. M. Worden, W. C. Hou, C. Corredor, B. Y. Moghadam, C. Musolff, L. Li, W. Walker, P. Westerhoff, A. J. Mason, P. Duxbury, J. D. Posner and R. M. Worden, Investigation of Peroxynitrite–Biomembrane Interactions Using Biomimetic Interfaces, copyright 2013 with permission from Interscience.

integrity. Statistical techniques can be used to analyze the dynamic current fluctuation patterns such as those shown in Figure 8.2 and could, in principle, be used to distinguish between different molecular mechanisms of PON-induced BLM disruption.[57,58]

While the pBLM method offers the advantages of a high dynamic range and single pore sensitivity, it also has disadvantages. A pBLM is inherently mechanically fragile and exhibits a relatively short lifetime (typically a few hours). Furthermore, the formation of tBLM and data interpretation require a highly trained specialist. Also, to achieve good signal-to-noise ratios with the extremely low baseline currents (sub-picoampere), expensive planar bilayer workstations and Faraday cages are typically required.

8.4.2 Tethered Bilayer Methods

To avoid the mechanical fragility characteristic of pBLM, a tBLM is tethered to an electrode, which provides physical reinforcement and increases mechanical stability.[57] The presence of an electrolyte layer between the tBLM and electrode allows the tBLM's impedance to be measured using EIS. As shown schematically in Figure 8.3, in EIS, the electrode is subjected to an oscillating electrical potential. The resulting charge transfer is analyzed using an equivalent circuit model to determine parameters such as the tBLM's capacitance and R_m. This information can then be used to characterize the effect of membrane-active agents such as oxidants, ionophores, channel-forming membrane proteins and a variety of chemical agents.[59–64]

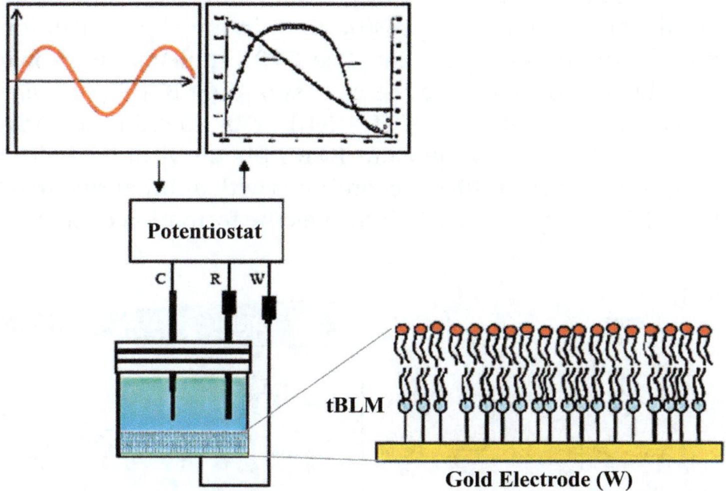

Figure 8.3   Schematic diagram of EIS characterization of the interaction between tBLM and PON. C: counter electrode; R: reference electrode; W: working electrode. Reprinted with permission from reference A. Vasilescu, A. Vezeanu, Y. Liu, L. Hosu, R. M. Worden and S. F. Peteu, *ACS Symposium Series: Instrumental Methods for the Analysis and Identification of Bioactive Molecules*, 2014, 303. Copyright 2014 American Chemical Society.

The EIS method is much less sensitive to electrical noise than electrophysiology methods and has been adapted to microelectrode arrays that can be used for high-throughput operations.[65–67]

8.5 Optical Methods to Measure PON Effects on Biomembrane Lipids

Optical methods to study the effect of PON on biomembranes provide complementary information to the electrochemical methods described above. In addition, optical and electrochemical methods can, in principle, be conducted simultaneously to provide synergistic information about lipid peroxidation and/or nitration rates, and the resulting changes in biomembrane properties.

Common optical approaches to quantify PON effects in biological fluids and biomembranes are based on fluorescent probes (dihydrorhodamine and dichlorodihydrofluorescein). The detection of 3-nitrotyrosine is still utilized widely because this is the major modification derived from the reaction of PON with proteins.[68] However, these assays have disadvantages, including the fact that they are indirect detection methods, requiring signal processing, and, thus, do not give real-time analyses.[69] To address these disadvantages, reliable, robust, and real-time detection methods are needed to detect PON and its effects.

Several optical probes for PON detection in both chemical and biological media were recently reviewed[69] and are also described in several chapters of this book. Hydrophobic tyrosine analogs, such as the previously mentioned BTBE, have been incorporated into biomimetic interfaces and used to study PON-induced, intra-membrane tyrosine protein and lipid nitration.[25] This approach is illustrated in Figure 8.4. The BTBE molecule is shown undergoing a one-electron oxidation to the corresponding BTBE phenoxyl radical either by PON-derived radicals ($^{\bullet}OH$, $^{\bullet}NO_2$) or by membrane-derived lipid peroxyl radicals ($ROO^{\bullet}$). The transient BTBE phenoxyl radical either reacts with $^{\bullet}NO_2$ to yield 3-nitro-BTBE or recombines with another phenoxyl radical to yield 3,3′di-BTBE. Figure 8.4 also indicates the formation of small amounts

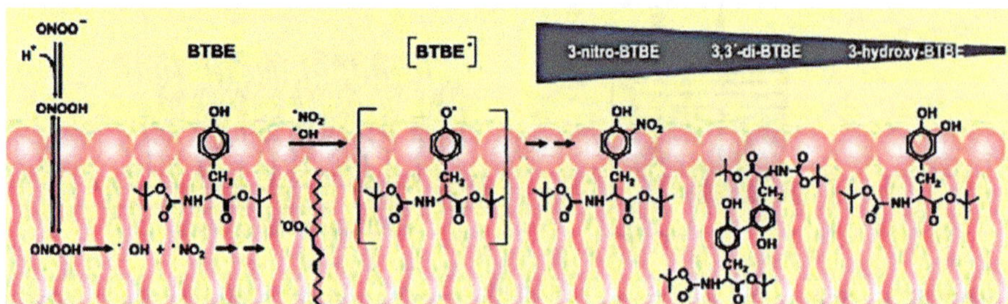

Figure 8.4 Tyrosine oxidation products in biomembranes induced by PON. Reprinted with permission from S. Bartesaghi, V. Valez, M. Trujillo, G. Peluffo, N. Romero, H. Zhang, B. Kalyanaraman and R. Radi, *Biochemistry*, 2006, **45**, 6813. Copyright 2006 American Chemical Society.

of the 3-hydroxy-BTBE from the addition reaction with hydroxyl radicals, and suggests the diffusion and homolysis of ONOOH within the BLM. The relative ease of BTBE synthesis, incorporation into model BLM, and subsequent extraction and quantitation of BTBE reaction products offers a unique opportunity to investigate PON-induced oxidation mechanisms. However, this and other similar optical methods lack biological relevance and clear applicability to disease models. Therefore, improved methods are still needed to study PON-induced oxidizing effects on *in vivo* cell membranes.

8.6 Example of the Measurement of PON Effects Using a Tethered Biomimetic Interface

In this section, we demonstrate the use of a biomimetic interface to electrochemically measure the effect of PON on a biomembrane. A tBLM was self-assembled on working electrodes consisting of a gold layer (100 nm thick) coated on a silicon wafer by chemical vapor deposition. The electrodes were first cleaned in fresh piranha solution for 30 s, washed with deionized water, and dried in nitrogen.[70,71] Next, a self-assembled monolayer of 1,2-dipalmitoyl-*sn*-glycero-phosphothioethanol (DPPTE; from Avanti Polar Lipids Inc., Alabaster, AL, USA) was chemically adsorbed onto the electrodes by dipping a freshly cleaned electrode into 1 mM ethanolic DPPTE solution for 1 h, followed by rinsing with ethanol, and drying with nitrogen. Small unilamellar liposomes composed of 1,2-dioleoyl-*sn*-glycero-phosphocholine (DOPC; Avanti Polar Lipids Inc.) were prepared by freeze-drying DOPC dissolved in chloroform at −47 °C for 2 h, followed by hydration in 10 mM HEPES buffer (pH 7.4) to a concentration of 1 mM. The DOPC liposome suspension was ultrasonicated in an ice bath for 20 min prior to use. Then, 300 μL of the liposome suspension was pipetted into a poly(dimethylsiloxane) (PDMS) reservoir and incubated at room temperature for 24 h. After tBLM formation, the liposome solution was replaced with fresh 0.1 M phosphate-buffered saline (PBS) buffer (pH 7.2). The PON-generating compound 3-morpholino-sydnonimine-*N*-ethylcarbamide (SIN-1; Cayman Chemicals, Ann Arbor, MI, USA) was then transferred to the PDMS reservoir and mixed by a pipette to achieve a final concentration of 30 mM SIN-1, which generates the equivalent of 300 μM PON.

The experimental system shown in Figure 8.5 was used to measure the effect of PON exposure on the tBLM's R_m value using an electrochemical workstation (CHI Instruments Inc., Austin, TX, USA) configured in a three-electrode setup. An AC perturbation of 5 mV was superimposed on a dc bias of 0 V over a frequency (ω) range of 0.01–10 000 Hz. Experiments were performed in triplicate for two cases: (1) exposure to SIN-1; and (2) exposure to both SIN-1 and 20 nm diameter COOH-functionalized polystyrene nanoparticles (PNP; obtained from Invitrogen, Eugene, OR, USA). The R_m values were determined at 10 min intervals by fitting a modified Randles equivalent circuit into the electrical impedance (Z) data using Zview software (Scribner Associates, Southern Pines, NC, USA).

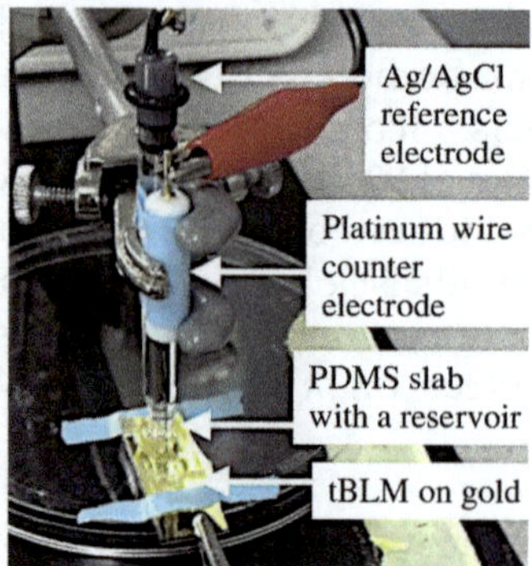

Figure 8.5 The tBLM device and the three-electrode electrochemical setup. Reprinted from *Biochim. Biophys. Acta, Biomembr.*, vol. 1838, 429, Y. Liu and R. M. Worden, Biomembrane disruption by silica-core nanoparticles: effect of surface functional group measured using a tethered bilayer lipid membrane, copyright 2014 with permission from Elsevier.

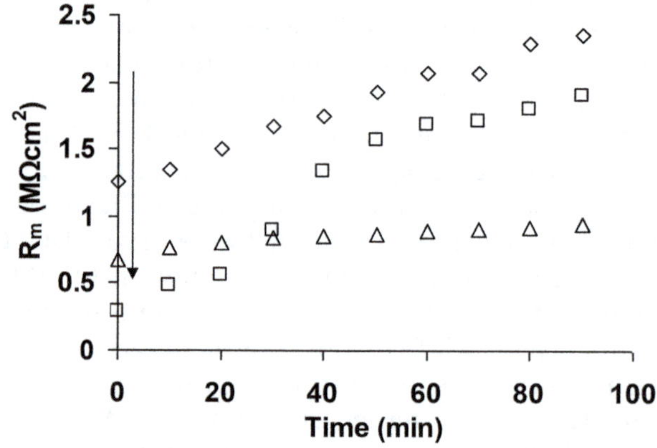

Figure 8.6 Triplicate R_m profiles recorded during the interaction of DPPTE/DOPC tBLM with 300 μM PON in PBS buffer (0.1 M PBS, 0.3 M NaCl, pH 7.2) over 90 min. The arrow indicates that tBLM were exposed to SIN-1 immediately after the first R_m value was measured.

In previous studies, we established that, in the absence of a biomembrane-active agent, a DOPC tBLM's R_m value remains constant over at least a 5 h incubation period.[70,71] However, exposure to SIN-1 significantly increased the tBLM's R_m (Figure 8.6). Because a BLM's electrical resistivity (~10^{12} Ω cm) is roughly ten orders of magnitude higher than that of the surrounding buffer (~70 Ω cm),[72] a tBLM's R_m value is extremely sensitive to small changes in the area of microdefects that expose the underlying electrode.[70,71] For example, an increase in the defect area fraction of only 1.8×10^{-11}% could give a 50% decrease in R_m.[73] Thus, the increase in R_m observed in Figure 8.6 suggests that PON-induced peroxidation caused the expansion of the tBLM into defects, thereby decreasing the defect area. This proposed mechanism is

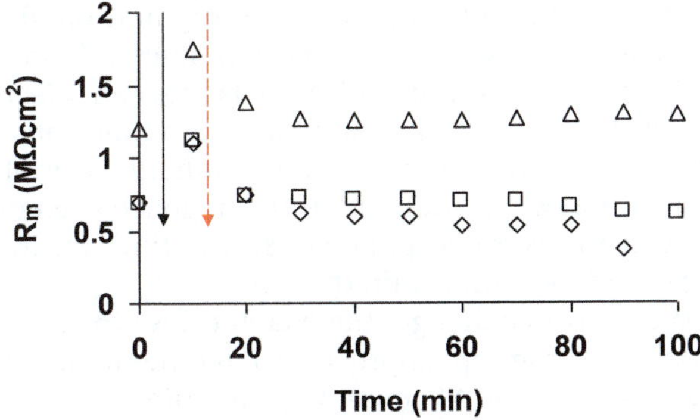

Figure 8.7　Triplicate R_m profiles recorded during the interaction of DPPTE/DOPC tBLM with 300 μM PON and 100 μg mL^{-1} 20 nm COOH-functionalized PNP in PBS buffer (0.1 M PBS, 0.3 M NaCl, pH 7.2) over 90 min. The black (solid) arrow indicates that tBLM were exposed to SIN-1 immediately after the first R_m value was measured, and the red (dashed) arrow indicates that 20 nm COOH-functionalized PNP were added after the second R_m value was measured.

consistent with reported changes in biomembrane lipid localization caused by peroxidation.[74] Such changes alter a biomembrane's phospholipid composition and key physical properties, including its packing density, surface area per lipid molecule, and lateral homogeneity.[75]

In a previous study, when we exposed a DOPC tBLM to 20 nm COOH-functionalized PNP, the R_m continuously declined along a trajectory that could be empirically modeled by a decaying exponential.[70] This decline was attributed to the removal of DOPC lipids by the PNP, thereby increasing the defect area. This result suggested the need for a second experiment, in which a tBLM was exposed to both PON (which caused an increase in R_m) and PNP (which caused a decrease in R_m). The results of the second experiment (Figure 8.7) show intermediate behavior between the continuous increase seen with PON alone and the continuous decrease seen with PNP alone. The initial increase in R_m triggered by PON was terminated by the addition of PNP, resulting in an essentially flat R_m during the remainder of the experiment.

These preliminary results establish that the tBLM biomimetic interface platform is able to measure the significant effect of PON when a DOPC tBLM is exposed to the PON-generating compound SIN-1. In addition, Figure 8.7 suggests that PON-induced peroxidation of a tBLM may influence how nanoparticles interact with the model biomembrane. Further research using biomimetic interfaces are needed to elucidate the mechanisms underlying these interesting observations.

8.7　Conclusions

This chapter reviewed the research challenges associated with elucidating the role of PON in biomembrane-associated disease processes. The challenges include: (1) the complex web of reactions interconnecting PON and several associated RNOS and ROS; (2) the high reactivity and rapid reaction rates of PON and the associated RNOS and ROS with biomembrane components; (3) the lack of robust, label-free analytical tools that can selectively measure PON and the associated species in real-time with a time constant sufficient to capture the rapid reaction dynamics; and (4) inherent complexities associated with biomembranes, including their heterogeneous compositions and properties that vary in a complicated manner with time, environmental conditions, cell type, and location within the cell.

To address the fourth challenge, the chapter described how biomimetic interfaces allow a bottom-up approach to biomembrane research. The approach uses a synthetic BLM containing the minimum number of components needed to mimic desired biomembrane properties. An overview of the literature describing molecular interactions between PON and biomembrane components was presented, with a focus on membrane lipids. The effect of PON and associated radical species on biomembrane properties, ion leakage, fluidity, molecular diffusion rates, sensitivity to oxidative stress, *etc.* was described.

The chapter then presented two major biomimetic interface platforms that are especially well suited for research on PON-related biomembrane phenomena (the pBLM and tBLM), as well as complementary optical methods suitable to measure PON–biomembrane interactions. A novel application of a biomimetic interface to characterize the influence of PON on biomembranes was then presented. A DOPC tBLM was exposed to PON-generating SIN-1, and a significant increase in tBLM's R_m value was measured over a 100 min period. To explore possible interactions between the effects of PON and nanoparticles on biomembranes, COOH-functionalized PNP were added immediately after SIN-1 addition. The addition of the nanoparticles negated the increase in R_m seen with SIN-1 alone. A possible mechanism to explain these novel experimental trends was suggested, setting the stage for further biomimetic interface research to further elucidate the underlying mechanism.

References

1. V. R. de Lima, M. P. Morfim, A. Teixeira and T. B. Creczynski-Pasa, *Chem. Phys. Lipids*, 2004, **132**, 197–208.
2. A. M. Cassina, R. Hodara, J. M. Souza, L. Thomson, L. Castro, H. Ischiropoulos, B. A. Freeman and R. Radi, *J. Biol. Chem.*, 2000, **275**, 21409–21415.
3. C. S. Cobbs, M. Samanta, L. E. Harkins, G. Y. Gillespie, B. A. Merrick and L. A. MacMillan-Crow, *Arch. Biochem. Biophys.*, 2001, **394**, 167–172.
4. M. Ramalingam and S. J. Kim, *J. Neural Transm.*, 2012, **119**, 891–910.
5. M. R. Reynolds, R. W. Berry and L. I. Binder, *Biochemistry*, 2005, **44**, 1690–1700.
6. C. Szabo, H. Ischiropoulos and R. Radi, *Nat. Rev. Drug Discovery*, 2007, **6**, 662–680.

7. N. Abello, H. A. M. Kerstjens, D. S. Postma and R. Bischoff, *J. Proteome Res.*, 2009, **8**, 3222–3238.
8. B. Alvarez and R. Radi, *Amino Acids*, 2003, **25**, 295–311.
9. B. Alvarez, H. Rubbo, M. Kirk, S. Barnes, B. A. Freeman and R. Radi, *Chem. Res. Toxicol.*, 1996, **9**, 390–396.
10. J. Ara, S. Przedborski, A. B. Naini, V. Jackson-Lewis, R. R. Trifiletti, J. Horwitz and H. Ischiropoulos, *Proc. Natl. Acad. Sci. U. S. A.*, 1998, **95**, 7659–7663.
11. P. Calcerrada, G. Peluffo and R. Radi, *Curr. Pharm. Des.*, 2011, **17**, 3905–3932.
12. C. Quijano, B. Alvarez, R. M. Gatti, O. Augusto and R. Radi, *Biochem. J.*, 1997, **322**, 167–173.
13. S. N. Savvides, M. Scheiwein, C. C. Bohme, G. E. Arteel, P. A. Karplus, K. Becker and R. H. Schirmer, *J. Biol. Chem.*, 2002, **277**, 2779–2784.
14. J. M. Souza, E. Daikhin, M. Yudkoff, C. S. Raman and H. Ischiropoulos, *Arch. Biochem. Biophys.*, 1999, **371**, 169–178.
15. J. M. Souza, G. Peluffo and R. Radi, *Free Radical Biol. Med.*, 2008, **45**, 357–366.
16. Y. H. Wang, Y. J. Luo and R. G. Zhong, *Prog. Chem.*, 2007, **19**, 893–901.
17. R. Radi, *Proc. Natl. Acad. Sci. U. S. A.*, 2004, **101**, 4003–4008.
18. G. Ferrer-Sueta and R. Radi, *ACS Chem. Biol.*, 2009, **4**, 161–177.
19. R. Radi, *J. Biol. Chem.*, 2013, **288**, 26464–26472.
20. R. Radi, J. S. Beckman, K. M. Bush and B. A. Freeman, *Arch. Biochem. Biophys.*, 1991, **288**, 481–487.
21. P. Pacher, J. S. Beckman and L. Liaudet, *Physiol. Rev.*, 2007, **87**, 315–424.
22. A. Vasilescu, A. Vezeanu, Y. Liu, L. Hosu, R. M. Worden and S. F. Peteu, in *ACS Symposium Series: Instrumental Methods for the Analysis and Identification of Bioactive Molecules*, ed. G. K. Jayprakasha, B. Patil and F. Pellati, American Chemical Society, 2014, pp. 303–332.
23. A. E.-D. A. Bekhit, D. L. Hopkins, F. T. Fahri and E. N. Ponnampalam, *Compr. Rev. Food Sci. Food Saf.*, 2013, **12**, 565–597.
24. Y. R. Chen, C. L. Chen, X. P. Liu, G. L. He and J. L. Zweler, *Arch. Biochem. Biophys.*, 2005, **439**, 200–210.
25. S. Bartesaghi, V. Valez, M. Trujillo, G. Peluffo, N. Romero, H. Zhang, B. Kalyanaraman and R. Radi, *Biochemistry*, 2006, **45**, 6813–6825.
26. R. S. Deeb, G. Hao, S. S. Gross, M. Laine, J. H. Qiu, B. Resnick, E. J. Barbar, D. P. Hajjar and R. K. Upmacis, *J. Lipid Res.*, 2006, **47**, 898–911.
27. J. A. Hunt, J. B. Lee and J. T. Groves, *Chem. Biol.*, 1997, **4**, 845–858.
28. S. F. Peteu and S. Szunerits, in *Detection Challenges in Clinical Diagnostics*, The Royal Society of Chemistry, 2013, pp. 156–181.
29. N. Hogg and B. Kalyanaraman, *Biochim. Biophys. Acta, Bioenerg.*, 1999, **1411**, 378–384.
30. P. S. Brookes, J. M. Land, J. B. Clark and S. J. R. Heales, *J. Neurochem.*, 1998, **70**, 2195–2202.
31. K. Jain, A. Siddam, A. Marathi, U. Roy, J. R. Falck and M. Balazy, *Free Radical Biol. Med.*, 2008, **45**, 269–283.

32. M. Shirai, J. H. Moon, T. Tsushida and J. Terao, *J. Agric. Food Chem.*, 2001, **49**, 5602–5608.

33. H. Zhang, J. Joseph, J. Feix, N. Hogg and B. Kalyanaraman, *Biochemistry*, 2001, **40**, 7675–7686.

34. H. Rubbo, A. Trostchansky and V. B. O'Donnell, *Arch. Biochem. Biophys.*, 2009, **484**, 167–172.

35. C. Richter, *Chem. Phys. Lipids*, 1987, **44**, 175–189.

36. V. R. de Lima, M. P. Morfim, A. Teixeira and T. B. Creczynski-Pasa, *Chem. Phys. Lipids*, 2004, **132**, 197–208.

37. J. Brzeszczynska and K. Gwozdzinski, *Cell Biol. Int.*, 2008, **32**, 114–120.

38. M. N. Starodubtseva, A. L. Tattersall, T. G. Kuznetsova, N. I. Yegorenkov and J. C. Ellory, *Bioelectrochemistry*, 2008, **73**, 155–162.

39. S. J. Marrink, A. H. de Vries and D. P. Tieleman, *Biochim. Biophys. Acta, Biomembr.*, 2009, **1788**, 149–168.

40. R. F. Khairutdinov, J. W. Coddington and J. K. Hurst, *Biochemistry*, 2000, **39**, 14238–14249.

41. S. S. Marla, J. Lee and J. T. Groves, *Proc. Natl. Acad. Sci. U. S. A.*, 1997, **94**, 14243–14248.

42. P. Schroeder, L. O. Klotz and H. Sies, *Biochem. Biophys. Res. Commun.*, 2003, **307**, 69–73.

43. H. L. Shi, N. Noguchi, Y. X. Xu and E. Niki, *Biochem. Biophys. Res. Commun.*, 1999, **257**, 651–656.

44. F. Castelli, D. Trombetta, A. Tomaino, F. Bonina, G. Romeo, N. Uccella and A. Saija, *J. Pharmacol. Toxicol. Methods*, 1997, **37**, 135–141.

45. Y. Yoshida, N. Ito, S. Shimakawa and E. Niki, *Biochem. Biophys. Res. Commun.*, 2003, **305**, 747–753.

46. Y. T. Li, S. H. Zhang, L. Wang, R. R. Xiao, W. Liu, X. W. Zhang, Z. Zhou, C. Amatore and W. H. Huang, *Angew. Chem., Int. Ed.*, 2014, **53**, 12456–12460.

47. D. Emerson, S. F. Peteu and R. M. Worden, *Biotechnol. Tech.*, 1996, **10**, 673–678.

48. S. F. Peteu, M. T. Widman and R. M. Worden, *Biosens. Bioelectron.*, 1998, **13**, 1197–1203.

49. B. Hille, *Ionic Channels of Excitable Membranes*, Sinauer Associates Inc., Sunderland, 2nd edn, 1992.

50. V. F. Antonov, A. A. Anosov, V. P. Norik and E. Y. Smirnova, *Eur. Biophys. J.*, 2005, **34**, 155–162.

51. R. Schonherr, M. Hilger, S. Broer, R. Benz and V. Braun, *Eur. J. Biochem.*, 1994, **223**, 655–663.

52. S. Matile, N. Berova and K. Nakanishi, *Chem. Biol.*, 1996, **3**, 379–392.

53. R. L. Baldwin, T. Mirzabekov, B. L. Kagan and B. J. Wisnieski, *J. Immunol.*, 1995, **154**, 790–798.

54. R. Margalit and A. Shanzer, *Biochim. Biophys. Acta*, 1981, **649**, 441–448.

55. R. L. Robinson and A. Strickholm, *Biochim. Biophys. Acta*, 1978, **509**, 9–20.

56. J. S. Shim, J. Geng, C. H. Ahn and P. X. Guo, *Biomed. Microdevices*, 2012, **14**, 921–928.

57. A. Negoda, Y. Liu, R. M. Worden, W. C. Hou, C. Corredor, B. Y. Moghadam, C. Musolff, L. Li, W. Walker, P. Westerhoff, A. J. Mason, P. Duxbury,

J. D. Posner and R. M. Worden, *Int. J. Biomed. Nanosci. Nanotechnol.*, 2013, **3**, 52–82.

58. A. Negoda, K. J. Kim and E. D. Crandall, *Biochim. Biophys. Acta, Biomembr.*, 2013, **1828**, 2215–2222.

59. C. Rossi and J. Chopineau, *Eur. Biophys. J.*, 2007, **36**, 955–965.

60. S. R. Jadhav, D. X. Sui, R. M. Garavito and R. M. Worden, *J. Colloid Interface Sci.*, 2008, **322**, 465–472.

61. S. R. Jadhav, Y. Zheng, R. M. Garavito and R. M. Worden, *Biosens. Bioelectron.*, 2008, **24**, 831–835.

62. J. W. F. Robertson, M. G. Friedrich, A. Kibrom, W. Knoll, R. L. C. Naumann and D. Walz, *J. Phys. Chem. B*, 2008, **112**, 10475–10482.

63. S. Terrettaz, M. Mayer and H. Vogel, *Langmuir*, 2003, **19**, 5567–5569.

64. T. N. Tun and A. T. A. Jenkins, *Electrochem. Commun.*, 2010, **12**, 1411–1415.

65. A. Negoda, Y. Liu, W. C. Hou, C. Corredor, B. Y. Moghadam, C. Musolff, L. Li, W. Walker, P. Westerhoff, A. J. Mason, P. Duxbury, J. D. Posner and R. M. Worden, *Int. J. Biomed. Nanosci. Nanotechnol.*, 2013, **3**, 52–83.

66. H. Yue, L. Ying, B. L. Hassler, R. M. Worden and A. J. Mason, *IEEE Trans. Biomed. Circuits Syst.*, 2013, **7**, 43–51.

67. Y. Liu, Z. Zhang, Q. Zhang, G. L. Baker and R. M. Worden, *Biochim. Biophys. Acta*, 2014, **1**, 429–437.

68. S. Malcolm, R. Foust, III, C. Hertkorn and H. Ischiropoulos, in *Septic Shock Methods and Protocols*, ed. T. Evans, Humana Press, 2000, vol. 36, pp. 171–177.

69. X. Chen, H. Chen, R. Deng and J. Shen, *Biomed. J.*, 2014, **37**, 120–126.

70. Y. Liu and R. M. Worden, *Biochim. Biophys. Acta, Biomembr.*, 2014, **1848**, 67–75.

71. Y. Liu and R. M. Worden, *Biochim. Biophys. Acta, Biomembr.*, 2014, **1838**, 429–437.

72. T. Cassier, A. Sinner, A. Offenhäuser and H. Möhwald, *Colloids Surf., B*, 1999, **15**, 215–225.

73. Y. Liu, Z. Zhang, Q. Zhang, G. L. Baker and R. M. Worden, *Biochim. Biophys. Acta, Biomembr.*, 2014, **1838**, 429–437.

74. E. Schnitzer, I. Pinchuk and D. Lichtenberg, *Eur. Biophys. J. Biophys. Lett.*, 2007, **36**, 499–515.

75. J. Wong-ekkabut, Z. Xu, W. Triampo, I. M. Tang, D. P. Tieleman and L. Monticelli, *Biophys. J.*, 2007, **93**, 4225–4236.

Recent Approaches to Enhance the Selectivity of Peroxynitrite Detection

ALINA VASILESCU[†b], VALENTINA DINCA[†a],
MIHAELA FILIPESCU[a], LAURENTIU RUSEN[a],
IOANA S. HOSU[c], RABAH BOUKHERROUB[d],
SABINE SZUNERITS[d], MARIA DINESCU[a],
AND SERBAN F. PETEU*[c,e]

[a]National Institute for Laser, Plasma and Radiation Physics (INFLPR),
Str Atomistilor 409, PO Box MG-36, 077125, Magurele, Bucharest, Romania;
[b]International Centre of Biodynamics, 1 B Intrarea Portocalelor, 060101,
Bucharest, Romania; [c]National Institute for R&D in Chemistry and
Petrochemistry, ICECHIM, Splaiul Independentei 202, 060021 Bucharest,
Romania; [d]Institut de Recherche Interdisciplinaire (IRI, USR 3078),
Université Lille 1, Parc de la Haute Borne, 50 Avenue de Halley, 59658
Villeneuve d'Ascq, France; [e]Chemical Engineering & Materials Science,
Michigan State University, 3523, Engineering Building, 428 S. Shaw Ln,
East Lansing, MI 48824-1226, USA
*E-mail: serbanfpeteu@gmail.com, peteu@msu.edu

[†]The first two authors have contributed equally to this work.

RSC Detection Science Series No. 7
Peroxynitrite Detection in Biological Media: Challenges and Advances
Edited by Serban F. Peteu, Sabine Szunerits, and Mekki Bayachou
© The Royal Society of Chemistry 2016
Published by the Royal Society of Chemistry, www.rsc.org

9.1 Introduction

Monitoring changes in peroxynitrite (PON) concentration in biological systems in real time represents a challenging endeavor. PON is a short-lived reactive species (half-life <1 s at physiological pH) and the concentration levels reported in cellular cultures and *in vivo* range from femtomolars to hundreds of nanomolars. Most of the time, a compromise is reached between maximum sensitivity, selectivity and temporal resolution. While there are currently no biosensors that respond reliably to PON *in vivo*, as is the case for other reactive species,[1,2] electrocatalytic materials and electrochemical approaches in general have been shown to drastically enhance the PON signal over other interferences such as superoxide (O_2^-), nitric oxide (NO) or hydrogen peroxide (H_2O_2).[3] Direct amperometric PON detection based on its oxidation using carbon-based and platinized electrodes or electrochemical reduction on gold electrodes at low negative potentials where interferences are at minimal levels has been proposed.[4,5]

Separation of PON from biological samples prior to its detection has been considered recently.[6] The approach is based on the use of a microchip electrophoresis system, coupled with amperometric detection where PON separation and detection occur in less than 1 min. However, such an approach does not allow the monitoring of rapid events, such as PON bursts. The release of PON from a single stimulated macrophage[4,7] or the measurement of the intracellular concentration of PON in neonatal myocardial cells[8] and in single aortic endothelial cells[9] have been successfully addressed by considering "fast" detection by two means:[10] chemical resolution and selective chemical sensitivity.

9.2 Approaches to Enhance the Selectivity of PON Detection

9.2.1 Selectivity *via* Chemical Resolution

Achievement of chemical resolution relies on detecting the electrochemical "signature" of the analyte of interest. Each analyte presents a specific current potential curve depending on its formal potential (Table 9.1) and electron transfer kinetics.[10] In some cases, techniques such as fast scan cyclic voltammetry allow the signals of multiple analytes to be resolved. While frequently used to monitor events associated with neurotransmitters, fast scan cyclic voltammetry has not been applied for the detection of reactive oxygen species (ROS)/reactive nitrogen oxide species (RNOS) due to the high capacitive currents associated with this technique—hence low sensitivity—combined with the lower concentrations of ROS/RNOS expected *in vivo*.

The monitoring of PON released by myocardial cells was reported by Jin using differential pulse amperometry (DPA).[8] The principle of DPA is illustrated in Figure 9.1.[8] In this procedure, the potential is kept at one value for a short time, and then switched to another value for the same length

Table 9.1   The formal potentials $E^{o\prime}$ of selected ROS and RNOS *vs.* the normal hydrogen electrode (NHE). Reproduced with kind permission from Springer Science and Business Media.[11]

Redox couple of ROS or RNOS	Formal potential $E^{o\prime}$ *vs.* NHE (V)
$O_2/O_2^{\cdot-}$	−0.33
H_2O_2/O_2	−0.15
$ONOO^{\cdot}/ONOO^-$	+0.20
$NO_2^{\cdot}/NO_2^-$	+0.99
$NO^+/NO^{\cdot}$	+1.21
$NO_2^+/NO_2^{\cdot}$	+1.56
$NO_3^{\cdot}/NO_3^-$	+2.50

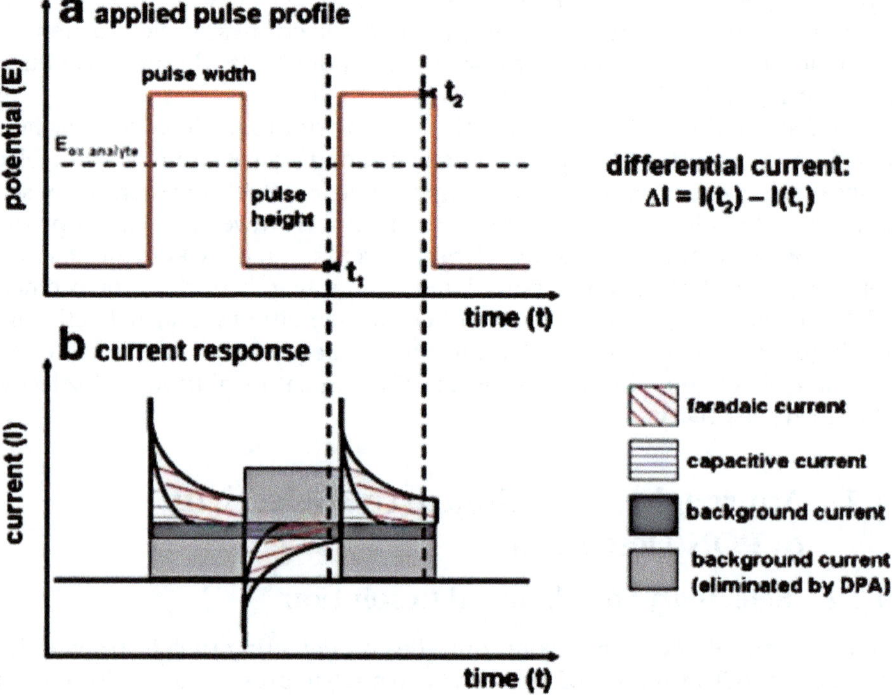

Figure 9.1   (a) Applied pulses and (b) current response in DPA. With kind permission from Springer Science and Business Media: *Anal. Bioanal. Chem.*, 2009, **394**, 95, S. Borgmann, Electrochemical quantification of reactive oxygen and nitrogen: challenges and opportunities, Figure 4. © Springer-Verlag 2009.

of time. The two potentials are chosen so they correspond to one value where the oxidation/reduction of the analyte of interest takes place, while at the other value only the background current caused by other electroactive species in the solution is measured. The difference in currents at the first and second potential is then measured. By taking the measurements at the end of each potential step, the contribution of capacitive current is

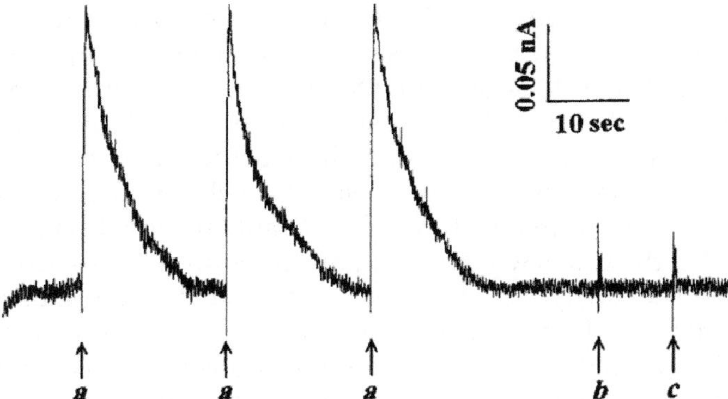

Figure 9.2 DPA responses to injection of (a) 1.0×10^{-7} mol L^{-1} PON, (b) 1.0×10^{-6} mol L^{-1} H_2O_2 and (c) 1.0×10^{-6} mol L^{-1} NO_2^{-}. Reprinted with permission from J. Xue, X. Ying, J. Chen, Y. Xian, L. Jin and J. Jin, *Anal. Chem.*, 2000, 72, 5313. Copyright 2000 American Chemical Society.

drastically diminished. To monitor the PON released by myocardial cells, the potential was kept at 0.2 V *vs.* reference for 100 ms, then switched to 0.0 V for 100 ms.[8]

Many interfering agents, including neurotransmitters, their metabolites and other species such as ascorbic acid, uric acid, O_2^{-} and nitrite (NO_2^{-}), have been investigated and found not to interfere with PON detection by DPA (Figure 9.2). The successful monitoring of PON with no interference was not due exclusively to the use of DPA, but to a combination of the electrochemical technique chosen, the low operating potential (0 V *vs.* the reference electrode), measurement in a region with minimum electrochemical interference and coating the sensor with a permselective membrane.

9.2.2 Selectivity *via* Chemical Sensitivity

In biological media, electroactive species give mostly overlapping electrochemical signals. Constant potential amperometry provides good temporal resolution, appropriate for monitoring oxidative bursts, but little chemical resolution. Selective chemical sensitivity approaches aim to maintain a high sensitivity for the species of interest while diminishing the response for possible interfering species. This can be achieved in a number of ways, including: (a) developing materials displaying high activity towards PON but negligible response towards interferents; (b) tuning the electrochemical techniques, sometimes combined with mathematical calculations, to isolate the PON signal from that of the other interfering species in the medium; and (c) using permeable membranes, limiting the access of possible interfering species.

The development of better electrocatalytic materials typically involves screening new molecules and combinations of the catalyst and support. This is more efficiently done by high-throughput screening, such as using electrochemical robotic systems.[11,12] Currently, there are a series of materials that have proven to be efficient electrocatalysts for PON, such as organometallic complexes of iron, manganese (Mn) and cobalt (porphyrins, phthalocyanines), polymer hybrids and reduced graphene oxide (rGO) based nanocomposites. The PON chemical sensors based on electrocatalytic materials have recently been reviewed.[3] Metallo-macrocycle-based electrodes, for example, were shown to allow sensitive electrochemical detection of PON in a variety of sensor architectures.[3,13–16] Graphene-based nanocomposites investigated in the past few years have shown promise, in particular, for augmenting the sensitivity of PON electrocatalytic detection. Electrocatalysts such as hemin or cobalt phthalocyanine carboxylic acid were supported on rGO nanocomposites *via* π–π stacking interactions.[17–24] Sensitivity enhancements were also obtained by utilizing conductive polymers (CPs) such as polyethylenedioxythiophene (PEDOT).[15,25] The use of CPs for sensing can be combined with their utility as soft actuators in methods for the responsive release of therapeutic agents[26–29] or the delivery of decomposition catalysts that could counteract the toxic effects of PON.[30]

9.2.3 Selectivity *via* Detection Method

Selective detection of PON can also be achieved by the appropriate choice of parameters of the detection method, combined sometimes with mathematical calculations. The group of Amatore pioneered electrochemical assessment of the relative concentrations of ROS/RNOS in synthetic mixtures and biological systems by an innovative approach.[4,7,31–35] The method, supported by both theoretical models and experimental validation is based on amperometric detection of ROS/RNOS species at selected potentials, using platinized carbon microelectrodes. Amatore's group reported that a single-cell oxidative burst could, in fact, be decomposed into each of the corresponding species, specifically, reduced oxygen ($O_2^{\cdot-}$, H_2O_2) and reduced nitrogen (NO, NO_2^-, $ONOO^-$) species. It was shown that NO, NO_2^-, $ONOO^-$ and H_2O_2 could be selectively quantified based on the contribution of each reactive species to the total current recorded at each of the three potentials. Figure 9.3 illustrates the steady-state voltammograms for NO, NO_2^-, $ONOO^-$ and H_2O_2, showing the partial overlap between the current potential traces for $ONOO^-$ and H_2O_2, and how the three potential values were chosen to allow the best combination of sensitivity and selectivity for the amperometric detection of PON.

Moreover, the authors reconstructed the original fluxes of NO and $O_2^{\cdot-}$ (Figure 9.4) using a system of mathematical equations and the fluxes of the four reactive species from Figure 9.3.[4] While this experimental algorithm allowed new investigations in controlled bio-environments, monitoring simultaneously the temporal evolution of individual RNOS species in single cells was not possible using this approach. The same group later reported the simultaneous monitoring of NO•, $ONOO^-$ and H_2O_2 released by single macrophages, using triple-potential step amperometry at platinized carbon microelectrodes.

Instead of carrying out individual experiments at single potentials and repeating them for a large number of cells to account for cellular variability,

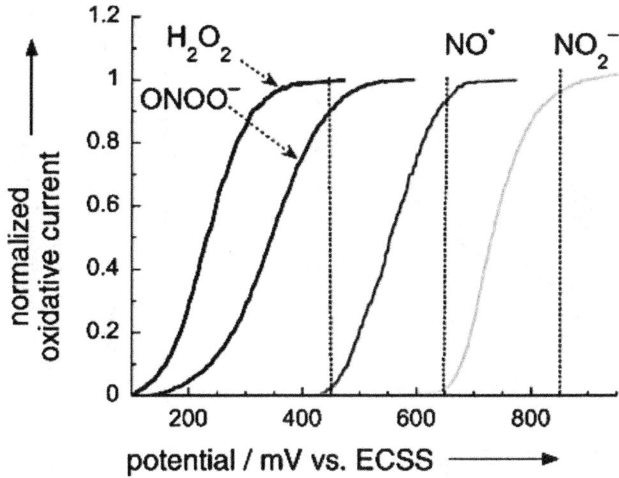

Figure 9.3   Normalized steady-state voltammograms (20 mV s^{-1}) obtained for the electrochemical oxidations of H$_2$O$_2$, ONOO$^-$, NO$^{\bullet}$ and NO$_2^-$ solutions (each at 1 mM in phosphate-buffered saline) at platinized carbon fiber microelectrodes. Dotted lines define the potentials offering the best sensitivity and selectivity of detection for each species by amperometry. Reprinted with permission from *ChemBioChem*, 7, 653, C. Amatore, S. Arbault, C. Bouton, K. Coffi, J. C. Drapier, H. Ghandour and Y. Tong, Monitoring in real time with a microelectrode the release of reactive oxygen and nitrogen species by a single macrophage stimulated by its membrane mechanical depolarization. Copyright © 2006 WILEY-VCH Verlag GmbH & Co. KGaA, Weinheim.

in this new method, the potential was stepped from 300 to 450 and 650 mV, and back to 300 mV in the span of 60 s. The current was measured at the end of each 20 s potential step to minimize the contribution of the capacitive current and increase the signal-to-noise ratio (Figure 9.5). Although limited by the current sampling ratio—at every 60 s for each potential—the method allowed the fluxes of NO$^{\bullet}$, ONOO$^-$ and H$_2$O$_2$ released by a single stimulated macrophage to be recorded simultaneously. By comparing the amperograms obtained for a single cell, the study also allowed the differences in the release profiles over time to be emphasized for the various reactive species.

9.2.4 Selectivity *via* Permeable Membranes

Polyethyleneimine (PEI) was used by Koh and colleagues[36] as an outer coating for modified microelectrodes used for PON detection. The platinum microelectrodes were modified by electrodeposition with a Mn–polymer complex film with poly-2,5-di-(2-thienyl)-1*H*-pyrrole)-1-(*p*-benzoic acid) (pDPB) and decorated with Au nanoparticles, providing enhanced electrochemical PON reduction. The PON electroreduction was monitored amperometrically at +0.2 V. Under these conditions, several electroactive species and PON decomposition molecules[36] were investigated as interferents and found to give negligible signals, with the exception of oxygen, which showed a small signal

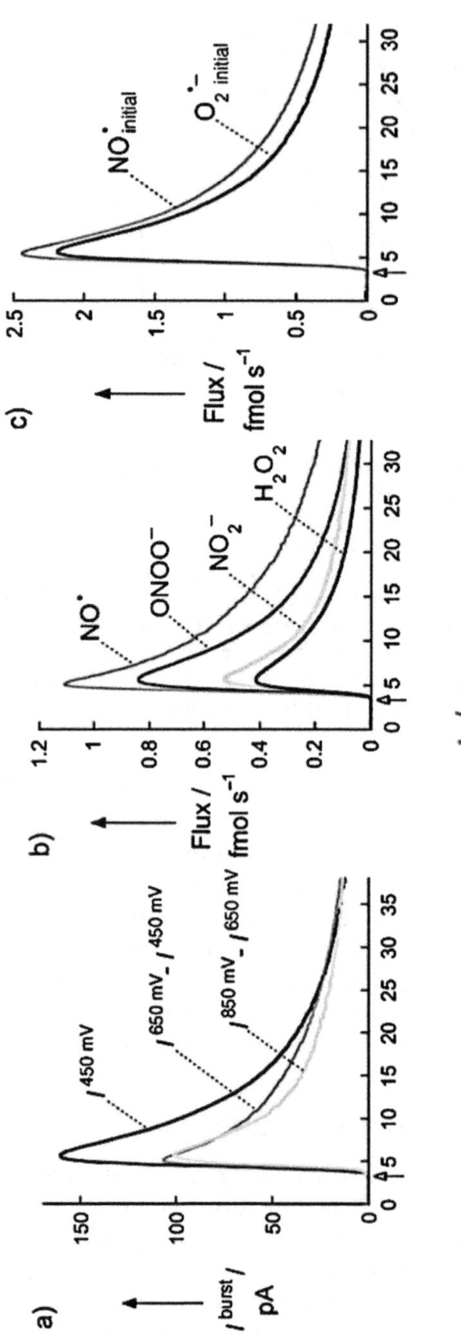

Figure 9.4 (a) Amperometric response at +450 mV *vs.* silver/silver chloride electrode and subtracted curves representative of the contribution of nitrogen-containing organic compounds ($I_{650mv} - I_{450mv}$) and NO_2^- ($I_{850mv} - I_{650mv}$) in the oxidative burst current detected on macrophages. (b) Time variations of the individual fluxes for the different species emitted by macrophages. Fluxes were determined according to Faraday's law from subtracted curves for NO⋅ and NO_2^- and from the current at +450 mV for H_2O_2 and $ONOO^-$ through the application of the procedure described by Amatore's group.[4] (c) Reconstruction of the fluxes of the primary species NO⋅ and $O_2^{\cdot-}$ that are responsible for the final cocktail of species in the oxidative burst of the macrophages. Reprinted with permission from *ChemBioChem*, **7**, 653, C. Amatore, S. Arbault, C. Bouton, K. Coffi, J. C. Drapier, H. Ghandour and Y. Tong, Monitoring in real time with a microelectrode the release of reactive oxygen and nitrogen species by a single macrophage stimulated by its membrane mechanical depolarization. Copyright © 2006 WILEY-VCH Verlag GmbH & Co. KGaA, Weinheim.

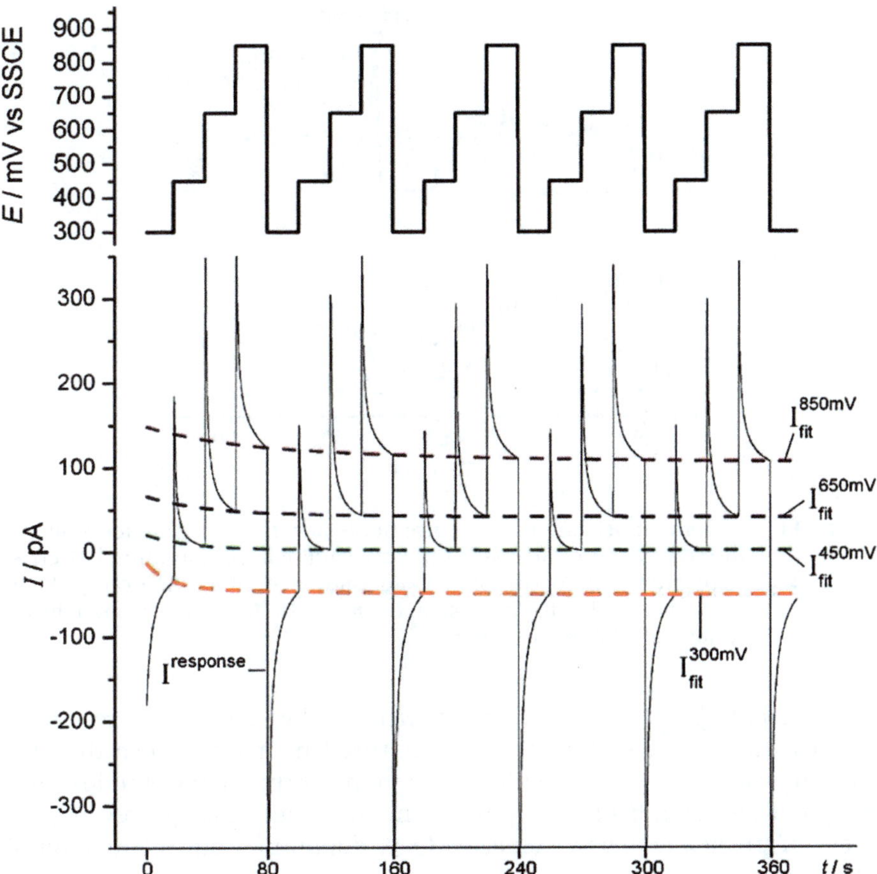

Figure 9.5 Measured current (bottom, solid line) resulting from a repetitive sequence of potential steps (top) applied to a platinized carbon microelectrode. The series of four dashed lines are the best fits taking into account the mean current near the end of each step. From bottom to top: +300, +450, +650 and +850 mV *vs.* silver/silver chloride electrode (SSCE), respectively. Reprinted with permission from C. Amatore, S. Arbault and A. C. W. Koh, *Anal. Chem.*, 2010, **82**, 1411. Copyright 2010 American Chemical Society.

(Figure 9.6). For this particular sensor architecture, the authors reported a response time of 15 s, which might limit the sensor's application for monitoring rapid events. However, the sensor was proven appropriate for PON detection in a rat spiked plasma sample, as well as for the measurement of PON generated by cultured rat glioma cells stimulated with phorbol myristate acetate.

Another preferred cationic polymer used for coating PON ultramicrosensors is poly(4-vinylpyridine). In two studies where this polymer was used as a coating, the detection was based on peroxynitrate reduction catalyzed by electropolymerized films of tetra-aminophthalocyanine Mn(II)[8]

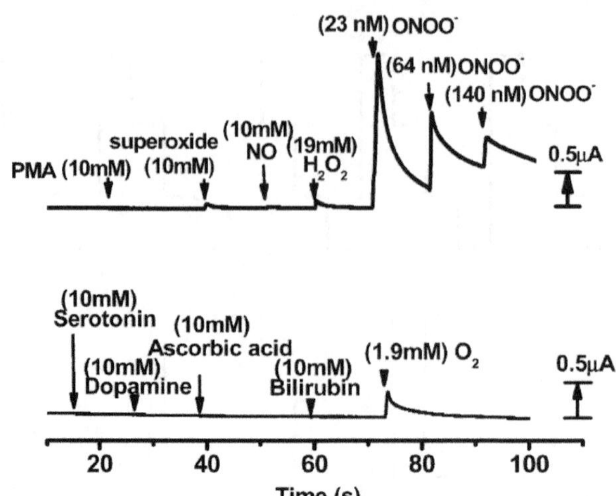

Figure 9.6 Chronoamperometric measurements for the interference effects of different compounds with the Mn–pDPB complex-modified electrode. Reprinted with permission from reference W. C. A. Koh, J. I. Son, E. S. Choe and Y. B. Shim, *Anal. Chem.*, 2010, **82**, 10075. Copyright 2010 American Chemical Society.

or Mn(III)-[2,2]*para*cyclophenyl-porphyrin.[37] The sensors were applied to study the release of ONOO⁻ from cultured neonatal myocardial cells, induced by ischemia/reperfusion and to monitor the NO/PON redox imbalance in endothelial cells. The poly(4-vinylpyridine) coating prevented the access of positively charged species (*e.g.*, dopamine) and of macromolecules such as proteins.[8]

9.2.5 Selectivity *via* Deposition Technique

Drop casting is extensively used for the integration of graphene-based composites into the sensing layer of the electrode. Although very simple, such an approach leads to less reproducible and sometimes quite fragile films with poor stability. Our group investigated an alternative approach, the use of pulsed laser evaporation (PLE), to prepare hybrid films of rGO, iron protoporphyrin (hemin) and rGO–hemin complexes from solution. These active films were prepared by PLE on glassy carbon electrodes (GCEs) and were characterized and subsequently utilized as electrochemical PON-sensitive interfaces.

PLE was reported to ensure clean processing of natural or synthetic materials, complex hybrids and composite materials with minimal damage to the deposited molecules.[38–41] PLE typically involves a relatively low fluence (radiative flux per unit area) laser applied in vacuum, to interact with a frozen mixture of material and solvent.[42–44] During this process, the solute molecules are gently propelled from the target and then deposited onto a substrate (such as a glassy carbon surface) located a few centimeters away. Solute desorbs intact during the process, without any significant decomposition and it is often uniformly deposited on the substrate. Other potential PLE advantages include the use of small amounts of precursor materials, the control of architecture in terms of the number of layers, improved adherence on the substrate and between layers, controlled morphology, homogeneity, and uniformity of the surfaces. Owing to these potential advantages, in recent years PLE has been successfully used[45–58] as an alternative to more conventional deposition techniques to obtain thin films from active materials such as polymers, proteins or enzymes on various substrates from planar sensors[38–50] to 3D substrates.[45] Our group made PON-sensitive films by depositing rGO–hemin films by PLE on GCEs. The nano- and microscale morphology and roughness of these rGO–hemin films prepared by PLE were imaged by atomic force microscopy (AFM), while their electrochemical interaction with PON was evaluated by cyclic voltammetry and amperometry. A schematic of the PLE experimental deposition system used in the study is shown in Figure 9.7. The general PLE setup has been described elsewhere.[49] Briefly, the PLE targets were prepared by suspending rGO, hemin and rGO–hemin complexes in 1% dimethylformamide solutions. The prepared solutions were gently homogenized and rapidly frozen by pouring in the liquid nitrogen-cooled copper holder target as illustrated in Figure 9.7A. The nano-imaging of surface morphology of the PLE films was performed by AFM. The general aspects and roughness, and the typical 3D features of the film surfaces are shown in Figure 9.7B. The morphology of the coatings differed substantially by compound type. More specifically, smooth films were obtained from hemin, while the presence of rGO compounds appeared to generate a conglomerate with sporadic pores. The differences could perhaps be correlated to the imperfect miscibility of the starting rGO or rGO–hemin solution compared with those containing hemin alone.

3-Morpholinosydnonimine-*N*-ethylcarbamide (SIN-1) was utilized as a PON donor[59,60] to evaluate its interaction with the PLE-prepared rGO–hemin films. Electrocatalytic oxidation in the presence of PON is mediated by the hemin iron center within the rGO–hemin film, showing an oxidative wave at 1.17 V *vs.* Ag/AgCl, as illustrated in Figure 9.8.[13–16]

The calibration curve is illustrated in Figure 9.9 with the oxidation current scaling linearly with the increasing $ONOO^-$ concentration. From the slope of the calibration curve, the sensitivity of the rGO–hemin GCE towards PON was evaluated at 4 nA nM^{-1} and the limit of detection (LoD) was 25 nM. As shown in Figure 9.9, the amperometric response is reversible after re-immersion in fresh phosphate-buffered saline. However, both the LoD and response time of 10 s need further improvement, and miniaturization is one way to achieve this.[61,62] Meanwhile, these preliminary results proved the feasibility of the PLE method to form stable films from solutions of rGO–hemin complexes.

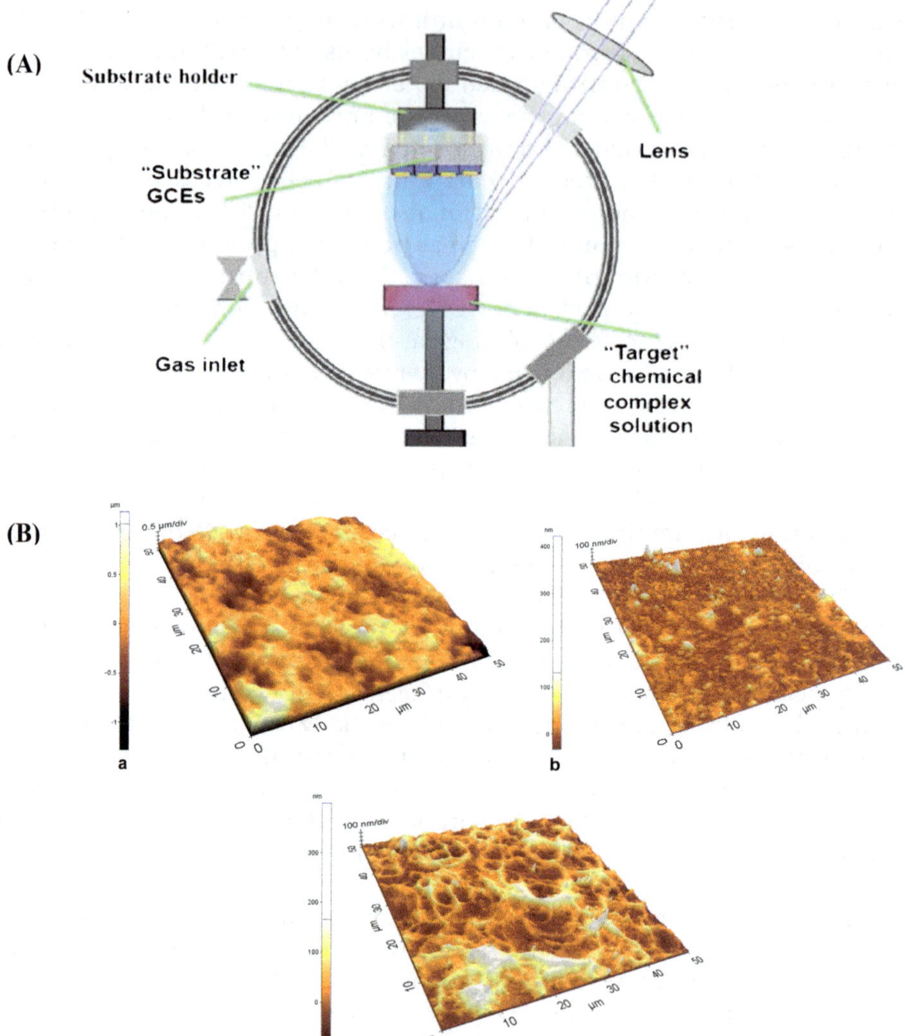

Figure 9.7 (A) Experimental setup of the PLE-assisted deposition of rGO–hemin films. A pulsed Nd:YAG laser system irradiated the frozen target, prepared by mixing the rGO, hemin or rGO–hemin in dimethylformamide solution to 1% final weight, and gently homogenized and then maintained frozen with liquid nitrogen. Thus, films of hemin and rGO–hemin were deposited, by PLE of the target solvent, onto the freshly cleaned GCE substrates. (B) 3D AFM images of (a) rGO, (b) hemin and (c) rGO–hemin films formed by PLE. The surfaces of the rGO films obtained had one order of magnitude higher roughness of 447 nm compared with 23 nm for just hemin and 50 nm for the rGO–hemin complex.

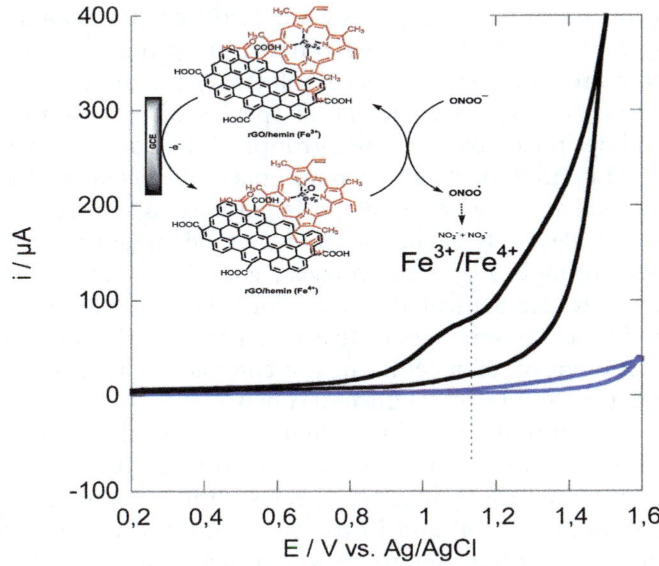

Figure 9.8　Electrocatalytic oxidation in the presence of ONOO⁻ is mediated by hemin centers within the rGO–hemin film. Reproduced from ref. 13 with permission from The Royal Society of Chemistry.

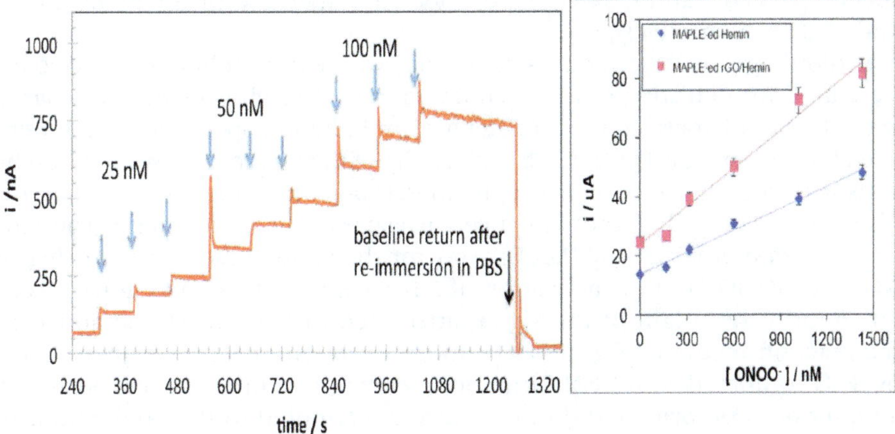

Figure 9.9　Typical amperometry response to PON obtained using rGO–hemin GCEs polarized at +1.1 V *vs.* Ag/AgCl for subsequent analyte aliquots (left) and calibration graphs of rGO–hemin and hemin modified GCEs (right).

9.3　Selective PON Detection with Simultaneous Detection of Interfering Species

To understand the mechanism by which cellular events trigger PON release it is important to monitor simultaneously, in addition to PON, the fluxes of other reactive species such as $NO^{\bullet}$, H_2O_2 and NO_2^{-}. PON is generated by the reaction of $NO^{\bullet}$ and $O_2^{\bullet-}$. H_2O_2 is generated from the spontaneous evolution of O_2^- and NO_2^- and it is a decomposition product of both $NO^{\bullet}$ and $ONOO^-$. There are two ways to ensure selectivity for the detection of each analyte. One elegant way is to detect all species with the same sensor, as described and optimized by Amatore's group.[31–35] By polarizing the sensor at four different potentials and using a mathematical calculation based on the currents recorded at these potentials, it is possible to extract the individual fluxes of $ONOO^-$, H_2O_2, $NO^{\bullet}$ and NO_2^-, and to calculate the original fluxes of NO and $O_2^{\bullet-}$ produced by macrophages, as explained above. The alternative approach is to use dedicated sensors for each of the reactive species of interest by using sensor arrays. Selective detection of each analyte is achieved through a combination of specific surface chemistry and a specific value of the applied potential for each dedicated sensor.

A module containing three individually addressable micrometric sensors, optimized for the amperometric detection of NO, $O_2^{\bullet-}$ and $ONOO^-$ was used to monitor these reactive species simultaneously, both *in vitro* for a single endothelial cell and in the vasculature of rats *in vivo*.[9] In this work, carbon fiber working electrodes were functionalized with a polymeric nickel(II)tetrakis(3-methoxy-4-hydroxyphenyl)porphyrin covered with Nafion for NO detection, with immobilized polypyrrole/horseradish peroxidase and superoxide dismutase (SOD) for the $O_2^{\bullet-}$ sensor, and a polymeric film of Mn(III)-[2,2]*para*cyclophenyl-porphyrin for $ONOO^-$ detection. The reactive species NO, $O_2^{\bullet-}$ and $ONOO^-$ were detected amperometrically at 0.67, 0.35 and −0.35 V, respectively.

Bedioui's group described a sensor array where bare gold microelectrodes were used for detecting $ONOO^-$ reduction at −0.1 V, while electrodes coated with electropolymerized films of eugenol and phenol, operated at 0.8 V were used for the detection of NO.[5] The selectivity of the two sensors was proven by tests with the interferents nitrite, hydrogen peroxide, dopamine and ascorbic acid. Recently, the proof of concept was presented by the same group for a new ultramicroelectrode (UME) array for the simultaneous monitoring of $NO^{\bullet}$ and $ONOO^-$ and the activation of HL60 cells as shown in Figure 9.10.[63] To confirm the origin of the signal attributed to PON, SOD was added in the reaction medium (Figure 9.11). In these conditions, scavenging of $O_2^{\bullet-}$ by SOD competed with PON formation, leading to a drastic decrease in the amount of PON formed and in the current attributed to PON reduction. In this study, however, NO and PON were formed *in situ* from an exogenous NO donor molecule ((Z)-1-[N-(2-aminopropyl)-N-(2-ammoniopropyl)amino]-diazen-1-ium-1,2-diolate) and its reaction with the superoxide released by phorbol-12-myristate-13-acetate-stimulated HL60 cells. Demonstration of the feasibility of sensor arrays for monitoring PON generated exclusively by cellular events remains to be done. As emphasized in a recent review,[64] electrochemical sensor arrays are particularly interesting not only for the simultaneous detection of several species but also for their potential use in mapping the spatio-temporal diffusion or transport of a single analyte. With the current technological advances related to fabrication of miniaturized sensors and arrays, advances can be anticipated in the near future.

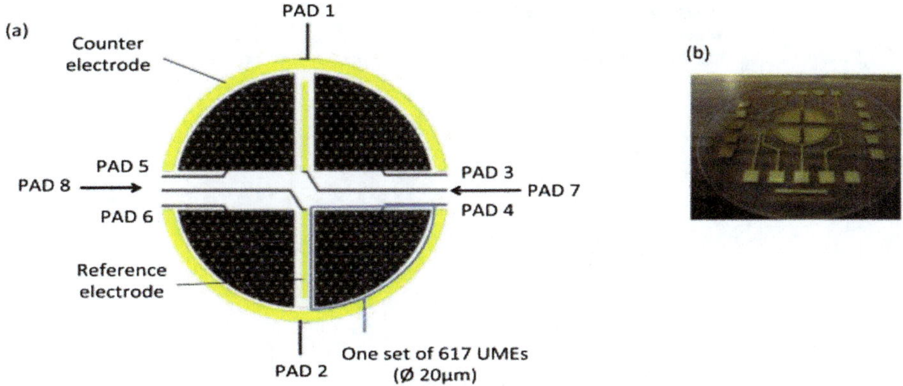

Figure 9.10 (a) Layout and design, and (b) picture of the electrochemical sensor array contained on a 50 mm circular wafer. The electrical connections are permitted through 20 PADs arranged around the periphery. The arrayed UMEs are contained within the confines of the counter electrodes (PADs 1 and 2), which are two half circles with a 7.5 mm radius. There are two reference electrodes, which are centered vertical bars (PADs 7 and 8). The working UMEs are defined by the etching of the parylene insulating layer on the gold electrode areas. They can be addressed in four sets of 617 electrodes of 20 µm diameter each (PADs 3–6). Reprinted from *Electrochim. Acta*, **140**, 33, L. T. O. Kim, V. Escriou, S. Griveau, A. Girard, L. Griscom, F. Razan, F. Bedioui, Array of ultramicroelectrodes for the simultaneous detection of nitric oxide and peroxynitrite in biological systems, copyright 2014 with permission from Elsevier.

9.4 Confirming the Identity of the PON Electrochemical Signal

Ensuring selectivity and correct identification of PON or other reactive species of interest based on their electrochemical signal requires a progressive approach. First, efforts are put into developing sensors displaying the optimal sensitivity for PON. Initial studies of selectivity performed with individual standards and synthetic mixtures in laboratory conditions target known decomposition products of PON, nitrite and nitrate, as well as other electroactive species typically found in biological systems. Both freshly synthesized and *in situ* produced PON should be used to establish sensor sensitivity and selectivity at physiological pH. Before applying the sensor for measuring dynamic concentration changes in biological systems, DPA is useful to confirm the PON identity and to measure basal levels in the reaction medium.[9]

Next, to confirm that the signals recorded with the sensor/sensor array are correctly attributed to PON or other reactive species of interest, exogenous donors, and known triggers or inhibitors of cellular events leading to the formation of relevant reactive species are added to the reaction medium. For example, Burewicz *et al.*[37] described a two-sensor system for the simultaneous monitoring of NO and ONOO⁻ in endothelial cells and validated the selectivity of their method by intervening in the pathway of NO and ONOO⁻ production. First, NO/ONOO⁻ release was stimulated by a calcium ionophore. Next,

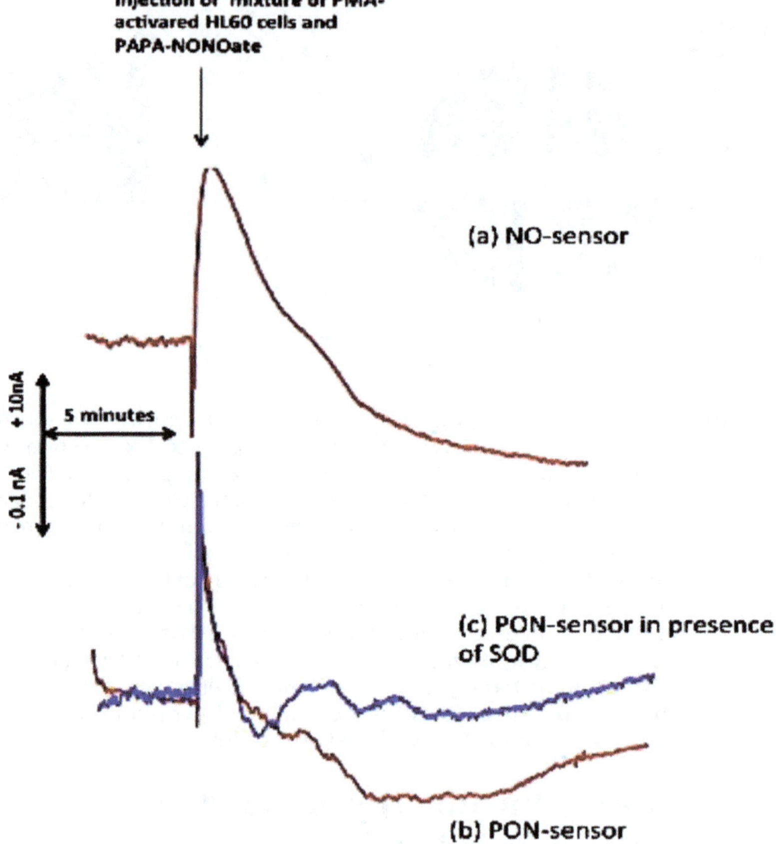

Figure 9.11   Simultaneous amperometric measurements at +0.8 V *vs.* Ag/AgCl at (a) the gold UME array coated by poly(eugenol)/poly(phenol) and (b) −0.1 V *vs.* Ag/AgCl at the bare gold UMEs upon injection of an aliquot of 500 μL of HEPES buffered solution containing 3 million phorbol ester myristate (PMA)-activated HL60 cells and alkaline propylamine propylamine NONOate (PAPA-NONOate) stock solution (100 μM) to the electrochemical cell containing 500 μL of HEPES buffer. (c) Curve obtained at −0.1 V *vs.* Ag/AgCl at the bare gold UMEs upon injection of an aliquot of 500 μL of HEPES buffered solution containing 3 million PMA-activated HL60 cells, alkaline PAPA-NONOate stock solution (100 μM) and SOD (600 U mL^{-1}) to the electrochemical cell containing 500 μL of HEPES buffer. Reprinted from *Electrochim. Acta*, **140**, 33, L. T. O. Kim, V. Escriou, S. Griveau, A. Girard, L. Griscom, F. Razan, F. Bedioui, Array of ultramicroelectrodes for the simultaneous detection of nitric oxide and peroxynitrite in biological systems, copyright 2014 with permission from Elsevier.

polyethylene glycol (PEG)-SOD was added to convert superoxide into H_2O_2, a specific inhibitor of NADPH oxidase (VAS2870) was added to inhibit superoxide production, a Mn(III)porphyrin was used as a ONOO$^-$ scavenger and an L-arginine analog (L-NMMA) was used to inhibit NO synthase in endothelial cells (Figure 9.12). As emphasized in Figure 9.12A, the addition of PEG-SOD

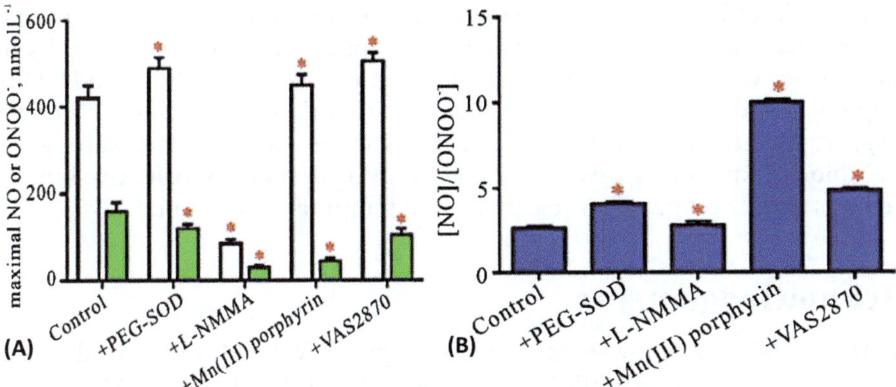

Figure 9.12 (A) Maximal NO (open bars) and ONOO⁻ (solid bars) concentrations released from HUVECs in the absence or presence of PEG-SOD, L-NMMA, VAS 2870, and Mn(III) porphyrin. (B) Ratio of the maximal NO concentration to maximal ONOO⁻ concentration measured from HUVEC cells in the absence or presence of PEG-SOD, L-NMMA, Mn(III) porphyrin, and VAS 2870. $*p < 0.05$ *vs.* control, $n = 3$. A. Burewicz, H. Dawoud, L. Jiang and T. Malinski, Nitric Oxide/Peroxynitrite Redox Imbalance in Endothelial Cells Measured with Amperometric Nanosensors, *Am. J. Anal. Chem.*, **4**, 2013, 30. Copyright 2013 Scientific Research Publishing.

and VAS2870 leads to an increase in the quantity of NO and a decrease in the concentration of PON compared with the control due to the fact that less superoxide was available to react with NO. The Mn(III)porphyrin efficiently scavenged PON, while L-NMMA inhibited the production of NO and, consequently, that of PON. Figure 9.12B presents the ratio between the maximum levels of NO/ONOO⁻ after the addition of different compounds. This provides more evidence that the two nanosensors do indeed capture variations in the concentrations of NO and ONOO⁻ in biological systems. Moreover, this ratio of NO : ONOO⁻ is considered an indicator of the nitroxidative stress caused by ONOO⁻ in the endothelium.[65]

In other studies, stimulation of cultured cells with phorbol myristate acetate[66] was utilized to induce superoxide production. Further addition of an exogenous NO donor led to *in situ* PON formation, thus confirming that the electrochemical signal recorded with the respective sensor was correctly attributed to PON.[63]

9.5 Conclusions

The efforts to develop better electrocatalysts for PON detection are constantly growing and various aspects related to this topic are currently being investigated, including new molecules, new supports, and better ways to deposit electrocatalytic films to improve activity and stability. In this context, we have described a novel PLE method to deposit a hemin–rGO hybrid film. The specificity of the PON detection was mainly evaluated *in vitro*. The adequate ratios of interferent:PON and concentration ranges utilized should mimic those expected in real situations *in vivo*. To ensure selectivity, the identity of the species of interest should be confirmed with additional methods. One approach is to generate it *in situ*, by using either donor molecules or known triggers of cellular events that produce the species. Often, for measurements in biological media, especially when attempting to monitor oxidative bursts, there is a trade off between selectivity and temporal resolution.

Acknowledgements

V. D. and M. D. acknowledge financial support from Ministerul Educatiei Nationale and CNCS-UEFISCDI *via* project PN-II-RU-TE-2011-3-0289. R. B. and S. S. gratefully acknowledge financial support from the Centre National de Recherche Scientifique (CNRS), the University Lille 1, the Nord Pas de Calais region and the Institut Universitaire de France (IUF). S. F. P. and I. S. H. kindly acknowledge funding by CNCS-UEFISCDI for the project PN-II-ID-PCE-2011-3-1076. A. V. acknowledges financial support from CNCS-UEFISCDI for the project PN-II-PT-PCCA-2011-3.1-1809. Also, S. S., R. B. and S. F. P. are grateful to the Institut Français de Roumanie and UEFISCDI for the Partenariat BHC bilateral project 28684 VM. Bogdan Dimitriu, Raluca Oprea and Ana Popescu are credited with preliminary assistance in the initial phase of this work.

References

1. C. Calas-Blanchard, G. Catanante and T. Noguer, *Electroanalysis*, 2014, **26**(6), 1277–1286.
2. S. Gaspar, Biosensors for detection of superoxide and hydrogen peroxide from living cells using electrochemical sensors, in *Oxydative stress: diagnostics, prevention and therapy*, ed. S. Andreescu and M. Hepel, ACS, 2011, pp. 289–309.
3. S. F. Peteu, R. Boukherroub and S. Szunerits, *Biosens. Bioelectron.*, 2014, **58**, 359–373.
4. C. Amatore, S. Arbault, C. Bouton, K. Coff, J. C. Drapier, H. Ghandour and Y. Tong, *ChemBioChem*, 2006, **7**, 653–661.
5. D. Quinton, A. Girard, L. T. T. Kim, V. Raimbault, L. Griscom, F. Razan, S. Griveau and F. Bedioui, *Lab Chip*, 2011, **11**(7), 1342–1350.
6. M. K. Hulvey, C. N. Frankenfeld and S. M. Lunte, *Anal. Chem.*, 2010, **82**(5), 1608–1611.
7. C. Amatore, S. Arbault, C. Bouton, J. C. Drapier, H. Ghandour and A. C. W. Koh, *ChemBioChem*, 2008, **9**, 1472–1480.
8. J. Xue, X. Y. Ying, J. S. Chen, Y. H. Xian, L. T. Jin and J. Jin, *Anal. Chem.*, 2000, **72**, 5313–5321.
9. R. Kubant, C. Malinski, A. Burewicz and T. Malinski, *Electroanal.*, 2006, **18**, 410–416.
10. P. E. M. Phillips and R. M. Wightman, *TRAC*, 2003, **22**, 509–514.

11. S. Borgmann, *Anal. Bioanal. Chem.*, 2009, **394**, 95–105.
12. A. S. Bandarenka, E. Ventosa, A. Maljusch, J. Masa and W. Schuhmann, *Analyst*, 2014, **139**, 1274–1291.
13. R. Oprea, S. F. Peteu, P. Subramanian, W. Qi, E. Pichonat, H. Happy, M. Bayachou, R. Boukherroub and S. Szunerits, *Analyst*, 2013, **138**, 4345–4352.
14. I. S. Hosu, Q. Wang, A. Vasilescu, S. F. Peteu, V. Raditoiu, S. Railian, V. Zaitsev, K. Turcheniuk, Q. Wang, M. Li, R. Boukherroub and S. Szunerits, *RSC Adv.*, 2015, **5**(2), 1474–1484.
15. S. F. Peteu, P. Peiris, E. Gebremichael and M. Bayachou, *Biosens. Bioelectron.*, 2010, **25**, 1914–1921.
16. S. F. Peteu, T. Bose and M. Bayachou, *Anal. Chim. Acta*, 2013, **780**, 81–88.
17. T. Xue, S. Jiang, Y. Q. Qu, Q. Su, R. Cheng, S. Dubin, C. Y. Chiu, R. Kaner, Y. Huang and X. F. Duan, *Angew. Chem., Int. Ed.*, 2012, **51**, 3822–3825.
18. Y. J. Guo, J. Li and S. J. Dong, *Sens. Actuators, B*, 2011, **160**, 295–300.
19. C. X. Guo, Y. Lei and C. M. Li, *Electroanalysis*, 2011, **23**, 885–893.
20. C. H. Xu, J. Li, X. B. Wang, J. C. Wang, L. Wan, Y. Y. Li, M. Zhang, X. P. Shang and Y. K. Yang, *Mater. Chem. Phys.*, 2012, **132**, 858–864.
21. Y. J. Guo, L. Deng, J. Li, S. J. Guo, E. K. Wang and S. J. Dong, *ACS Nano*, 2011, **5**(2), 1282–1290.
22. A. A. Vernekar and G. Mugesh, *Chem.–Eur. J.*, 2012, **18**(47), 15122–15132.
23. I. Kaminska, M. R. Das, Y. Coffinier, J. Niedziolka-Jonsson, J. Sobczak, P. Woisel, J. Lyskawa, M. Opallo, R. Boukherroub and S. Szunerits, *ACS Appl. Mater. Interfaces*, 2012, **4**(2), 1016–1020.
24. I. Kaminska, M. R. Das, Y. Coffinier, J. Niedziolka-Jonsson, P. Woisel, M. Opallo, S. Szunerits and R. Boukherroub, *Chem. Commun.*, 2012, **48**(9), 1221–1223.
25. C. Zanardi, F. Terzi and R. Seeber, *Anal. Bioanal. Chem.*, 2013, **405**(2–3), 509–531.
26. R. Balint, N. J. Cassidy and S. H. Cartwell, *Acta Biomater.*, 2014, **10**(6), 2341–2353.
27. S. F. Peteu, F. Oancea, O. A. Sicuia and F. Constantinescu, *Polymers*, 2010, **2**(3), 229–251.
28. S. F. Peteu, *J. Intell. Mater. Syst. Struct.*, 2007, **18**(2), 147–152.
29. S. Deo, E. Moschou, S. F. Peteu, P. Eisenhardt, L. Bachas, M. Madou and S. Daunert, *Anal. Chem.*, 2003, **75**, 207A–213A.
30. L. M. Sloski and T. W. Vanderah, *Expert Opin. Ther. Pat.*, 2015, **25**, 1.
31. C. Amatore, S. Arbault, D. Bruce, P. De Oliveira, M. Erard and M. Vuillaume, *Faraday Discuss.*, 2000, **116**, 319–333.
32. C. Amatore, S. Arbault, D. Bruce, P. De Oliveira, L. M. Erard and M. Vuillaume, *Chem.–Eur. J.*, 2001, **7**, 4171–4179.
33. C. Amatore, S. Arbault, Y. Chen, C. Crozatier and I. Tapsoba, *Lab Chip*, 2007, **7**, 233–238.
34. C. Amatore, S. Arbault and A. C. W. Koh, *Anal. Chem.*, 2010, **82**, 1411–1419.
35. M. R. Filipović, A. C. W. Koh, S. Arbault, V. Niketić, A. Debus, U. Schleicher, C. Bogdan, M. Guille, F. Lemaître, C. Amatore and I. Ivanović-Burmazović, *Angew. Chem., Int. Ed. Engl.*, 2010, **49**, 4228–4232.

36. W. C. A. Koh, J. I. Son, E. S. Choe and Y. B. Shim, *Anal. Chem.*, 2010, **82**(24), 10075–10082.
37. A. Burewicz, H. Dawoud, L.-L. Jiang and T. Malinski, *Am. J. Anal. Chem.*, 2013, **4**(10A), 30–36.
38. A. Piqué, P. Wu, B. R. Ringeisen, D. M. Bubb, J. S. Melinger, R. A. McGill and D. B. Chrisey, *App. Surf. Sci.*, 2002, **186**(1–4), 408–415.
39. A. Piqué, R. C. Y. Auyeung, R. C. Y. Auyeung, J. L. Stepnowski, D. W. Weir, C. B. Arnold, R. A. McGill and D. B. Chrisey, *Surf. Coat. Technol.*, 2003, **16**, 293–299.
40. A. Purice, J. Schou, P. Kingshott and M. Dinescu, *Chem. Phys. Lett.*, 2007, **435**(4–6), 350–353.
41. K. Wu, B. R. Ringeisen, D. B. Krizman, C. G. Frondoza, M. Brooks, D. M. Bubb, R. C. Y. Auyeung, A. Piqué, B. Spargo, R. A. McGill and D. B. Chrisey, *Rev. Sci. Instrum.*, 2003, **74**(4), 2546–2557.
42. V. Dinca, P. E. Florian, L. E. Sima, L. Rusen, C. Constantinescu, R. W. Evans, M. Dinescu and A. Roseanu, *Biomed. Microdevices*, 2014, **16**(1), 11–21.
43. L. Rusen, C. Mustaciosu, B. Mitu, M. Filipescu, M. Dinescu and V. Dinca, *Appl. Surf. Sci.*, 2013, **278**, 198–202.
44. L. Rusen, V. Dinca, B. Mitu, C. Mustaciosu and M. Dinescu, *Appl. Surf. Sci.*, 2014, **302**, 134–140.
45. V. Califano, F. Bloisi, L. R. M. Vicari, D. M. Yunos, X. Chatzistavrou and A. R. Boccaccini, *Adv. Eng. Mater.*, 2009, **11**, 685–689.
46. C. N. Hunter, M. H. Check, J. E. Bultman and A. A. Voevodin, *Surf. Coat. Technol.*, 2008, **203**, 300–306.
47. A. P. Caricato, A. Luches and R. Rella, *Sensors*, 2009, **9**(4), 2682–2696.
48. V. Dinca, R. Oprea, A. M. Popescu, M. Dinescu, S. F. Peteu, R. Boukherroub and S. Szunerits, COLL. Symposium, ACS National Meeting 2013, Indianapolis IN, United States, 2013.
49. A. Palla-Papavlu, V. Dinca, M. Dinescu, F. Di Pietrantonio, D. Cannata, M. Benetti and E. Verona, *Appl. Phys. A*, 2011, **105**, 651–659.
50. F. Di Pietrantonio, M. Benetti, V. Dinca, D. Cannatà, E. Verona, S. D'Auria and M. Dinescu, *Appl. Surf. Sci.*, 2014, **302**, 250–255.
51. A. T. Sellinger, E. M. Leveugle, K. Gogick, L. V. Zhigilei and J. M. Fitz-Gerald, *J. Vac. Sci. Technol. A*, 2006, **24**(4), 1618–1622.
52. B. R. Ringeisen, J. Callahan, P. K. Wu, A. Piqué, B. Spargo, R. A. McGill, M. Bucaro, H. Kim, D. M. Bubb and D. B. Chrisey, *Langmuir*, 2001, **17**(11), 3472–3479.
53. A. Purice, J. Schou, P. Kingshott and M. Dinescu, *Chem. Phys. Lett.*, 2007, **435**(4–6), 350–353.
54. P. K. Wu, B. R. Ringeisen, D. B. Krizman, C. G. Frondoza, M. Brooks, D. M. Dubb, R. C. Y. Auyeung, A. Pique, B. Spargo, R. A. McGill and D. B. Crisey, *Rev. Sci. Instrum.*, 2003, **74**, 2546–2557.
55. R. C. Y. Auyeung, A. Piqué, B. Spargo, R. A. McGill and D. B. Chrisey, *Rev. Sci. Instrum.*, 2003, **74**(4), 2546–2557.

56. A. P. Caricato, M. G. Manera, M. Martino, R. Rella, F. Romano, J. Spadavecchia, T. Tunno and D. Valerini, *Appl. Surf. Sci.*, 2007, **253**(15), 6471–6475.
57. D. M. Bubb, P. K. Wu, J. S. Horowitz, J. H. Callahan, M. Galicia, A. Vertes, R. A. McGill, E. J. Houser, B. R. Ringeisen and D. B. Chrisey, *J. Appl. Phys.*, 2002, **91**(4), 2055–2058.
58. A. P. Caricato, M. Cesaria, G. Gigli, A. Loiudice, A. Luches, M. Martino, V. V. Resta, A. Rizzo and A. Taurino, *Appl. Phys. Lett.*, 2012, **100**, 073306.
59. Y. Asahi, K. Shinozaki and M. Nagaka, *Chem. Pharm. Bull.*, 1971, **19**, 1079–1088.
60. M. Feelisch, J. Ostrowski and E. Noack, *J. Cardiovasc. Pharmacol.*, 1989, **14**, S13–S22.
61. S. F. Peteu, M. T. Widman and R. M. Worden, *Biosens. Bioelectron.*, 1998, **13**, 1197–1203.
62. D. Emerson, S. F. Peteu and R. M. Worden, *Biotechnol. Tech.*, 1996, **10**, 673–678.
63. L. T. O. T. Kim, V. Escriou, S. Griveau, A. Girard, L. Griscom, F. Razan and F. Bedioui, *Electrochim. Acta*, 2014, **140**, 33–36.
64. S. Griveau and F. Bedioui, *Anal. Bioanal. Chem.*, 2013, **405**, 3475–3488.
65. R. P. Mason, R. Kubant, R. F. Jacob, M. F. Walter, B. Boychuk and T. Malinski, *J. Cardiovasc. Pharmacol.*, 2006, **48**(1), 862–869.
66. A. Karlsson, J. B. Nixon and L. C. McPhail, *J. Leukocyte Biol.*, 2000, **67**(3), 396–404.

Development of Fluorescent Probes for the Detection of Peroxynitrite

ZHIJIE CHEN[a], TAN M. TRUONG[b], AND HUI-WANG AI*[a,b]

[a]Department of Chemistry, University of California Riverside, 501 Big Springs Road, Riverside, CA 92521, USA; [b]Cell, Molecular and Developmental Biology Graduate Program, University of California Riverside, Riverside, CA 92521, USA
*E-mail: huiwang.ai@ucr.edu

10.1   Introduction to Peroxynitrite

10.1.1   Production of Peroxynitrite in Biological Systems

In 1990, Beckman *et al.* observed the production of peroxynitrite while investigating local tissue perfusions of nitric oxide (NO) and superoxide species as potential sources of endothelial injury.[1,2] They were able to show that peroxynitrite generates hydroxyl radical ($^{\bullet}$OH) more efficiently than the Fenton reaction [reaction of reduced iron with hydrogen peroxide (H_2O_2)] in biological systems.[3,4] Their findings were soon confirmed by several independent studies using co-generated NO and superoxide.[5] It was these initial observations that compelled a review of the role peroxynitrite played in oxidative[6] and nitrosative stresses,[7,8] which were previously attributed to the actions of NO or $^{\bullet}$OH.[9] Although no enzymes dedicated to metabolizing peroxynitrite

RSC Detection Science Series No. 7
Peroxynitrite Detection in Biological Media: Challenges and Advances
Edited by Serban F. Peteu, Sabine Szunerits, and Mekki Bayachou
© The Royal Society of Chemistry 2016
Published by the Royal Society of Chemistry, www.rsc.org

have yet been discovered, the discovery of NO synthase (NOS),[10,11] NADPH oxidase (NOX),[11] superoxide dismutase (SOD),[12] and other peroxidases[13] put further weight on the argument for the biological relevance of peroxynitrite. Produced almost spontaneously from the diffusion-limited reaction of NO and superoxide radical ($k = 6.7 \times 10^9$ M^{-1} s^{-1}), peroxynitrite exerts most of its biological effects through direct, or radical-mediated, oxidation and nitration of biomolecules, such as proteins, nucleic acids, and lipids.[14-17] These reactions and their associated products have significant consequences on the function and fate of cells and tissues. Deregulation of such processes can lead to cardiac,[18,19] vascular,[20,21] circulatory,[22] inflammatory,[23,24] diabetic,[7,16] and neurodegenerative diseases.[25,26]

10.1.2 Biological Chemistry of Peroxynitrite

As a short-lived oxidant (half life ~10 ms at physiological pH), peroxynitrite, once produced, can have a multitude of effects on biological systems.[14] Both peroxynitrite anion (ONOO$^-$) and peroxynitrous acid (ONOOH; pK_a = 6.8) can undergo direct one- or two-electron oxidation reactions with redox-active molecules, such as glutathione (GSH),[27,28] metalloproteins,[29] hemeproteins,[30-32] and thiol-containing proteins,[32,33] to name just a few. In addition to direct oxidation of its targets, peroxynitrite can trigger more complex reactions *via* radical-mediated mechanisms.[17] Homolytic fission of peroxynitrite produces ˙OH and nitric dioxide (˙NO$_2$) radicals, whereas the fast reaction of peroxynitrite with carbon dioxide readily generates carbonate (CO$_3$˙$^-$) and ˙NO$_2$ radicals. These radicals can propagate and further induce DNA damage, lipid peroxidation,[34] and protein nitration.[35] In particular, the production of 3-nitrotyrosine is frequently regarded as a marker of peroxynitrite-induced nitrosative stress,[36] despite the fact that 3-nitrotyrosine can also be generated *via* alternative routes.[37] Depending on the local concentration and exposure time, these peroxynitrite-induced reactions can lead to alterations in protein structure and function, inappropriate activation/inactivation of signal transduction pathways, impairment of mitochondrial function, and initiation of cell death (apoptosis or necrosis).[16] However, there are peroxynitrite scavengers and neutralizers, such as uric acid, peroxiredoxins, and selenium-containing GSH peroxidases, that allay the potentially deleterious effects of mild peroxynitrite production.[38] In cases with toxic overproduction of peroxynitrite, treatment with peroxynitrite decomposition catalysts, notably metalloporphyrins, has shown protective antioxidant-like effects in various disease models (Figure 10.1).[39-43]

10.1.3 Peroxynitrite-Induced Cytotoxicity

Due to its high reactivity, both exogenous and endogenous peroxynitrite is capable of conferring deleterious and cytotoxic effects on cells and tissues. For example, peroxynitrite-dependent oxidation of mitochondrial electron transport complexes and mitochondrial permeability transition

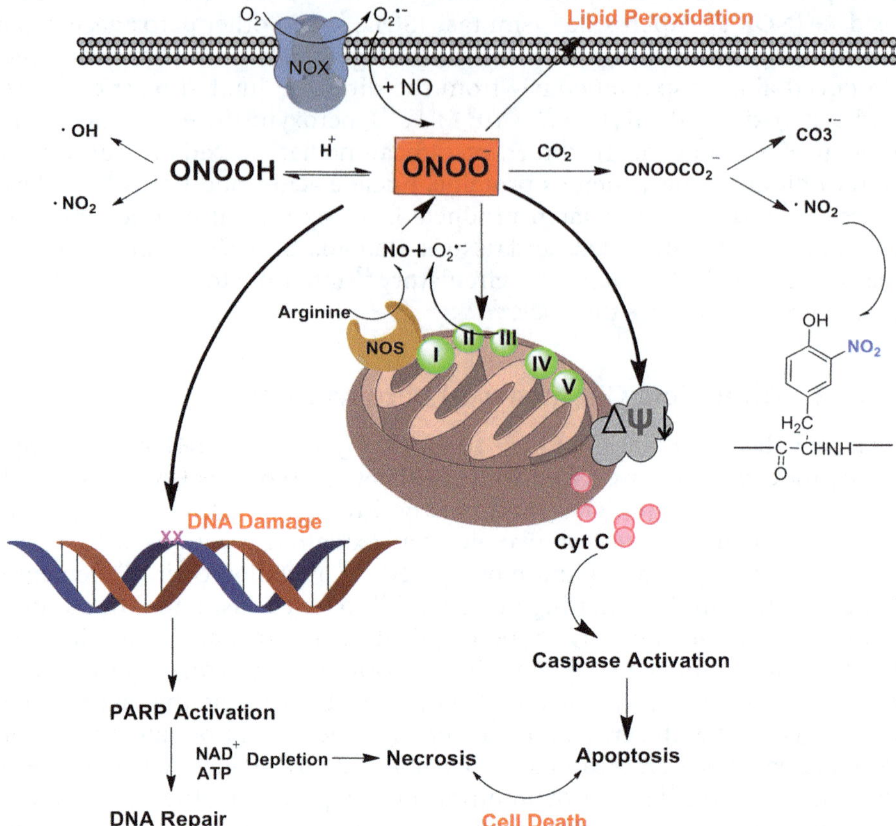

Figure 10.1   Sources and targets of cellular peroxynitrite. Peroxynitrite is a transient cellular oxidant that can diffuse and interact with the immediate and adjacent cells (~5–20 μm). ONOO⁻ is formed from a reaction between $O_2^{\cdot-}$ and NO, which can originate from the mitochondrial respiratory chain or NOX, and from arginine-derived NO synthesis *via* NOS, respectively. Within the cells, ONOO⁻ is in equilibrium with ONOOH, which can decompose into ˙OH and ˙NO₂ radicals. Moreover, peroxynitrite preferentially reacts with carbon dioxide (CO_2) to form nitrosoperoxycarbonate ($ONOOCO_2^-$), which can decompose into $CO_3^{\cdot-}$ and ˙NO₂ radicals. Accumulation of ˙NO₂ radicals can lead to tyrosine nitration on proteins—*i.e.* nitrotyrosine formation—which is indicative of oxidative cellular stress. Indeed, peroxynitrite accumulation confers cellular stress in a number of ways, such as DNA damage from purine oxidation, lipid peroxidation from free radical propagation of fatty acids, and cellular respiration dysfunction from an inappropriately decreased mitochondrial membrane potential ($\Delta\Psi$). These cellular defects can be circularly propagated if the mitochondria continuously produce ROS, thereby ultimately leading to cell death as the oxidative damage incurred exceeds the capacity of the cellular antioxidant and repair machineries. Cyt C: cytochrome c; PARP: poly (ADP-ribose) polymerase.

A Oxidation-based Peroxynitrite Probes

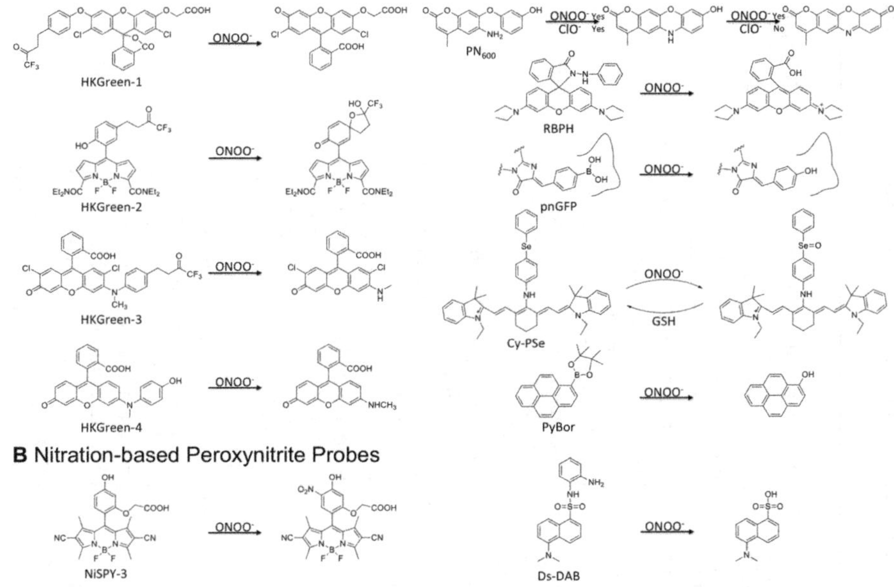

B Nitration-based Peroxynitrite Probes

Figure 10.2 Fluorescent peroxynitrite detection probes. Chemical structures of (A) oxidation- and (B) nitration-based peroxynitrite probes before and after reaction with peroxynitrite, as outlined in Table 10.1.

pores can significantly impair cellular respiration and induce apoptosis, respectively.[44,45] Furthermore, peroxynitrite-induced DNA oxidation and double-strand breaks activate the poly(ADP-ribose) polymerase (PARP) DNA repair pathway, which triggers cell necrosis by depleting the cellular NAD^+ and ATP stores.[22] Peroxynitrite-mediated cytotoxicity can also come about through a positive-feedback cycle, whereby peroxynitrite could inhibit the activity of antioxidant enzymes, such as manganese SOD (MnSOD),[46,47] and also induce inflammatory and apoptotic responses that further stimulate additional peroxynitrite production.[48,49]

10.2 Principles of Fluorescence and Fluorescent Probes

10.2.1 The Phenomenon of Fluorescence

The process of fluorescence involves the absorption of relatively higher energy, shorter wavelength light to enable emission of lower energy, longer wavelength light. This process is observed with fluorophores, which typically have distinct molecular structures that are commonly polyaromatic hydrocarbons or heterocycles.[50,51] Except for these highly conjugated organic

Table 10.1 Summary of fluorescent probes for peroxynitrite detection. The chemical structures and chemical transformations of the indicated probes are illustrated in Figure 10.2.

Probes	λ_{ex} (nm)	λ_{em} (nm)	Selectivity	Detection limit	Response *in vitro*	Response in cells or tissues	Reference
HKGreen-1	490	521	Reacts with OH· and ONOO⁻	0–10 μM	>9-Fold fluorescence turn on in 1 h (20 μM probe, 300 μM ONOO⁻, pH 7.3)	Primary cultured neuronal cells; fluorescence turn on by 10 μM SIN-1 (20 μM probe loaded)	98
HKGreen-2	520	539	Reacts with OH· and ONOO⁻	0–10 μM	21-Fold fluorescence turn on in 30 min (10 μM probe, 10 μM ONOO⁻, pH 7.4)	Murine macrophage cell line J774.1; fluorescence turn on by LPS (1 μg mL⁻¹) and IFN-γ (50 ng mL⁻¹) for 4 h, then PMA (10 nM) for 0.5 h	100
HKGreen-3	520	535	Reacts with OH·, OCl⁻, and ONOO⁻	<50 nM	140-Fold fluorescence turn on (5 μM probe, 5 μM ONOO⁻, pH 7.4)	Murine macrophage cell line RAW264.7; fluorescence turn on by LPS (1 μg mL⁻¹) and IFN-γ (50 ng mL⁻¹) for 4 h, then PMA (10 nM) for 0.5 h	101
HKGreen-4	517	535	Reacts with ONOO⁻	10 nM	290-Fold fluorescence turn on in 5 s (1 μM probe, 1 μM ONOO⁻, pH 7.4)	Respond to SIN-1 generated ONOO⁻ in RAW264.7, C17.2, CHO and BV-2 cells; respond to LPS/IFN-γ or heat-killed *E. coli* stimulation in RAW264.7 cells; imaging of ONOO⁻ generated in live smooth muscles from a mouse model of atherosclerosis	102
BzSe-Cy	770	800	Reacts with ONOO⁻	0–5 μM	>5-Fold fluorescence decrease in 5 min (5 μM probe, 5 μM ONOO⁻, pH 7.4)	Murine macrophage cell line RAW264.7; fluorescence decrease by SIN-1 (20 μM) in 1 h (20 μM probe loaded for 30 min)	106
Cy-PSe	758	775	Reacts with ONOO⁻	0–10 μM	23.3-Fold fluorescence turn on (10 μM probe, 10 μM ONOO⁻, pH 7.4)	Murine macrophage cell line RAW264.7; fluorescence turn on by LPS (1 μg mL⁻¹) and IFN-γ (50 ng mL⁻¹) for 4 h, then PMA (10 nM) for 0.5 h (10 μM probe loaded for 5 min)	107

Cy-NTe	793	820	Reacts with ONOO$^-$	917 nM	>10-Fold fluorescence turn on in 10 min (10 μM probe, 10 μM ONOO$^-$, pH 7.4)	Murine macrophage cell line RAW264.7; fluorescence turn on by LPS (1 μg mL^{-1}) and IFN-γ (50 ng mL^{-1}) for 4 h, then PMA (10 nM) for 0.5 h (1 μM probe loaded for 15 min)	108
PN$_{600}$	355 (576)	525 (595)	Reacts with ONOO$^-$ and O$_2^{\cdot-}$	>10 μM	>5-Fold fluorescence turn on at 595 nm (20 μM probe, 300 μM ONOO$^-$, pH 7.4)	Human glioma cell line U87; fluorescence turn on at 595 nm induced by 200 μM SIN-1 (20 μM probe loaded)	109
PyBor	347	410	Reacts with ONOO$^-$, ClO$^-$, BrO$^-$, and H$_2$O$_2$	100 nM	>10-Fold fluorescence turn on in 1 min (10 μM probe, 10 μM ONOO$^-$, pH 7.4)	Murine macrophage cell line RAW264.7; fluorescence turn on by LPS (1 μg mL^{-1}) and IFN-γ (50 ng mL^{-1}) for 4 h, then PMA (10 nM) for 0.5 h (10 μM probe loaded for 30 min)	119
pnGFP	484	508	Reacts with ONOO$^-$	553 nM	25-Fold fluorescence turn on in 1 min (0.5 μM probe, 100 μM ONOO$^-$, pH 7.4)	HEK 293T cells; probe genetically expressed; fluorescence turn on by SIN-1 (60 μM)	121
NiSPY-3	505	518	Reacts with ONOO$^-$ and other nitrating reagents	0–5 μM	>10-Fold fluorescence turn on in 3 min (10 μM probe, 10 μM ONOO$^-$, pH 7.4)	HeLa cells; fluorescence turn on by ONOO$^-$ (2 μM probe loaded)	126
Ds-DAB	350	505	Reacts with ONOO$^-$	52.4 nM	6.1-Fold fluorescence turn on in 2 h (10 μM probe, 50 μM ONOO$^-$, pH 7.4)	Murine macrophage cell line RAW264.7; fluorescence turn on by LPS (1 μg mL^{-1}) for 4 h, then PMA (10 nM) for 0.5 h (10 μM probe loaded for 15 min to 2 h)	127

molecules, other common fluorophores include fluorescent proteins (FPs), semiconductor crystals, and lanthanides.[52,53] The mechanism underlying the fluorescence process can be interpreted as three stages. In the first stage, a photon from an excitation light source is absorbed by the fluorophore, which increases the electron's energy level and brings it from the ground state to an excited state. In the second stage, the excited state electron interacts with its molecular environment and quickly dissipates some of this energy into heat. Finally, the electron relaxes back to the ground state by emitting the remaining energy as a photon.[51] Because energy is partially converted to heat during the transition, the emitted light is always of a lower energy (longer wavelength) compared with the excitation light. This wavelength difference is known as the Stokes shift. Most fluorophores can be repeatedly excited many times before being irreversibly damaged or photobleached—an intrinsic trait that depends on a fluorophore's lifetime. Another defining feature of fluorescence technology is its high sensitivity, which is attributable to the fact that thousands of photons can be generated from a single fluorophore, and the excitation and emission light can be efficiently separated and collected by using optical fluorescence instrumentations.[50]

10.2.2   Applications of Fluorescence Technology

The discovery of fluorescence and the development of a wide variety of useful fluorophores have fueled astonishing research in fluorescence instrumentation and applications for biological chemistry and cell biology. Organic dyes were among the first fluorophores to be studied and applied into biological research.[54] Their small molecular weight, reasonable quantum yield and photostability, and customizable photophysical performances have paved the way for organic fluorophores to pioneer the use of fluorescence in biochemical assays.[55] Many organic dye derivatives, such as fluorescein isothiocyanate (FITC), can be conjugated to macromolecules—such as antibodies, for example.[56] Most notably, the bioconjugation of organic dyes with primary or secondary antibodies has laid the foundation for immunofluorescence, which is now an indispensible technique in most biology laboratories.[57] In contrast to being synthesized *in vitro*, biological fluorophores can be genetically encoded in virtually any type of aerobic cells, tissues, or organisms.[58] Green fluorescent protein (GFP), originally isolated from the jellyfish *Aequorea victoria*, is most widely used as a biological marker for gene expression and protein localization.[59,60] Tsien and colleagues further engineered GFP derivatives with favorable photophysical properties for monitoring biological events.[61] These derivatives varied in hues that spanned the entire visible spectrum, thereby enabling multicolored imaging for simultaneously tracking multiple biological parameters. This is of particular importance because, for the first time, it allowed scientists to interpret cellular behaviors in an integrated perspective *via* direct monitoring of multiple labeled events.[62] In addition to being capable of fusion to proteins of interest as a potent protein-labeling reagent, FPs can be engineered into genetically encoded biosensors that function beyond emitting fluorescence.[58,63,64] These biosensors, when carefully designed and systematically optimized, have proven to be invaluable for monitoring a multitude of cellular parameters, such as pH,[65,66] temperature,[67] voltage,[68] metal ions,[69] reactive oxygen species (ROS),[70] protein post-translational modifications,[71] and enzymatic activities.[72] The third type of fluorophore that is increasingly being employed in biological applications is the quantum dot.[73] Developed in the 1980s, quantum dots are nanoscale-size (2–50 nm) crystals that serve to complement other types of fluorophore.[74] A unique feature of quantum dots is the proportionality of their emission properties relative to the particle size—*i.e.* the larger the particle, the longer the emission wavelength. By controlling the particle size, therefore, one can diversify a range of fluorescence emission particles.[75] Quantum dots are becoming increasingly popular due to their tunable fluorescence range and exceptional photostability, which allows fluorescent samples to be analyzed over an extended period of time.[75–77]

The vast and versatile panel of fluorophores will allow researchers to inevitably continue to revolutionize biotechnology platforms. For example, organic fluorophores have been extensively used in gene sequencing projects to decode the genome of thousands of organisms.[78] Flow cytometry, a high throughput, laser-based biophysical technology is another example that benefits from fluorescence principles. In flow cytometry experiments, a diverse group of cells are diluted in a stream of fluid and passed through an electronic detection apparatus that counts and sorts cells based on their fluorescence properties. Nowadays, flow cytometry is routinely used in basic research as well as clinical practice.[79] These biotechnological applications of fluorescence technology have allowed research to be conducted in a more confident, efficient, and high-throughput manner.

10.2.3 Fluorescent Probes

One of the most fascinating applications of fluorescence technology is the emerging development of fluorescent probes for the detection of biologically important molecules.[58] Fluorescent probes are bioengineered fluorophores that report on the cellular environmental status or specific analytes of interest. Since their initial development, fluorescent probes have gained extraordinary popularity among the biological community because information on invisible molecules can now be interpreted directly from the fluorescence of the corresponding fluorescent probes, which are so carefully designed that even nuanced alterations in target concentrations or conformations could trigger fluorescence changes.[53] This presents a stark contrast to traditional biochemical assays that are generally more time consuming, laborious, and expensive. Moreover, fluorescent probes can be observed in live cells by fluorescence microscopy imaging, where high spatio-temporal resolution can be obtained without having to sacrifice samples of interest.[80] Furthermore, because the use of fluorescent probes introduces only minimal cellular perturbations, their use preserves the target molecules in their native environment to better reflect the real underlying biology.[81] Fluorescent probes can be categorized into several groups, depending on the nature of their targets:

probes for cellular structures (ion channels, organelles, *etc.*), cell function (viability, pH, membrane potential, *etc.*), nucleic acids, proteins, lipids, and small molecules (metal ions, metabolites, ROS, *etc.*). Among them, fluorescent probes for small molecules are of particular interest because very few other methods are able to detect, track, and quantify these molecules.[82] Small molecule targets can change a probe's fluorescence properties *via* two discrete mechanisms: ligand binding or chemical reactions.[83] Prototypical and pioneering examples of binding-based small molecule fluorescent probes are fluorescent calcium ion indicators that report on the concentration changes of calcium ion, which is an important secondary messenger for signal transduction pathways.[69] Most organic dye-based calcium indicators contain derivatives of metal ion chelator motifs that display unique spectral features upon calcium binding. For example, fura and indo fluorescent calcium indicators, upon calcium binding, shift their excitation and emission spectra, respectively, and are hence considered excitation ratiometric. By contrast, rhod and calcium green family indicators are intensiometric.[84] An exciting advancement of fluorescent calcium indicators is the development of a family of genetically encoded FP-based calcium probes, such as fluorescence resonance energy transfer (FRET)-based cameleon indicators[85] and single FP-based GCaMP-type indicators.[86] These FP-based indicators integrate the calcium binding domain, calmodulin (CaM), into FPs, thereby allowing the resulting indicators to be genetically encoded into target cellular organelles or tissues of interest, which has broad applications in cell biology and neuroscience.[87,88] In contrast to binding small molecules, fluorescent probes for redox active small molecules, such as hydrogen sulfide (H_2S), rely on a reaction-based detection mechanism.[89] Chang and coworker were among the first researchers to incorporate an azide reactive functional group into their designs for H_2S-responsive fluorescent probes. By masking a rhodamine with an azide group, they generated the first-generation H_2S turn-on probes—Sulfidefluor-1 (SF1) and Sulfidefluor-2 (SF2).[90] H_2S-mediated reduction of aryl azide to a corresponding aniline elicits a dramatic fluorescence turn-on response in both aqueous solution and living HEK 293T cells. Akin to azide reduction for H_2S detection, chemospecific boronate and keto-acid oxidation reactions were widely adopted for the design of an assortment of fluorescent probes for H_2O_2.[90–92] These reaction-based probes combine various fluorophore scaffolds with reactive functional groups, resulting in far-ranging types of fluorescent probes with high reactivity and sensitivity. Taken together, progress in the development of various types of fluorescent probe offers a promising approach for interrogating biological puzzles.[82] Together with seminal advancements in optical spectroscopy and fluorescence microscopy, the development of fluorescent probes shall continue to open new doors and opportunities for both industrial innovations and scientific breakthroughs. Herein, we will provide a brief survey of recent developments of fluorescent probes for detecting and imaging peroxynitrite in biological systems. We will emphasize general probe design strategies and performances in biological media.

10.3 Development of Fluorescent Probes for Peroxynitrite

The detection of peroxynitrite, a prerequisite for understanding its production and for dissecting its signal/stress dichotomy in biological samples, has never been an easy task—given its low steady-state concentration, short physiological half life (~10 ms), and multiplex reaction pathways.[93] From a historic perspective, many different analytical methods, including chromatographical,[94] electrophoretic,[95] and electrochemical assays[96] (reviewed in other chapters) have been employed in peroxynitrite detection and quantification. In addition, immunohistological detection of 3-nitrotyrosine still continues to be extensively used as a marker for biological peroxynitrite formation despite numerous controversies regarding the ambiguous interpretation surrounding this detection method.[37] Most of these methods suffer from poor sensitivity, limited spatial and temporal resolution, and the requisite for cell lysis, which consequently precludes them from reliable live cell or *in vivo* detection, real-time monitoring, and imaging of peroxynitrite. The advent of fluorescent probes has made such a daunting task possible, in part due to their high sensitivity, tunable selectivity, and ability to complement modern spectroscopic and microscopic platforms. In particular, the successful development of a diverse array of small molecule fluorescent probes for metal ions, hydrogen sulfide, and H_2O_2—among many others—has drastically advanced our understanding of cellular homeostasis, signal transduction, and pathophysiology in a manner that was previously unattainable from using traditional biochemical methods.[82,83] Similarly, the development of fluorescent probes for peroxynitrite has hence been the major focus in the field during the past decade to meet the pressing need for tools to explore peroxynitrite biology.[97] Among the many challenges facing the probe designs, the unambiguous detection of peroxynitrite amidst other ROS and reactive nitrogen species (RNS) in a complex cellular milieu has motivated the use of a number of different chemical principles.

10.3.1 Oxidation-Based Probes

10.3.1.1 Oxidation of Activated Ketone

The first peroxynitrite-specific fluorescent probe, HKGreen-1,[98] was devised when Yang and co-workers examined the similar chemical behavior of $ONOO^-$ and peroxymonosulfate ($HOOSO_3^-$) towards activated ketones. In this regard, they found that peroxynitrite could efficiently oxidize an anisole-derived ketone to dienone *via* a dioxirane intermediate.[99] Such a reaction does not occur when the same ketone is incubated with other ROS and RNS, such as H_2O_2, $^1O_2$, NO, superoxide ($O_2^{\cdot-}$), ^-OCl, and peroxyl radical ($ROO^\cdot$), suggesting that this reaction scheme could be harnessed to develop fluorescent probes selective towards $ONOO^-$. Following this reaction principle, HKGreen-1 was synthesized by replacing the methoxy group of the anisole-derived ketone with dichlorofluorescein (DCF). Upon reaction with $ONOO^-$, the masked DCF is released while the phenyl ring is oxidized to produce a dienone product, and consequently results in a significantly augmented fluorescence. The fluorescence intensity of HKGreen-1 increases 8-fold and 3-fold, respectively, when incubated with 15 equiv. of $ONOO^-$ and $^{\cdot}OH$, whereas NO, $^1O_2$, $O_2^{\cdot -}$, and H_2O_2 trigger less than 1-fold fluorescence enhancement. More importantly, HKGreen-1 can be loaded into primary cultured neuronal cells to image 3-Morpholinosydnonimine-*N*-ethylcarbamide (SIN-1) generated peroxynitrite, demonstrating the potential to employ fluorescent probes to detect and image $ONOO^-$ in living cells.[98] Utilizing a similar peroxynitrite-mediated ketone oxidation reaction, Yang's group developed the second-generation probe for peroxynitrite, HKGreen-2, based on the photoinduced electron transfer (PET) mechanism of a boron-dipyrromethene (BODIPY)-type fluorescent probe.[100] Compared with HKGreen-1, HKGreen-2 has increased reactivity and dynamic range for $ONOO^-$, and as a result, is able to detect endogenously generated $ONOO^-$ in J774.1 macrophage cells stimulated with a combination of lipopolysaccharide (LPS), interferon-γ (IFN-γ), and phorbol 12-myristate 13-acetate (PMA).[100] Subsequently, by incorporating a diarylamine-derived ketone into the rhodol scaffold, Yang's laboratory was able to synthesize a new probe, HKGreen-3, for highly selective and sensitive detection of peroxynitrite.[101] HKGreen-3 exhibited negligible background fluorescence and peroxynitrite could readily mediate *N*-dearylation to release a highly fluorescent fluorophore, resulting in a 140-fold fluorescence turn-on response with 1 equiv. of $ONOO^-$. The detection of endogenous $ONOO^-$ with HKGreen-3 was further demonstrated by immune stimulation of both RAW264.7 and J774.1 macrophage cell lines.[101] Similarly, HKGreen-4 has recently been reported for visualizing peroxynitrite endogenously generated in *Escherichia coli*-treated macrophages and in live atherosclerotic mouse tissues. HKGreen-4 contains no ketone functional group but instead has a phenol group that is cleavable by peroxynitrite.[102]

10.3.1.2   *Oxidation of Organoselenium Fluorophores*

Selenium has long been established as an important cofactor for many redox cycling enzymes owing to its essential roles in intracellular redox homeostasis.[103,104] Indeed, many organoselenium compounds have been found to have applications as anti-oxidant agents that demonstrate protective benefits against oxidative species—such as $ONOO^-$.[105] Inspired by these selenium-mediated redox cycling events, Xu *et al.* reported a near-infrared organoselenium probe, benzylselenide-tricarbocyanine (BzSe-Cy), for the selective detection of $ONOO^-$ and reversible monitoring of redox cycles in live cells.[106] Due to its unusually high oxidation potential (0.622 V), BzSe-Cy can only be oxidized by $ONOO^-$, and not by other ROS or RNS, to yield an oxidized BzSe-Cy (BzSeO-Cy) that is weakly fluorescent due to a PET process. Remarkably, the oxidized probe can be recycled back to BzSe-Cy by reduced ascorbate ($ASCH_2$), and the $ONOO^-/ASCH_2$ mediated redox cycles can be repeated at least eight times without significant loss of fluorescence. In live RAW264.7 cells, BzSe-Cy preferentially localizes to mitochondria without apparent permeability and cytotoxicity issues, and is able to undergo multiple ONOO⁻ and ASCH₂ stimulated oxidation–reduction reaction cycles. BzSe-Cy represents the first ONOO⁻ selective probe that is reversible, and establishes a general selenium oxidation strategy for ONOO⁻ probe design. To further explore the effectiveness of the probe in ONOO⁻ imaging, Tang and co-workers reported, in a follow-up study, a novel fluorescent nanoprobe that incorporated BzSe-Cy and a reference dye, isopropylrhodamine B, for the ratiometric detection of ONOO⁻.[97] Coated with polymeric micelle and cell-penetrating peptide (TAT), this nanoprobe bestows attractive biophysical properties, including enhanced solubility, permeability, photostability, biocompatibility, and near-infrared excitation/emission profiles. Furthermore, the remarkable diffusibility of ONOO⁻ across membranes (permeability coefficient 8.0×10^4 cm s⁻¹) enables the nanomatrix of the nanoprobe to act as a shield that effectively blocks interferences from macromolecules and other oxidizing species, such as ˙OH and ClO⁻—even at high concentrations. It is thus conceivable that strategies using hybrid polypeptide–polymer–organic dye based nanoprobes should be generally applicable and beneficial for other fluorescent probes as well, provided that an extra layer of specificity and permeability is desired. Almost concomitantly, the Han laboratory reported another near-infrared, reversible fluorescent probe for peroxynitrite, Cy-PSe.[107] Cyanine (Cy) and 4-phenylselenyl-aniline (PSe) were selected as the signal transducer and modulator, respectively. Selective oxidation of selenium by ONOO⁻ disrupts the PET process between the signal transducer and the modulator, resulting in a fluorescence turn-on response. Similar to BzSe-Cy, the oxidized Cy-PSe (Cy-PSeO) can be repeatedly recycled with reducing agents, such as GSH and L-cysteine, allowing prolonged live cell monitoring of ONOO⁻ mediated oxidation–reduction events.[107] To improve the activity of the probe, they reported another similar fluorescent probe, Cy-NTe, based on heptamethine Cy and a telluroenzyme mimic, for reversible imaging of ONOO⁻/GSH redox cycles *in vivo*.[108] Compared with organoselenium analogs, GSH peroxidase (GPx)-like telluroenzyme mimics display much higher activity, thereby making Cy-NTe a more sensitive probe for imaging ONOO⁻/GSH redox cycles both in living cell cultures and mouse models. Owing to their reversibility, sensitivity, selectivity, and near-infrared emission profiles, these GPx-like analog-based fluorescent probes have proven to be invaluable for studying ONOO⁻ mediated redox homeostasis as well as antioxidant GSH repair.

10.3.1.3 Coupled Phenol Oxidation

One of the challenging issues of ONOO⁻ specific detection is the interference from other highly oxidizing species—particularly, ˙OH and hypochlorite (ClO⁻). To partially overcome this problem, Yang's group designed a three-channel fluorescent probe, PN₆₀₀, which distinguishes ONOO⁻ from ClO⁻.[109]

While both ONOO⁻ and ClO⁻ can oxidize PN$_{600}$ (green) into an aminophenol intermediate (orange), only ONOO⁻ could further mediate the two-electron oxidation of aminophenol into an iminoquinone end product (red). This unique three-color coupled fluorophore transformation of PN$_{600}$ spectrally separates ONOO⁻ from other oxidizing species, including highly oxidative ·OH and ClO⁻. Moreover, the fluorescence of the end product from ONOO⁻ oxidation is compatible with common tetramethylrhodamine filter sets, further facilitating live cell detection of ONOO⁻ with fluorescence microscopy.

10.3.1.4 Oxidation and Cleavage of Spiro Xanthene Derivatives

Due to their long emission wavelengths, high quantum efficiency, and water solubility, xanthene derivatives have found broad applications in fluorescent probe designs. Typically, these probes are non-fluorescent in their spiro closed forms and are converted, upon bond rearrangement, to their fluorescent spiro open forms.[110] This feature has been utilized by Malingappa's group to develop a new fluorescent probe, rhodamine B phenyl hydrazide (RBPH), for peroxynitrite imaging in MCF-7 cells.[111] The presence of a phenylhydrazone and a spirolactam group has made RBPH a very sensitive probe towards ONOO⁻, with a detection limit of 1.4 nM. RBHP is cell permeable and selective to ONOO⁻ over other oxidants and metals in the pH range of 6–8. Hence, oxidation and ring opening of xanthene derivatives by ONOO⁻ represents a new strategy for the development of sensitive fluorescent peroxynitrite probes.

10.3.1.5 Boronate Oxidation

A growing number of fluorescent probes targeting the boronate site have recently emerged and proven to be effective for the detection of oxidative species, such as H_2O_2 and ONOO⁻.[112] In particular, due to its mild reaction conditions and potential to modulate fluorescence, the oxidation of aryl boronates to phenols has gained unparalleled popularity in fluorescent probe designs. The reaction is derived from the famous hydroboration–oxidation reaction, whose inventor, Herbert C. Brown, was awarded the Nobel Prize in Chemistry in 1979.[113] Decades later, the group led by Chang fully exploited the chemical principles of boronate oxidation for biological applications.[90,111] As a bio-orthogonal reaction approach, boronate oxidation has laid the foundation for the development of a family of fluorescent H_2O_2 probes, which have been demonstrated to be of tremendous usefulness in living systems.[114] However, recent studies have shown that boronate-based probes also react with peroxynitrite in a ~1:1 stoichiometry.[115–117] Despite the similar reaction mechanisms, which involve formation of a peroxyborate intermediate, aryl migration to give aryl borane, and subsequent hydrolysis into phenol, peroxynitrite reacts with boronates much faster than H_2O_2 (the bimolecular rate constant is ~10^6 M^{-1} s^{-1}), thereby making the reaction invaluable for peroxynitrite probe designs.[117,118] Inspired by this reaction,

Han's group reported the use of PyBor, which contains a pyrene fluorophore and dioxaborolane reaction group, for the detection of ONOO$^-$ in aqueous solutions and in RAW264.7 cell extracts.[119] Upon addition of ONOO$^-$, faintly fluorescence PyBor is oxidized into highly fluorescent PyOH, with a limit of detection of 0.1 μM. Similarly, Kim *et al.* reported another probe composed of a benzothiazolyl iminocoumarin scaffold and a *p*-dihydroxyborylbenzyloxy group. ONOO$^-$ specifically oxidizes the aryl-boronate group into the corresponding phenol, which undergoes spontaneous reactions to release a highly fluorescent benzothiazolyl iminocoumarin product.[120] Compared with PyBor, this probe displays higher selectivity for ONOO$^-$ than for H_2O_2 and OCl$^-$, and can be further utilized to visualize both exogenous and endogenous ONOO$^-$ in living J774A.1 macrophage cells. Yet, from a mechanistic perspective, these probes are unable to unambiguously distinguish ONOO$^-$ from H_2O_2. To address this challenge, Chen and co-workers have developed the first genetically encoded fluorescent probe, pnGFP, for the selective detection of peroxynitrite.[121] pnGFP contains a *p*-boronophenylalanine derived chromophore, whose fluorescence can be turned on by oxidation of aryl-boronate into the corresponding phenol. Directed evolution of the FP template was conducted to identify key mutations near the chromophore that confer a selective reaction for ONOO$^-$ over H_2O_2. Although the detailed mechanism behind the observed selectivity is still under investigation, this proof-of-principle study demonstrates that the selectivity of a protein-based probe could be fine tuned by evolution of the protein template. In addition to being selective for ONOO$^-$, pnGFP can be genetically introduced into specific cell types or subcellular localizations within a single cell, allowing for the detection of ONOO$^-$ with unprecedented spatial and temporal precision unattainable with small molecule probes. Based on a similar principle, Sun *et al.* developed a boronate-based synthetic fluorescent probe that utilizes D-fructose as an "insulator" to hinder the reaction between boron and H_2O_2.[122] In the presence of D-fructose, boronic ester formation induces enhanced N–B interactions between neighboring amine and the boron atoms, which protects the probe from being oxidized by H_2O_2. Being extremely nucleophilic, ONOO$^-$ readily reacts with boronic ester to selectively modulate the fluorescence of the probe.

10.3.2 Nitration-Based Probes

10.3.2.1 Aromatic Nitration

As a potent nitrating agent, ONOO$^-$ exerts most of its biological and pathological roles *via* nitration of protein tyrosine residues.[14,36] Nevertheless, the development of fluorescence probes for monitoring nitration processes has proven challenging because nitro groups are generally regarded as quenchers of most fluorescent dyes.[123,124] Ueno *et al.* reasoned that the quenching mechanism of the nitro group is simply due to its electron-withdrawing effects, and the fluorescence quantum efficiency (Φ_{fl}) of nitro group-substituted fluorescent dyes can be precisely predicted.[125] On the basis of this concept, they reported the development of NiSPYs, a group of BODIPY-derived fluorescent dyes for monitoring nitrosative stress.[126] In the absence of nitration, the probe fluorescence is quenched *via* a PET process from the benzene moiety to the excited fluorophore (a-PET). The probes become highly fluorescent after nitration of the aromatic ring because of the decreased HOMO energy and diminished PET. NiSPY-3 is water-soluble, selective for ONOO⁻ and other ROS, and applicable for imaging ONOO⁻ in living cells. Therefore, nitration of fluorescent probes could serve as a strategy to distinguish other ROS and as a general principle for designing ONOO⁻ probes.

10.3.2.2 *Nucleophilic Nitrosation of Secondary Amine*

Taking advantage of the nucleophilicity of ONOO⁻, another reaction-based fluorescent probe, *N*-(2-aminophenyl)-5-(dimethylamino)-1-naphthalenesulfonic amide (Ds-DAB), was synthesized by reacting *o*-phenylenediamine with dansyl chloride.[127] By direct nucleophilic nitrosation of the secondary amine, ONOO⁻ can selectively react with Ds-DAB to form nitrosamine, which is further protonated and hydrolyzed to yield dansyl acid along with the production of benzotriazole. Compared with Ds-DAB, dansyl acid is much more fluorescent. Thus, after reacting with ONOO⁻, the fluorescence intensity of Ds-DAB increases drastically. Furthermore, both Ds-DAB and dansyl acid show almost no pH-dependent fluorescence alternations over the pH range 5.8–11.9, making Ds-DAB an ideal probe for ONOO⁻ detection in biological media.

10.4 Conclusions and Future Perspectives

Progress in the development of novel fluorescent probes for peroxynitrite has patently revolutionized our ability to dissect the intricate role that peroxynitrite plays in a complex cellular milieu. However, our understanding of peroxynitrite biochemistry and its far-reaching contributions to cell physiology and pathology is still in its infancy.[14,17] The inextricably linked relationship between peroxynitrite production and a variety of disease phenotypes, such cancer, diabetes, stroke, chronic heart failure, and neurodegenerative disorders, further provides compelling motivation for devising novel tools to monitor its production, trafficking, transformation, and detoxification in live cells or disease models.[16] Fluorescent probes responsive to peroxynitrite offer an appealing, non-invasive approach to visualize and interrogate peroxynitrite biology because of their exceptional sensitivity, specificity, and biocompatibility. As the list of peroxynitrite-specific fluorescent probes continues to grow, we can envision that novel probes with enhanced properties will emerge and become widely adopted by the scientific community, facilitating deeper insights into peroxynitrite-mediated cell signaling/stress responses and inspiring new pharmacological strategies toward many related diseases.[128]

Despite the exciting progress accomplished thus far, novel fluorescent peroxynitrite probes with new and/or enhanced properties await further exploration. To make full use of these probes and achieve a greater biological impact, further developments in probe brightness, photostability, sensitivity, and biocompatibility are needed to enhance signal-to-noise ratios and spatio-temporal imaging resolution.[82,83] Equally important in terms of probe design is the probe's selectivity. Aside from tests conducted in aqueous solutions, probes are constantly challenged by a complex molecular environment (high ionic strength, neutral pH, *etc.*), as well as an overwhelming class of biological molecules, including ROS, RNS, sulfur species, millimolar GSHs, metal ions, proteins, lipids, and metabolites, *etc.*[129] Hence, probes need to be designed and tested against a wild array of biologically relevant molecules *in vitro* before being applied to live cell samples to ensure signal fidelity. In most cases, this can be achieved *via* exploration of novel chemoselective reaction schemes that exclude possible interfering molecules from reacting. Alternatively, molecular evolution efforts combined with customized screening assays might also be feasible to achieve the desired specificity, although such efforts are more formidable. Notwithstanding, relevant control experiments remain a crucial component in evaluating a probe's performance and the underlying biology its findings imply in order to achieve unambiguous conclusions.[123] In addition to good selectivity, probes that have distinct and orthogonal spectral properties will be of great value, since they complement current probe sets. For example, the majority of current peroxynitrite-specific probes are intensiometric and their measurements of peroxynitrite are qualitative or semi-quantitative at best. To gauge peroxynitrite levels in a more quantitative manner, probes that respond to ratiometric changes in excitation or emission profiles are likely to yield more precise measurements.[120] This can be accomplished by developing compound fluorescent probes that incorporate two spectrally compatible fluorophores. The peroxynitrite-induced changes in the interactions between the two fluorophores, either in the form of reciprocal quenching or FRET, can then be quantified and converted to concentration changes.[130] Furthermore, there are a host of opportunities for the development of genetically encoded fluorescent probes for peroxynitrite. These probes are desirable because they are genetically encodable, can be targeted toward specific cell compartments or tissue types, and are easily deliverable through DNA transfection. In addition, they hold the promise of being applied to transgenic animals and/or for long-term monitoring of peroxynitrite fluxes within desired cellular locals.[131] However, the engineering of such probes with high selectivity and sensitivity requires experience and painstaking efforts. Finally, fluorescent probes with reversible, yet chemoselective, responses to peroxynitrite should be emphasized because they offer a dynamic perspective that is not possible with irreversible probes based on stoichiometric reactions. In terms of future developments, novel signal amplification systems or targeted delivery technologies could also be employed to establish more robust detection and imaging platforms.[132,133] These enhanced probes and methodologies, together with our rapidly growing aspirations in peroxynitrite biology, will paint a colorful future for peroxynitrite detection and imaging.

References

1. J. S. Beckman, T. W. Beckman, J. Chen, P. A. Marshall and B. A. Freeman, *Proc. Natl. Acad. Sci. U. S. A.*, 1990, **87**, 1620–1624.
2. J. S. Beckman, *Nature*, 1990, **345**, 27–28.
3. J. S. Beckman and W. H. Koppenol, *Am. J. Physiol.*, 1996, **271**, C1424–C1437.
4. J. S. Beckman, M. Carson, C. D. Smith and W. H. Koppenol, *Nature*, 1993, **364**, 584.
5. V. M. Darleyusmar, N. Hogg, V. J. Oleary, M. T. Wilson and S. Moncada, *Free Radical Res. Commun.*, 1992, **17**, 9–20.
6. R. Radi, *J. Biol. Chem.*, 2013, **288**, 26464–26472.
7. M. H. Zou, R. Cohen and V. Ullrich, *Endothelium*, 2004, **11**, 89–97.
8. P. Pacher, I. G. Obrosova, J. G. Mabley and C. Szabo, *Curr. Med. Chem.*, 2005, **12**, 267–275.
9. E. O. Feigl, *Nature*, 1988, **331**, 490–491.
10. C. Nathan and Q. W. Xie, *Cell*, 1994, **78**, 915–918.
11. K. Bedard and K. H. Krause, *Physiol. Rev.*, 2007, **87**, 245–313.
12. C. Bowler, M. Vanmontagu and D. Inze, *Annu. Rev. Plant Physiol. Plant Mol. Biol.*, 1992, **43**, 83–116.
13. N. Fawal, Q. Li, B. Savelli, M. Brette, G. Passaia, M. Fabre, C. Mathe and C. Dunand, *Nucleic Acids Res.*, 2013, **41**, D441–D444.
14. G. Ferrer-Sueta and R. Radi, *ACS Chem. Biol.*, 2009, **4**, 161–177.
15. L. Liaudet, G. Vassalli and P. Pacher, *Front. Biosci., Landmark Ed.*, 2009, **14**, 4809–4814.
16. P. Pacher, J. S. Beckman and L. Liaudet, *Physiol. Rev.*, 2007, **87**, 315–424.
17. C. Szabo, H. Ischiropoulos and R. Radi, *Nat. Rev. Drug Discovery*, 2007, **6**, 662–680.
18. R. S. Ronson, M. Nakamura and J. Vinten-Johansen, *Cardiovasc. Res.*, 1999, **44**, 47–59.
19. I. N. Mungrue, R. Gros, X. You, A. Pirani, A. Azad, T. Csont, R. Schulz, J. Butany, D. J. Stewart and M. Husain, *J. Clin. Invest.*, 2002, **109**, 735–743.
20. F. Paneni, J. A. Beckman, M. A. Creager and F. Cosentino, *Eur. Heart J.*, 2013, **34**, 2436–2443.
21. U. Forstermann and T. Munzel, *Circulation*, 2006, **113**, 1708–1714.
22. C. Szabo, *Shock*, 1996, **6**, 79–88.
23. N. W. Kooy, S. J. Lewis, J. A. Royall, Y. Z. Ye, D. R. Kelly and J. S. Beckman, *Crit. Care Med.*, 1997, **25**, 812–819.
24. R. C. van der Veen, D. R. Hinton, F. Incardonna and F. M. Hofman, *J. Neuroimmunol.*, 1997, **77**, 1–7.
25. M. A. Smith, P. L. Richey Harris, L. M. Sayre, J. S. Beckman and G. Perry, *J. Neurosci.*, 1997, **17**, 2653–2657.
26. F. Torreilles, S. Salman-Tabcheh, M. Guerin and J. Torreilles, *Brain Res. Rev.*, 1999, **30**, 153–163.
27. T. Koppal, J. Drake and D. A. Butterfield, *Biochim. Biophys. Acta, Mol. Basis Dis.*, 1999, **1453**, 407–411.

28. J. E. Barker, J. P. Bolanos, J. M. Land, J. B. Clark and S. J. R. Heales, *Dev. Neurosci.*, 1996, **18**, 391–396.

29. C. Quijano, D. Hernandez-Saavedra, L. Castro, J. M. McCord, B. A. Freeman and R. Radi, *J. Biol. Chem.*, 2001, **276**, 11631–11638.

30. M. Mehl, A. Daiber, S. Herold, H. Shoun and V. Ullrich, *Nitric Oxide*, 1999, **3**, 142–152.

31. V. Ullrich, M. Zou and M. Mehl, *FASEB J.*, 1997, **11**, A784.

32. C. Quijano, B. Alvarez, R. M. Gatti, O. Augusto and R. Radi, *Biochem. J.*, 1997, **322**, 167–173.

33. K. Takakura, J. S. Beckman, L. A. MacMillan-Crow and J. P. Crow, *Arch. Biochem. Biophys.*, 1999, **369**, 197–207.

34. R. Radi, J. S. Beckman, K. M. Bush and B. A. Freeman, *Arch. Biochem. Biophys.*, 1991, **288**, 481–487.

35. M. Trujillo, M. Naviliat, M. N. Alvarez, G. Peluffo and R. Radi, *Analusis*, 2000, **28**, 518–527.

36. H. Ischiropoulos and A. B. Almehdi, *FEBS Lett.*, 1995, **364**, 279–282.

37. B. Halliwell, *FEBS Lett.*, 1997, **411**, 157–160.

38. I. N. Singh, P. G. Sullivan and E. D. Hall, *J. Neurosci. Res.*, 2007, **85**, 2216–2223.

39. P. Pacher, L. Liaudet, P. Bai, J. G. Mabley, P. M. Kaminski, L. Virag, A. Deb, E. Szabo, Z. Ungvari, M. S. Wolin, J. T. Groves and C. Szabo, *Circulation*, 2003, **107**, 896–904.

40. G. M. Pieper, V. Nilakantan, M. Chen, J. Zhou, A. K. Khanna, J. D. Henderson, C. P. Johnson, A. M. Roza and C. Szabo, *J. Pharmacol. Exp. Ther.*, 2005, **314**, 53–60.

41. G. B. Mackensen, M. Patel, H. X. Sheng, C. L. Calvi, I. Batinic-Haberle, B. J. Day, L. P. Liang, I. Fridovich, J. D. Crapo, R. D. Pearlstein and D. S. Warner, *J. Neurosci.*, 2001, **21**, 4582–4592.

42. H. X. Sheng, J. J. Enghild, R. Bowler, M. Patel, I. Batinic-Haberle, C. L. Calvi, B. J. Day, R. D. Pearlstein, J. D. Crapo and D. S. Warner, *Free Radical Biol. Med.*, 2002, **33**, 947–961.

43. G. Tumurkhuu, N. Koide, J. Dagvadorj, F. Hassan, S. Islam, Y. Naiki, I. Mori, T. Yoshida and T. Yokochi, *FEMS Immunol. Med. Microbiol.*, 2007, **49**, 304–311.

44. J. J. Shacka, M. A. Sahawneh, J. D. Gonzalez, Y. Z. Ye, T. L. D'Alessandro and A. G. Estevez, *Cell Death Differ.*, 2006, **13**, 1506–1514.

45. R. Radi, M. Rodriguez, L. Castro and R. Telleri, *Arch. Biochem. Biophys.*, 1994, **308**, 89–95.

46. L. A. MacMillanCrow, J. P. Crow, J. D. Kerby, J. S. Beckman and J. A. Thompson, *Proc. Natl. Acad. Sci. U. S. A.*, 1996, **93**, 11853–11858.

47. L. A. MacMillan-Crow, J. P. Crow and J. A. Thompson, *Biochemistry*, 1998, **37**, 1613–1622.

48. N. Nalwaya and W. M. Deen, *Chem. Res. Toxicol.*, 2005, **18**, 486–493.

49. C. Szabo and A. L. Salzman, *Biochem. Biophys. Res. Commun.*, 1995, **209**, 739–743.

50. J. R. Albani, *Principles and Applications of Fluorescence Spectroscopy*, Blackwell, Oxford, 2007.

51. F. W. J. Teale, *Nature*, 1984, **307**, 486.

52. A. Miyawaki, A. Sawano and T. Kogure, *Nat. Cell Biol.*, 2003, S1–S7.

53. B. N. G. Giepmans, S. R. Adams, M. H. Ellisman and R. Y. Tsien, *Science*, 2006, **312**, 217–224.

54. M. Sameiro and T. Goncalves, *Chem. Rev.*, 2009, **109**, 190–212.

55. S. van de Linde, S. Aufmkolk, C. Franke, T. Holm, T. Klein, A. Loschberger, S. Proppert, S. Wolter and M. Sauer, *Chem. Biol.*, 2013, **20**, 8–18.

56. A. N. Kapanidis and S. Weiss, *J. Chem. Phys.*, 2002, **117**, 10953–10964.

57. I. D. Odell and D. Cook, *J. Invest. Dermatol.*, 2013, **133**, E1–E4.

58. J. Zhang, R. E. Campbell, A. Y. Ting and R. Y. Tsien, *Nat. Rev. Mol. Cell Biol.*, 2002, **3**, 906–918.

59. M. Chalfie, Y. Tu, G. Euskirchen, W. W. Ward and D. C. Prasher, *Science*, 1994, **263**, 802–805.

60. R. Y. Tsien, *Annu. Rev. Biochem.*, 1998, **67**, 509–544.

61. N. C. Shaner, P. A. Steinbach and R. Y. Tsien, *Nat. Methods*, 2005, **2**, 905–909.

62. N. C. Shaner, G. H. Patterson and M. W. Davidson, *J. Cell Sci.*, 2007, **120**, 4247–4260.

63. A. Ibraheem and R. E. Campbell, *Curr. Opin. Chem. Biol.*, 2010, **14**, 30–36.

64. W. B. Frommer, M. W. Davidson and R. E. Campbell, *Chem. Soc. Rev.*, 2009, **38**, 2833–2841.

65. M. Tantama, Y. P. Hung and G. Yellen, *J. Am. Chem. Soc.*, 2011, **133**, 10034–10037.

66. D. E. Johnson, H. W. Ai, P. Wong, J. D. Young, R. E. Campbell and J. R. Casey, *J. Biol. Chem.*, 2009, **284**, 20499–20511.

67. S. Kiyonaka, T. Kajimoto, R. Sakaguchi, D. Shinmi, M. Omatsu-Kanbe, H. Matsuura, H. Imamura, T. Yoshizaki, I. Hamachi, T. Morii and Y. Mori, *Nat. Methods*, 2013, **10**, 1232–1238.

68. D. R. Hochbaum, Y. Zhao, S. L. Farhi, N. Klapoetke, C. A. Werley, V. Kapoor, P. Zou, J. M. Kralj, D. Maclaurin, N. Smedemark-Margulies, J. L. Saulnier, G. L. Boulting, C. Straub, Y. K. Cho, M. Melkonian, G. K. S. Wong, D. J. Harrison, V. N. Murthy, B. L. Sabatini, E. S. Boyden, R. E. Campbell and A. E. Cohen, *Nat. Methods*, 2014, **11**, 825–833.

69. K. P. Carter, A. M. Young and A. E. Palmer, *Chem. Rev.*, 2014, **114**, 4564–4601.

70. V. V. Belousov, A. F. Fradkov, K. A. Lukyanov, D. B. Staroverov, K. S. Shakhbazov, A. V. Terskikh and S. Lukyanov, *Nat. Methods*, 2006, **3**, 281–286.

71. H. Neumann, S. Y. Peak-Chew and J. W. Chin, *Nat. Chem. Biol.*, 2008, **4**, 232–234.

72. F. S. Wouters, P. J. Verveer and P. I. H. Bastiaens, *Trends Cell Biol.*, 2001, **11**, 203–211.

73. I. L. Medintz, H. T. Uyeda, E. R. Goldman and H. Mattoussi, *Nat. Mater.*, 2005, **4**, 435–446.

74. X. Michalet, F. F. Pinaud, L. A. Bentolila, J. M. Tsay, S. Doose, J. J. Li, G. Sundaresan, A. M. Wu, S. S. Gambhir and S. Weiss, *Science*, 2005, **307**, 538–544.

75. J. K. Jaiswal and S. M. Simon, *Trends Cell Biol.*, 2004, **14**, 497–504.

76. M. F. Frasco and N. Chaniotakis, *Anal. Bioanal. Chem.*, 2010, **396**, 229–240.

77. A. P. Alivisatos, W. W. Gu and C. Larabell, *Annu. Rev. Biomed. Eng.*, 2005, **7**, 55–76.

78. T. P. Niedringhaus, D. Milanova, M. B. Kerby, M. P. Snyder and A. E. Barron, *Anal. Chem.*, 2011, **83**, 4327–4341.

79. J. W. Tung, K. Heydari, R. Tirouvanziam, D. R. Parks, L. A. Herzenberg and L. A. Herzenberg, *Clin. Lab. Med.*, 2007, **27**, 453–468.

80. K. M. Dean and A. E. Palmer, *Nat. Chem. Biol.*, 2014, **10**, 512–523.

81. P. I. H. Bastiaens and R. Pepperkok, *Trends Biochem. Sci.*, 2000, **25**, 631–637.

82. J. Chan, S. C. Dodani and C. J. Chang, *Nat. Chem.*, 2012, **4**, 973–984.

83. L. M. Wysocki and L. D. Lavis, *Curr. Opin. Chem. Biol.*, 2011, **15**, 752–759.

84. K. M. Dean, Y. Qin and A. E. Palmer, *Biochim. Biophys. Acta, Mol. Cell Res.*, 2012, **1823**, 1406–1415.

85. A. Miyawaki, J. Llopis, R. Heim, J. M. McCaffery, J. A. Adams, M. Ikura and R. Y. Tsien, *Nature*, 1997, **388**, 882–887.

86. J. Nakai, M. Ohkura and K. Imoto, *Nat. Biotechnol.*, 2001, **19**, 137–141.

87. S. A. Hires, L. Tian and L. L. Looger, *Brain Cell Biol.*, 2008, **36**, 69–86.

88. M. Mank and O. Griesbeck, *Chem. Rev.*, 2008, **108**, 1550–1564.

89. V. S. Lin and C. J. Chang, *Curr. Opin. Chem. Biol.*, 2012, **16**, 595–601.

90. A. R. Lippert, G. C. V. De Bittner and C. J. Chang, *Acc. Chem. Res.*, 2011, **44**, 793–804.

91. E. W. Miller, A. E. Albers, A. Pralle, E. Y. Isacoff and C. J. Chang, *J. Am. Chem. Soc.*, 2005, **127**, 16652–16659.

92. E. W. Miller, O. Tulyanthan, E. Y. Isacoff and C. J. Chang, *Nat. Chem. Biol.*, 2007, **3**, 263–267.

93. A. Gomes, E. Fernandes and J. L. F. C. Lima, *J. Fluoresc.*, 2006, **16**, 119–139.

94. H. L. Jiang and M. Balazy, *Nitric Oxide*, 1998, **2**, 350–359.

95. M. K. Hulvey, C. N. Frankenfeld and S. M. Lunte, *Anal. Chem.*, 2010, **82**, 1608–1611.

96. J. Xue, X. Y. Ying, J. S. Chen, Y. H. Xian, L. T. Jin and J. Jin, *Anal. Chem.*, 2000, **72**, 5313–5321.

97. J. W. Tian, H. C. Chen, L. H. Zhuo, Y. X. Xie, N. Li and B. Tang, *Chem.–Eur. J.*, 2011, **17**, 6626–6634.

98. D. Yang, H. L. Wang, Z. N. Sun, N. W. Chung and J. G. Shen, *J. Am. Chem. Soc.*, 2006, **128**, 6004–6005.

99. D. Yang, Y. C. Tang, J. Chen, X. C. Wang, M. D. Bartberger, K. N. Houk and L. Olson, *J. Am. Chem. Soc.*, 1999, **121**, 11976–11983.

100. Z. N. Sun, H. L. Wang, F. Q. Liu, Y. Chen, P. K. H. Tam and D. Yang, *Org. Lett.*, 2009, **11**, 1887–1890.

101. T. Peng and D. Yang, *Org. Lett.*, 2010, **12**, 4932–4935.

102. T. Peng, N. K. Wong, X. Chen, Y. K. Chan, D. H. Ho, Z. Sun, J. J. Hu, J. Shen, H. El-Nezami and D. Yang, *J. Am. Chem. Soc.*, 2014, **136**, 11728–11734.

103. J. T. Rotruck, A. L. Pope, H. E. Ganther, A. B. Swanson, D. G. Hafeman and W. G. Hoekstra, *Science*, 1973, **179**, 588–590.

104. G. Mugesh, W. W. du Mont and H. Sies, *Chem. Rev.*, 2001, **101**, 2125–2179.
105. C. W. Nogueira, G. Zeni and J. B. T. Rocha, *Chem. Rev.*, 2004, **104**, 6255–6285.
106. K. H. Xu, H. C. Chen, J. W. Tian, B. Y. Ding, Y. X. Xie, M. M. Qiang and B. Tang, *Chem. Commun.*, 2011, **47**, 9468–9470.
107. F. Yu, P. Li, G. Li, G. Zhao, T. Chu and K. Han, *J. Am. Chem. Soc.*, 2011, **133**, 11030–11033.
108. F. B. Yu, P. Li, B. S. Wang and K. L. Han, *J. Am. Chem. Soc.*, 2013, **135**, 7674–7680.
109. Q. J. Zhang, Z. C. Zhu, Y. L. Zheng, J. G. Cheng, N. Zhang, Y. T. Long, J. Zheng, X. H. Qian and Y. J. Yang, *J. Am. Chem. Soc.*, 2012, **134**, 18479–18482.
110. H. N. Kim, M. H. Lee, H. J. Kim, J. S. Kim and J. Yoon, *Chem. Soc. Rev.*, 2008, **37**, 1465–1472.
111. V. S. Lin, B. C. Dickinson and C. J. Chang, *Meth. Enzymol.*, 2013, **526**, 19–43.
112. K. Lacina, P. Skladal and T. D. James, *Chem. Cent. J.*, 2014, **8**, 60.
113. H. C. Brown and G. Zweifel, *J. Am. Chem. Soc.*, 1959, **81**, 247.
114. B. C. Dickinson, C. Huynh and C. J. Chang, *J. Am. Chem. Soc.*, 2010, **132**, 5906–5915.
115. J. Zielonka, A. Sikora, J. Joseph and B. Kalyanaraman, *J. Biol. Chem.*, 2010, **285**, 14210–14216.
116. A. Sikora, J. Zielonka, M. Lopez, A. Dybala-Defratyka, J. Joseph, A. Marcinek and B. Kalyanaraman, *Chem. Res. Toxicol.*, 2011, **24**, 687–697.
117. A. Sikora, J. Zielonka, M. Lopez, J. Joseph and B. Kalyanaraman, *Free Radical Biol. Med.*, 2009, **47**, 1401–1407.
118. W. G. Keith and R. E. Powell, *J. Chem. Soc. A*, 1969, 90.
119. F. B. Yu, P. Song, P. Li, B. S. Wang and K. L. Han, *Analyst*, 2012, **137**, 3740–3749.
120. J. Kim, J. Park, H. Lee, Y. Choi and Y. Kim, *Chem. Commun.*, 2014, **50**, 9353–9356.
121. Z. J. Chen, W. Ren, Q. E. Wright and H. W. Ai, *J. Am. Chem. Soc.*, 2013, **135**, 14940–14943.
122. X. Sun, Q. Xu, G. Kim, S. E. Flower, J. P. Lowe, J. Yoon, J. S. Fossey, X. Qian, S. D. Bull and T. D. James, *Chem. Sci.*, 2014, **5**, 3368–3373.
123. P. Wardman, *Free Radical Biol. Med.*, 2007, **43**, 995–1022.
124. T. Nagano, *J. Clin. Biochem. Nutr.*, 2009, **45**, 111–124.
125. T. Ueno, Y. Urano, K. Setsukinai, H. Takakusa, H. Kojima, K. Kikuchi, K. Ohkubo, S. Fukuzumi and T. Nagano, *J. Am. Chem. Soc.*, 2004, **126**, 14079–14085.
126. T. Ueno, Y. Urano, H. Kojima and T. Nagano, *J. Am. Chem. Soc.*, 2006, **128**, 10640–10641.
127. K. K. Lin, S. C. Wu, K. M. Hsu, C. H. Hung, W. F. Liaw and Y. M. Wang, *Org. Lett.*, 2013, **15**, 4242–4245.
128. L. D. Lavis and R. T. Raines, *ACS Chem. Biol.*, 2008, **3**, 142–155.
129. X. Chen, H. Chen, R. Deng and J. Shen, *Biomed. J.*, 2014, **37**, 120–126.

130. L. E. McQuade and S. J. Lippard, *Curr. Opin. Chem. Biol.*, 2010, **14**, 43–49.
131. L. M. DiPilato and J. Zhang, *Curr. Opin. Chem. Biol.*, 2010, **14**, 37–42.
132. A. P. de Silva, H. Q. N. Gunaratne, T. Gunnlaugsson, A. J. M. Huxley, C. P. McCoy, J. T. Rademacher and T. E. Rice, *Chem. Rev.*, 1997, **97**, 1515–1566.
133. H. Kobayashi, M. Ogawa, R. Alford, P. L. Choyke and Y. Urano, *Chem. Rev.*, 2010, **110**, 2620–2640.

Reversible Near-Infrared Fluorescent Probes for Peroxynitrite Monitoring

PENG LI[a] AND KELI HAN*[a]

[a]State Key Laboratory of Molecular Reaction Dynamics, Dalian Institute of Chemical Physics, Chinese Academy of Sciences, 457 Zhongshan Road, Dalian 116023, P. R. China
*E-mail: klhan@dicp.ac.cn

11.1 Introduction

Peroxynitrite (ONOO$^-$, PN) is a biologically reactive species that is generated by the coupling reaction of nitric oxide ($^{\bullet}$NO) and superoxide ($O_2^{\bullet-}$) radicals at diffusion-controlled rates of ~10 M^{-1} s^{-1}.[1-3] *In vivo*, endogenous PN has long been recognized to act as a strong oxidant, and can oxidize various biomolecules such as proteins, nucleic acids and lipids. This chemical properties of PN make it a central biological pathogenic factor in a variety of diseases, including cancer, cardiovascular, neurodegenerative and inflammatory disorders.[4-6] PN has also been shown to be a nitrating agent that causes nitrative stress in cells, but more investigations on endogenous PN have been devoted to its ability to modulate signal transduction pathways *via* nitro compounds, including nitrated tyrosine residues, 8-nitroguanosine and nitro fatty acids, and thereby influencing cellular processes.[7-10] The complex oxidation and nitration biology of PN and its broad implications in human disease and

RSC Detection Science Series No. 7
Peroxynitrite Detection in Biological Media: Challenges and Advances
Edited by Serban F. Peteu, Sabine Szunerits, and Mekki Bayachou

Published by the Royal Society of Chemistry, www.rsc.org

health provide motivation for developing new methods to detect PN in living systems. In comparison with other technologies, fluorescence imaging with the assistance of a fluorescent probe offers advantages including sensitivity, selectivity and *in situ* observation.[11-14] Several fluorescent probes have been synthesized and applied to detect PN in living cells. Some of them are based on the oxidation of activated ketones and subsequent oxidation of the anisole derivatives by PN and the dioxirane intermediate, respectively. Fluorescein, boron-dipyrromethene (BODIPY) and rhodol green scaffolds have all been used as fluorescence reporters for PN.[15,16] Some other fluorescent probes selective for PN were designed using its unique nitrification properties.[17] The nitrification process dramatically influences the excited state dynamic of the dyes and thereby their fluorescence. Fluorescent probes for PN have also been designed by deprotection strategies. Recently, we developed PN-selective fluorescent probes based on deprotection of borate and amide groups, respectively.[18,19] It was demonstrated that the PN-induced hydrolysis of borates was very fast and thus escaped from the interference of hydrogen peroxide (H_2O_2).

Although these probes were proved selective for PN, they suffer two major drawbacks. One is their relatively short excitation and emission wavelength (~550 nm), which renders them difficult to employ for sensing and imaging PN in deep tissues and living animals. Light in the ultraviolet (UV)-visible (Vis) range is well absorbed and scattered by endogenous biomolecules and cannot penetrate tissues deeply. The short wavelength is also problematic when trying to avoid background noise because the inherent autofluorescence of biological samples also falls in this range.[20-26] The other major drawback of these probes is their irreversibility in responding to PN. Irreversible fluorescent probes respond to PN in an accumulative fashion, with their fluorescence always increasing over time in PN-generating systems, irrespective of PN fluctuations. Consequently, these probes are unable to report PN fluctuations or the related redox changes.

To solve these problems, we devoted our effort to near-infrared, reversible fluorescent probes that selectively respond to PN over other reactive oxygen species. Herein we would like to review the design, synthesis, characterization and imaging applications of our two cyanine-based, selenium/tellurium modulated fluorescent probes for PN.[27,28]

11.2 Design of the Probes

Near-infrared light (650–900 nm) is superior to UV-vis light in fluorescence imaging due to its minimal photo damage to biological systems, deep tissue penetration and minimal interference from background autofluorescence from biomolecules in living systems.[20-26] Thanks to great efforts by researchers in the past few decades, many near-infrared fluorescent dyes have been developed and applied in biological fluorescence imaging.[29,30] Of these near-infrared fluorescent dyes, heptamethine cyanines are the most used because of their high molar absorption coefficient and synthetic accessibility.

Traditionally, heptamethine cyanines usually suffered from relatively poor photo stability, low fluorescence quantum yields and undesired self-aggregation in aqueous solution. However, continuous efforts have achieved tremendous progress in this area and most of the above-mentioned limitations have been overcome. For example, it has been shown that the integration of a rigid hexatomic ring in the methine chain can dramatically increase the photo stability and fluorescence quantum yields;[31,32] indol rings also increase photo stability; and introduction of methyl groups on the indol rings can reduce self-aggregation,[33] *etc.* Consequently, heptamethine cyanines have become readily available near-infrared fluorescence dyes whose photophysical properties can be easily regulated. For these reasons, heptamethine cyanines were selected as the signal transducer for the PN-selective near-infrared fluorescent probes.

We next sought the recognition species that responded to PN not only selectively but also reversibly. Redox reversible fluorescent probes are much more valuable for visualizing cycles of redox signaling, stress or repair, and their dynamic interconversion induced by PN. However, it is not easy for a fluorescent probe to respond to the redox cycles reversibly because intracellular redox potentials are usually restricted in narrow ranges to maintain normal physiological functions.[34] Our strategy to design selective and reversible fluorescent probes relied on mimicking enzyme activity, because almost all enzymatic processes in biological systems are specific and reversible.

Intracellular PN is modulated by biological antioxidant defense systems, in which selenium plays an important role as the active site of the antioxidant enzyme glutathione peroxidase (GPx). GPx can catalyze the reduction of PN by glutathione (GSH) *via* a unique ping-pong mechanism in which reduced GPx reduces PN to give nitrite (NO_2^-) and oxidized GPx, which is then regenerated by GSH. In this process, the selenium center of GPx undergoes reversible redox changes in response to PN and GSH, respectively.[35–37] Taking advantage of this biological process, we mimicked the catalytic cycle and explored organic selenium/tellurium species as the selective and reversible recognition centers for PN in order to develop the near-infrared fluorescent probes. The selenium/tellurium based probes were also anticipated to play a role in monitoring the homeostasis between PN and GSH. It has been pointed out that PN can significantly perturb the mitochondrial ratio of reduced to oxidized glutathione (GSH:GSSG) and further cause irreversible damage to respiration.

Having designated the signal transducer and recognition species for the near-infrared fluorescent probes, the bridge that connected these modules was then considered. Generally, aromatic bridges are preferred over aliphatic ones because aliphatic selenium/tellurium compounds are prone to undergo β-elimination reactions upon oxidation, which would disable the reversible redox processes of selenium/tellurium compounds. Keeping this in mind, Cy-PSe was first designed as the near-infrared reversible fluorescent probe. According to the authors' hypotheses, the fluorescence of Cy-PSe should be quenched due to photo-induced electron transfer (PET) from the selenium center to the cyanine backbone in the excited state. However, the PET process should be prohibited when the selenium center is oxidized to selenoxide because of the loss of its electron-donating ability. As a result, the fluorescence is restored when Cy-PSe responds to PN. Under intracellular conditions, reduction of Cy-PSeO by GSH would restart the PET mechanism and quench the fluorescence again when PN concentrations decreased (Scheme 11.1).

To evaluate the feasibility of the design strategies, the photophysical properties were further investigated using time-dependent density functional theory calculations to confirm the PET mechanism. The results showed that the process of PET from the PSe moiety to the cyanine moiety occurs in the S1 state of Cy-PSe (Figure 11.1). The small oscillator strength of the S0 $\rightarrow$ S1 excitation ($f = 0.013$) implies a forbidden transition, indicating that the S1 state is a dark state rather than an emissive state. In other words, this state cannot be accessed directly by excitation from the S0 state. However, it can be populated by internal conversion from the S3 state due to the maximum oscillator strength of the S0 $\rightarrow$ S3 excitation ($f = 1.617$). In contrast to Cy-PSe, oxidized Cy-PSe (Cy-PSeO) is a more strongly fluorescent molecule because of the absence of the PET process. The calculated results revealed that the low-lying transition-dipole-allowed S0 $\rightarrow$ S1 excitation with an oscillator strength of $f = 1.014$ corresponds to the orbital transition from the highest occupied molecular orbital to the lowest unoccupied molecular orbital, both of which reside on the cyanine moiety. In other words, these calculations rationalized the redox controlled PET process and further demonstrated the feasibility of the design strategies.

In a similar manner to the design of Cy-PSe, a tellurium based near-infrared fluorescent probe, *i.e.* Cy-NTe, for selective and reversible detection of PN was designed. Cy-NTe also works in a ping-pong-like way: the weak fluorescent Cy-NTe and the strong fluorescent oxidized Cy-NTe (Cy-NTeO) converted to each other in the presence of PN and GSH, respectively, and thus reported

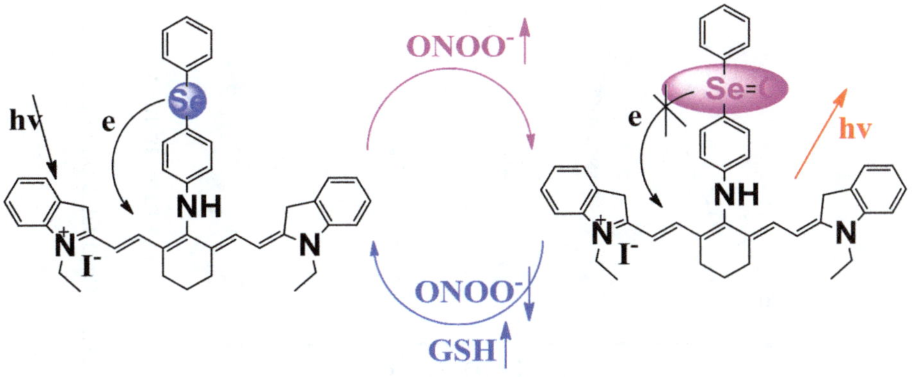

Scheme 11.1 Structures of Cy-PSe and Cy-PSeO, and the fluorescence switching mechanism.

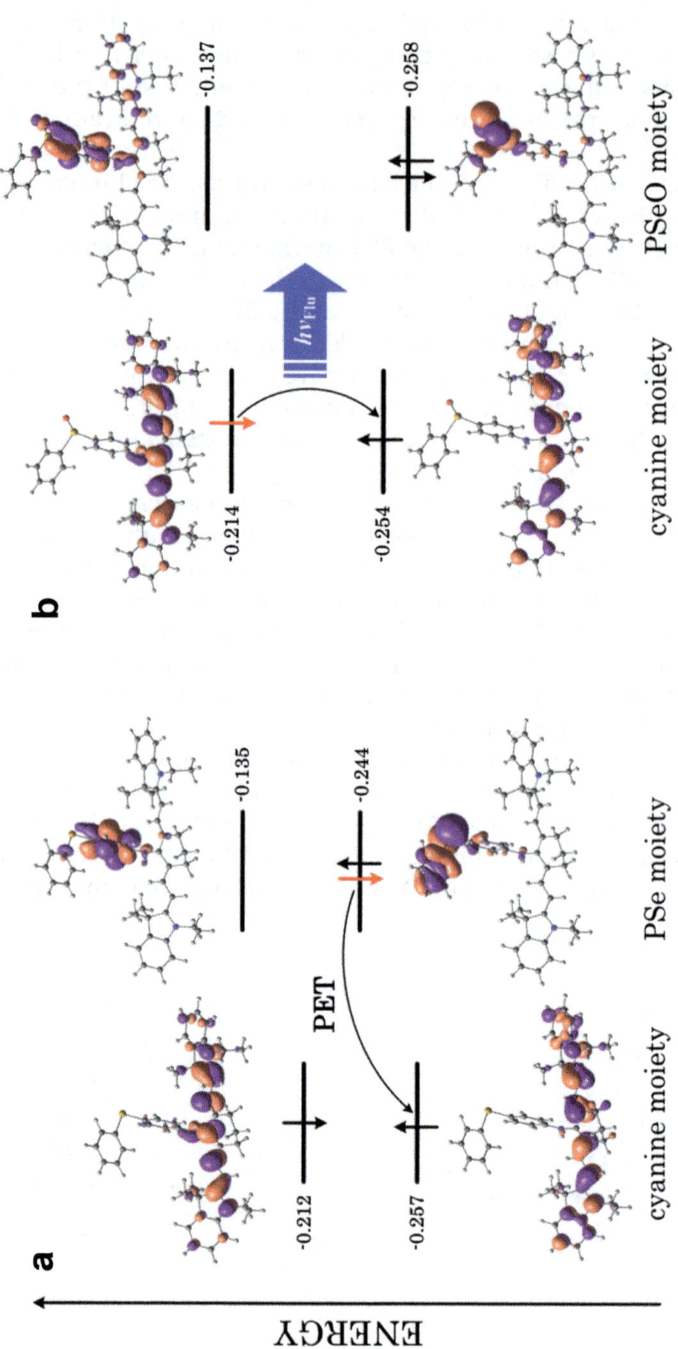

Figure 11.1 Calculated photophysical properties of (a) Cy-PSe and (b) Cy-PSeO.

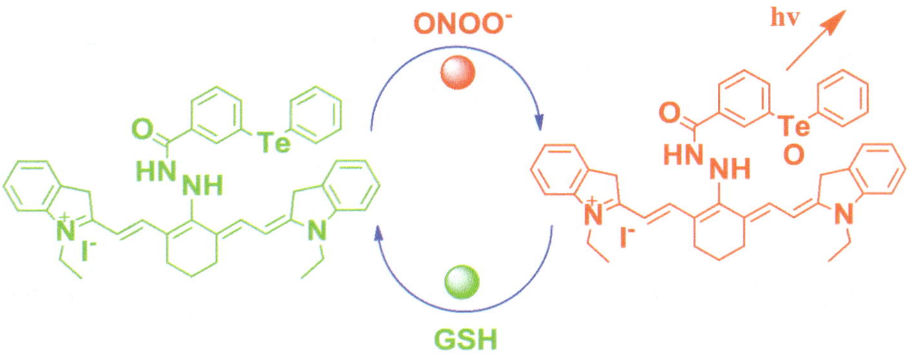

ONOO⁻

hv

GSH

Scheme 11.2 Structures of Cy-NTe and Cy-NTeO, and the sensing mechanism.

the PN dynamics. Changes in fluorescence were also controlled by PET. Unlike with Cy-PSe, however, an *ortho* benzohydrazide group was employed to bridge the phenyltellurium and cyanine to improve the performance of the probe. The 2-(phenyltellanyl)benzohydrazide structure was designed to allow for efficient reversible spirocyclization, which could protect the tellurium moiety from over-oxidation (Scheme 11.2).

11.3 Synthesis, Characterization and Spectroscopic Evaluation of the Probes

The synthesis of Cy-PSe and Cy-NTe are outlined in Schemes 11.3 and 11.4. These compounds were characterized by ^{1}HNMR, ^{13}CNMR, mass spectrometry and ^{77}SeNMR (for Cy-PSe), and ^{125}TeNMR (for Cy-NTe). The spectroscopic properties of both probes were evaluated under simulated physiological conditions [phosphate buffered saline (PBS), pH 7.4]. Cy-PSe showed λ_{max} for absorption at 758 nm ($\varepsilon_{758\,nm}$ = 203 700 M^{-1} cm^{-1}) and emission at 800 nm (Figure 11.2), while Cy-NTe showed λ_{max} for absorption at 793 nm ($\varepsilon_{793\,nm}$ = 95 390 M^{-1} cm^{-1}) and emission at 800 nm (Figure 11.3). All of these wavelengths lie in the near-infrared window. In response to PN, Cy-PSe showed an increase in fluorescence intensity by ~23.3 fold, corresponding to a quantum yield increase from Φ = 0.05 for Cy-PSe to Φ = 0.12 for the PN oxidized probe Cy-PSeO (Figure 11.2b). In the case of Cy-NTe, the fluorescence intensity increased by ~13.5 fold and the quantum yield increased from Φ = 0.0032 to 0.0431 when the probe was oxidized to Cy-NTeO by PN (Figure 11.3b). During the off–on fluorescence processes, neither absorbance spectra changed significantly (Figures 11.2a and 11.3a). These spectroscopic properties further verified the proposed PET mechanism in both probes. Both probes could detect PN quantitatively under simulated physiological conditions. When the probes were treated with different concentrations of PN, their fluorescence intensity increased linearly at λ_{max} with r = 0.990 and 0.963, respectively

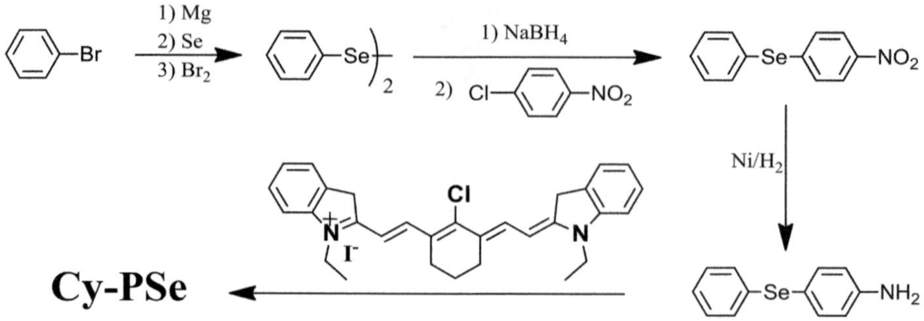

Cy-PSe

Scheme 11.3 Synthesis of Cy-PSe.

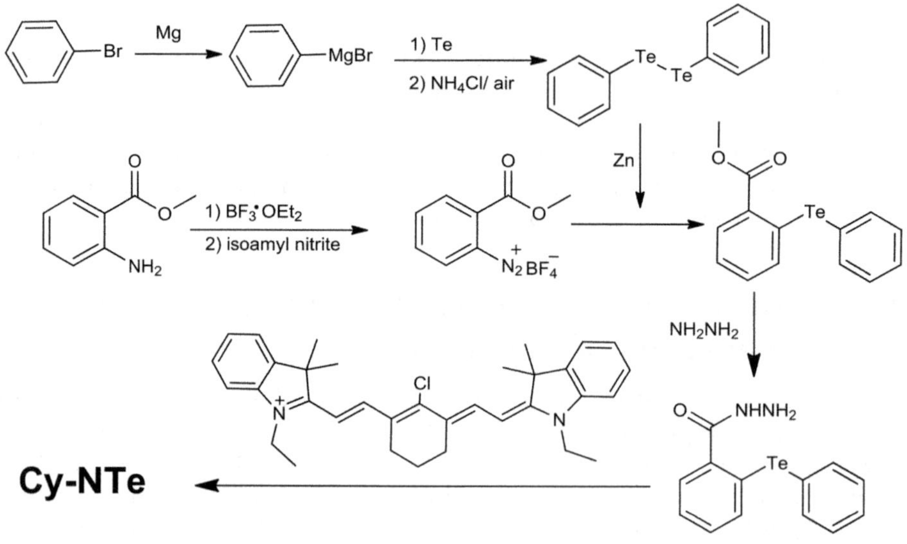

Cy-NTe

Scheme 11.4 Synthesis of Cy-NTe.

(Figure 11.2 inset and Figure 11.3 inset, respectively). As designed, both probes exhibited good selectivity towards PN. Other reactive oxygen species such as $O_2^{\bullet-}$, H_2O_2, hypochlorite (ClO^-), NO, hydroxyl radical ($^{\bullet}OH$) and $^1O_2$ that may be generated in association with PN in biological systems did not induce significant signal noise when incubated with the probes (Figures 11.4 and 11.5).

The authors then investigated the reversibility of the probes. As shown in Figure 11.6, the oxidized probes were easily reduced by various thiols, including GSH and cysteine. The fluorescence recovery rate was up to 97% for both probes. Other protein thiols, including thioredoxin and metallo-thionein, induced limited fluorescent responses, whereas the responses toward other antioxidants, such as vitamin C, vitamin E, uric acid and histamine were negligible during the 30 min experiment. Because the

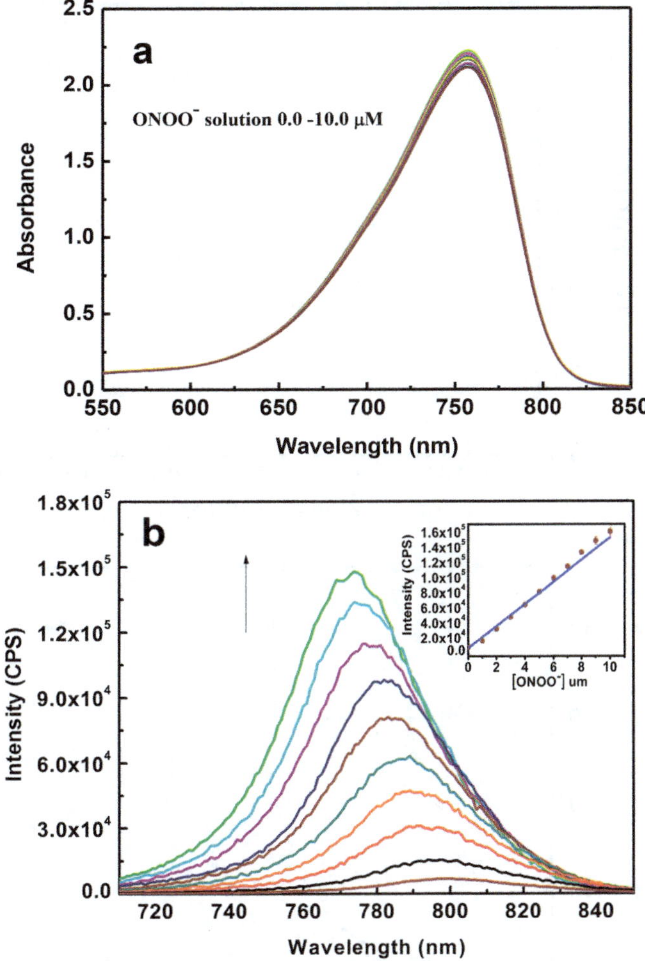

Figure 11.2 (a) Absorbance and (b) fluorescence spectra of Cy-PSe in the presence of ~0–10 µM PN.

concentration of intracellular GSH was far higher than that of cysteine, protein thiols and other endogenous antioxidants, the fluorescence signal could be regarded as a response of the oxidized probes to intracellular GSH. These results were consistent with the design strategies because the enzyme GPx is also recovered by GSH in nature. Importantly, the authors demonstrated that the probes could report on the redox cycles mediated by PN and GSH repeatedly at least five times with acceptable fluorescence decrements (Figure 11.6), indicating that continuous redox cycles induced by PN may be observed in living cells with these probes. A 2 h kinetic assay for Cy-PSe showed that the probe responded to PN in 8 min and was stable to water, air and light (Figure 11.7).

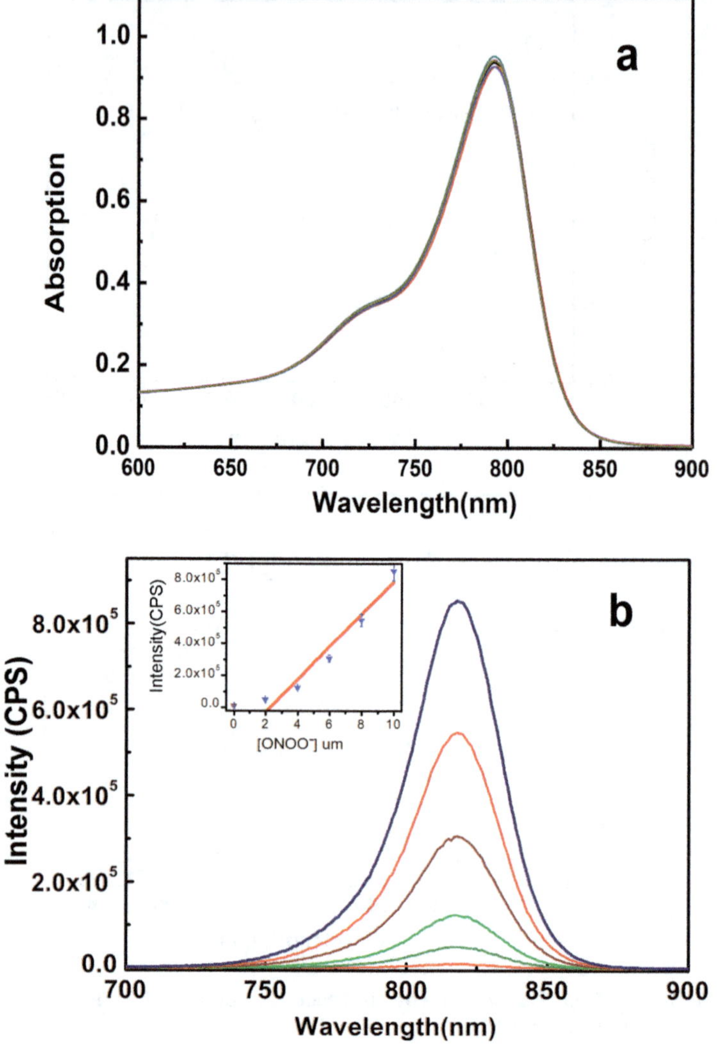

Figure 11.3 (a) Absorbance and (b) fluorescence spectra of Cy-NTe in the presence of ~0–10 µM PN.

11.4 Fluorescence Imaging of PN in Living Cells and Animals

When applied to image PN in living cells using confocal fluorescence microscopy, both probes showed dramatic fluorescence increases in response to endogenous PN generation. As shown in Figures 11.8 and 11.9, both Cy-PSe and Cy-NTe loaded cells showed faint fluorescence, and the fluorescence became strong when PN was generated by stimulating the cells with

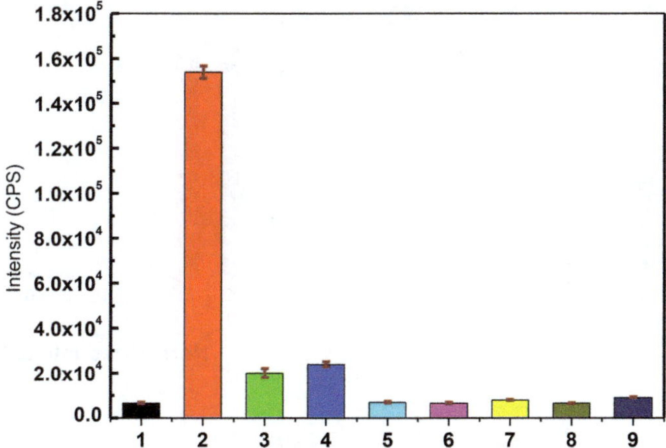

Figure 11.4 Fluorescence response of Cy-PSe (10 µM) to diverse ROS: 1, blank; 2, ONOO⁻ (10 µM); 3, NO (100 Mm); 4, H_2O_2 (200 µM); 5, ˙OH (200 µM); 6, $^1O_2$ (200 µM); 7, O_2^- (200 µM); 8, methyl linoleate hydroperoxide (200 µM); 9, *tert*-butyl hydroperoxide (200 µM).

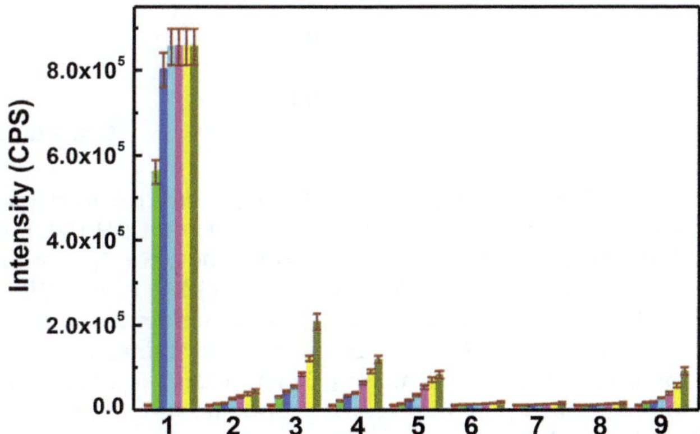

Figure 11.5 Time-dependent fluorescence response of Cy-NTe (10 µM) to diverse ROS: 1, ONOO⁻ (10 µM); 2, NO (300 µM); 3, H_2O_2 (200 µM); 4, *tert*-butyl hydroperoxide (250 µM); 5, cumene hydroperoxide (300 µM); 6, methyl linoleate hydroperoxide (400 µM); 7, ˙OH (200 µM); 8, O_2^- (200 µM); 9, ClO⁻ (200 µM).

lipopolysaccharide (LPS)/interferon-γ (IFN-γ) followed by phorbol myristate acetate (PMA). By contrast, when the probe loaded cells were pretreated with either aminoguanidine (AG), a NO synthesis inhibitor, or with TEMPO, a superoxide scavenger, only much weaker fluorescence was obtained in the stimulated cells because the generation of PN was blocked. To further demonstrate the selectivity of the probes toward PN in cells, the authors also treated the probe loaded cells with various other reactive oxygen species. The results showed that neither Cy-PSe loaded cells treated with H_2O_2 and NOC-5

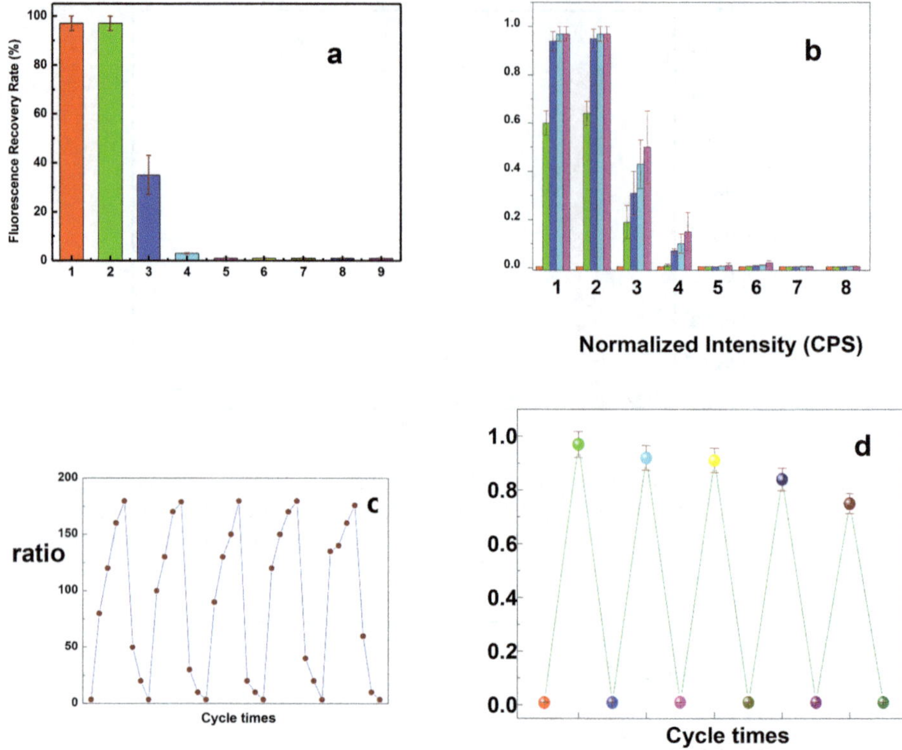

Figure 11.6 Reversibility of the probes. (a) Fluorescence recovery ratio of oxidized Cy-PSe (10 μM): 1, L-cysteine (20.0 μM); 2, GSH (20.0 μM); 3, metallothionein (100.0 μM); 4, vitamin C (100 μM); 5, vitamin E (100 μM); 6, uric acid (100 μM); 7, tyrosine (100 μM); 8, histidine (100 μM); 9, hydroquinone (100 μM). (b) Time-dependent fluorescence recovery ratio of oxidized Cy-NTe (10 μM): 1, L-cysteine; 2, glutathione; 3, thioredoxin; 4, metallothionein; 5, vitamin C; 6, vitamin E; 7, uric acid; 8, histamine. (c) Fluorescence responses of Cy-PSe (10.0 μM) to redox cycles. (d) Fluorescence responses of Cy-NTe (10.0 μM) to redox cycles. For examining the fluorescence response of Cy-PSe and Cy-NTe to redox cycles, the probes were first oxidized by 1 equiv. PN and then treated with 2 equiv. GSH. The redox cycle was repeated five times.

(NO precursor) nor Cy-NTe loaded cells treated with Paraquat (to generate H_2O_2), NOC-5 and sodium hypochlorite (NaClO) showed remarkable fluorescence. These results demonstrated that both probes can image PN specifically in living cells.

The reversibility of the probes in living cells was also verified by stimulating the cells with PN-generating and PN-reducing drugs alternately. As an example, the strong fluorescence in Cy-PSe loaded cells stimulated with a PN precursor 3-morpholinosydnonimine-*N*-ethylcarbamide (SIN-1) readily faded when the cells were subsequently treated with glutathione *S*-transferase (GST), because the enzyme effectively reduced PN generated

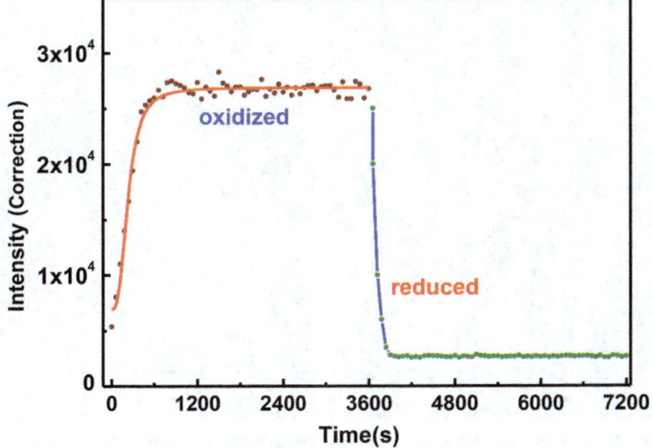

Figure 11.7   Time course of Cy-PSe (10.0 µM) as measured by a spectrofluorometer. Cy-PSe was oxidized by 1 equiv. of added ONOO⁻ for 1 h, after which the solution was treated with 2 equiv. of GSH for 1 h.

in cells (Figure 11.10a–c). However, when the reduced cells were treated with another dose of SIN-1, the faded fluorescence appeared again due to the re-oxidation of the probe by PN (Figure 11.10d). Similar fluorescence switches were also observed for Cy-NTe loaded cells, where stimulations of PMA/LPS, AG/GST, SIN-1 and L-cysteine/GST were successively executed.

Of the various subcellular compartments in the cell cytoplasm, mitochondria represent the cellular power plants. They are also involved in signaling, cellular differentiation, cell death, as well as the control of the cell cycle and cell growth. Endogenous PN are produced *via* the enzymatic activity of inducible NO synthase 2 (NOS2) and NADPH oxidase. NOS2 is expressed primarily in macrophages after induction by IFN-γ and LPS. In addition, mitochondrial electron leakage is the main pathway to produce superoxide anion *via* NADPH oxidase. Therefore, mitochondrial respiration is a major source of PN in macrophage cells. In a co-localization experiment, the authors co-stained the macrophage cells with Cy-PSe and rhodamine 6G, a nucleolus specific dye, and found that Cy-PSe was located mainly in the cytoplasm of living cells. When the co-stained cells were stimulated with LPS/PMA to generate PN, a great enhancement of fluorescence in the cytoplasm was observed, indicating an increase in the level of cytoplasmic PN. For the Cy-NTe probe, a more detailed co-localization experiment was carried out to assess whether the probe can respond to specific mitochondrial PN. By co-staining the stimulated cells with Cy-NTe and rhodamine 123, a mitochondrial tracker, the authors easily identified the precise location of Cy-NTe in mitochondria (Figure 11.11). The intensity profiles of the linear regions of interest across RAW264.7 cells co-stained with Cy-NTeO and rhodamine 123 (red arrow in Figure 11.11c) varied in close synchrony (Figure 11.11f). The Pearson's coefficient $r = 0.94$ and the Manders' coefficients $m_1 = 0.99$

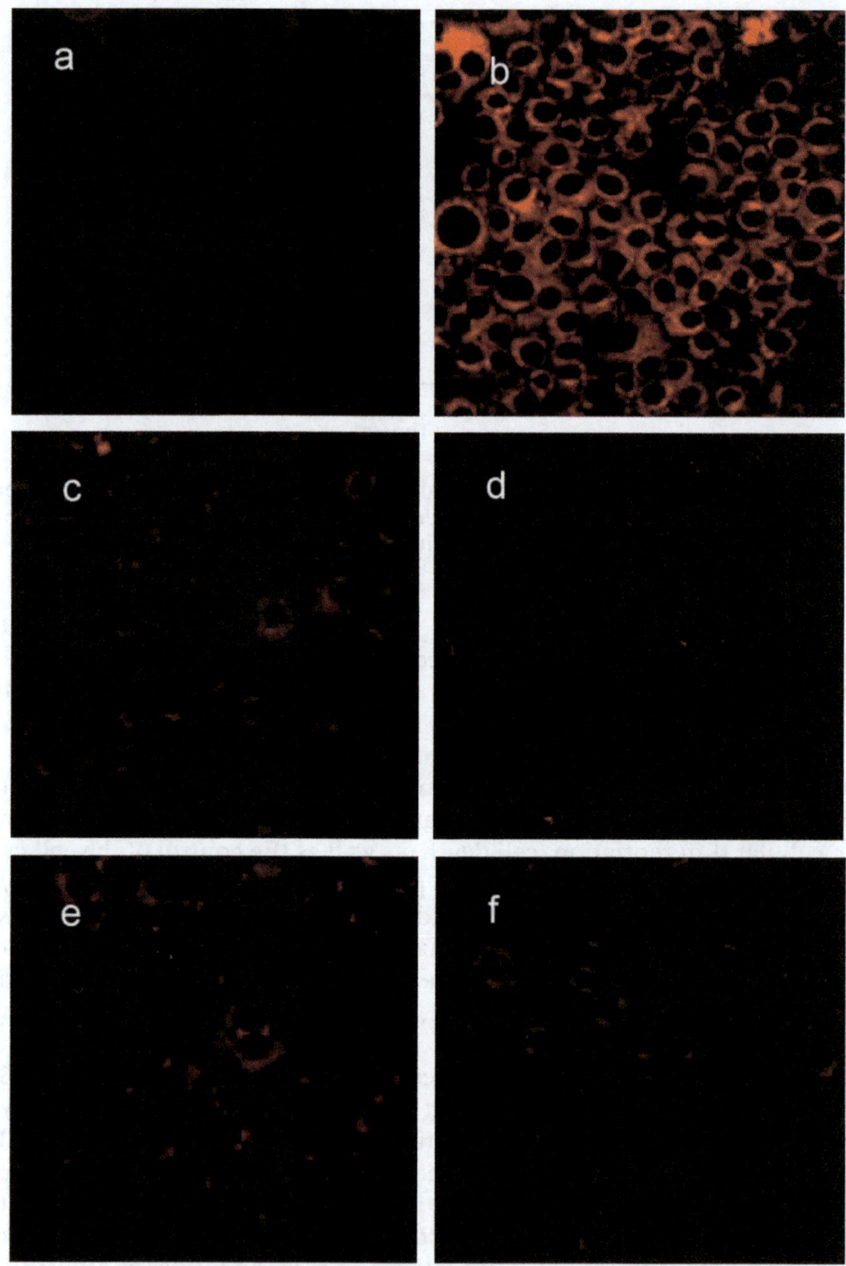

Figure 11.8 Confocal fluorescence images of oxidative stress RAW264.7 cells. Macrophage cells were incubated with Cy-PSe (10.0 μM) for 5 min after being treated with various stimulants: (a) control; (b) LPS (1 μg mL^{-1}) and IFN-γ (50 ng mL^{-1}) for 4 h then PMA (10 nM) for 0.5 h; (c) AG (1 mM), LPS (1 μg mL^{-1}) and IFN-γ (50 ng mL^{-1}) for 4 h then PMA (10 nM) for 0.5 h; (d) TEMPO (100 μM), LPS (1 μg mL^{-1}) and IFN-γ (50 ng mL^{-1}) for 4 h then PMA (10 nM) for 0.5 h; (e) H$_2$O$_2$ (50 μM) for 9 min; (f) NOC-5 (50 μM) for 3 min. Scale bar = 20 μm.

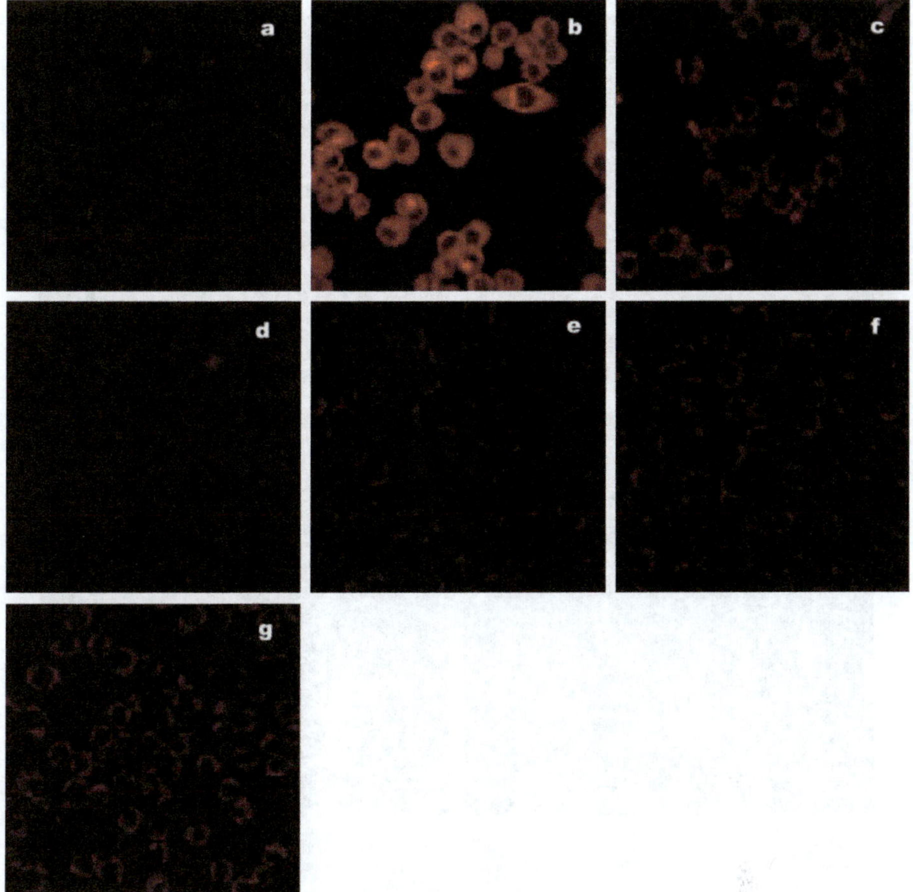

Figure 11.9 Fluorescence confocal microscopic images of RAW264.7 cells exposed to oxidative stress. Macrophage cells were loaded with 1 μM Cy-NTe for 15 min after being treated with various drugs: (a) control; (b) LPS (1 μg mL^{-1}) and IFN-γ (50 ng mL^{-1}) for 4 h then PMA (10 nM) for 0.5 h; (c) AG (1 mM), LPS (1 μg mL^{-1}) and IFN-γ (50 ng mL^{-1}) for 4 h then PMA (10 nM) for 0.5 h; (d) PMA (10 nM) for 0.5 h; (e) Paraquat (200 μM) for 8 h; (f) NOC-5 (50 μM) for 3 min. (g) NaClO (100 μM) for 0.5 h. Scale bar = 20 μm.

and m_2 = 0.98 were evaluated using ImagePro Plus software. A color intensity correlation analysis was carried out by plotting the intensity of the Cy-NTeO stain against that of rhodamine 123 for each pixel, and revealed that the dependent staining in Figure 11.11a and b resulted in highly correlated plots (Figure 11.11d and e). These results suggest that Cy-NTe is a strong candidate for imaging mitochondrial PN in living cells. The targeting ability of Cy-NTe to mitochondria was attributed to its ammonium cation on the cyanine backbone.

To demonstrate the great advantage of near-infrared fluorescent probes in deep tissue imaging, the authors further investigated the applicability of

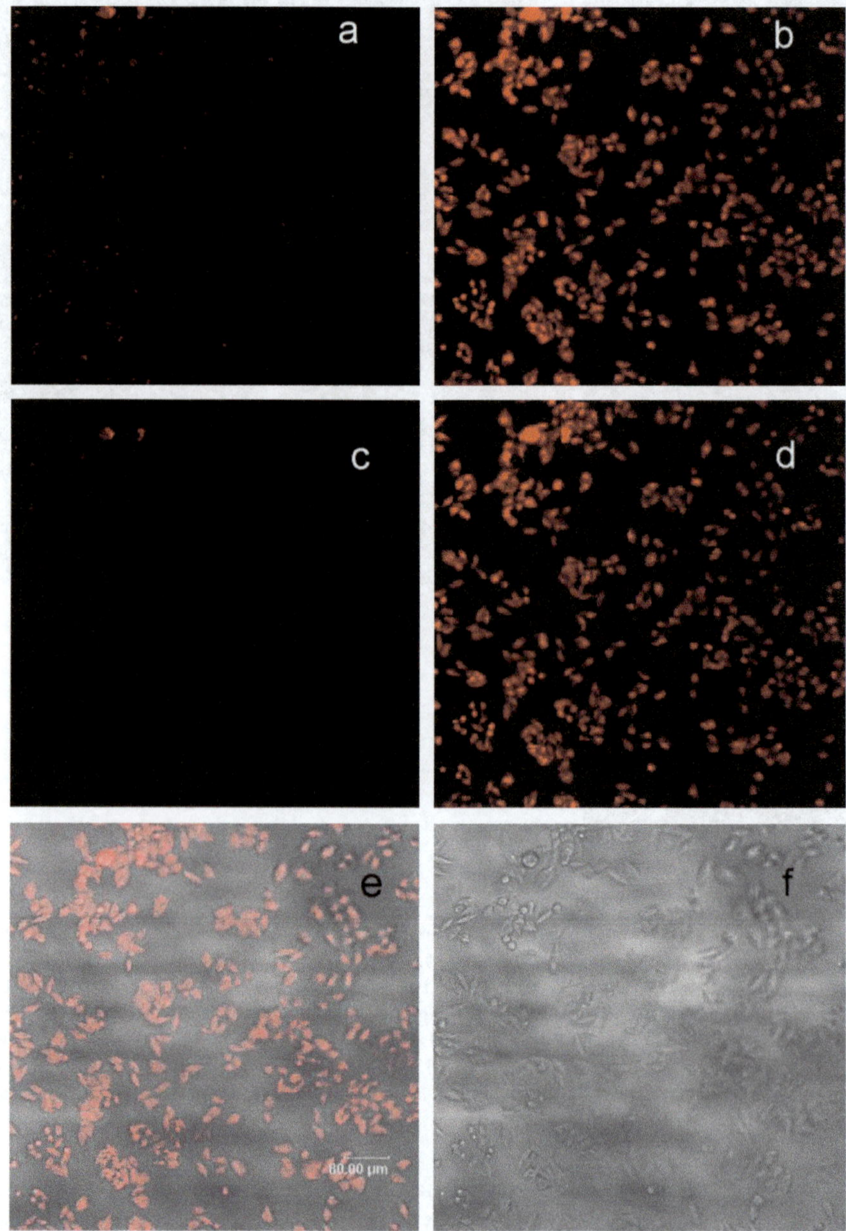

Figure 11.10 Confocal fluorescence images of reversible redox cycles in living RAW264.7 cells. (a) RAW264.7 cells loaded with 10.0 μM Cy-PSe for 5 min; (b) dye-loaded cells treated with 10.0 μM SIN-1 for 10 min; (c) dye-loaded, SIN-1-treated cells incubated with GST (125 U ml^{-1}) for 10 min; (d) cells exposed to a second dose of SIN-1 for an additional 10 min; (e) merged images of (b) red and (f) bright-field channels; (f) bright-field image of (a). Scale bar = 80 μm.

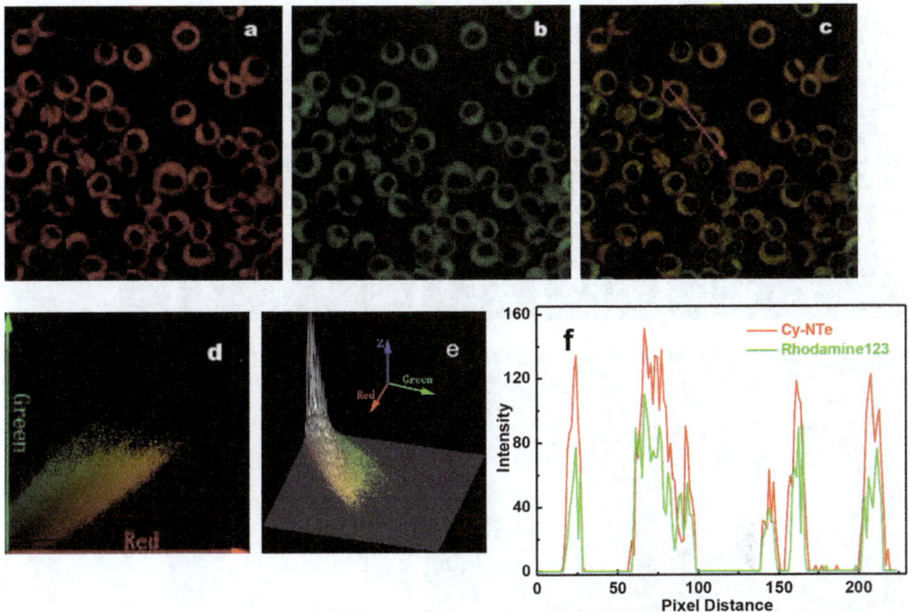

Figure 11.11 Cy-NTe and rhodamine 123 dye co-localization at mitochondria in RAW264.7 cells. Cells were treated with LPS (1 μg mL⁻¹) and IFN-γ (50 ng mL⁻¹) for 4 h, then with PMA (10 nM) for 0.5 h, and finally washed with RPMI-1640. The cells were loaded with (a) 1 μM Cy-NTe and (b) 5 μg mL⁻¹ rhodamine 123 for 15 min. Confocal fluorescence microscopic images were constructed from 790 to 860 nm for (a) and from 500 to 600 nm for (b); λ_{ex} = 780 and 488 nm, respectively. (c) Merged (a) red and (b) green channels. (d) Co-localization areas of the red and green channels selected. (e) Co-localization plot of (a) and (b); z-axis represents the frequency that color pair exists in (c), y- and x-axes represent the intensity of each pixel in (d). (f) Intensity profile of the region of interest [red arrow in (c)] across RAW264.7 cells.

Cy-NTe in imaging endogenous PN redox cycles in living animals. The results showed that fluorescence in mice injected with Cy-NTe and L-cysteine was weaker than those injected only with Cy-NTe (Figure 11.12f), indicating that the probe could be sensitive enough to detect basal levels of PN that were endogenously produced without stimulated PN production, and highlighting the potential utility of Cy-NTe for PN detection *in vivo*. The fluorescence in Cy-NTe loaded mice further enhanced greatly when the mice were pretreated with LPS/IFN-γ/PMA due to a burst of PN (Figure 11.12c and f). Inhibiting the inducible NOS activity by AG in the stimulated mice and then reducing the generated PN with GST quenched the probes' fluorescence (Figure 11.12d and f), which could rise again in response to another burst of PN induced by SIN-1 (Figure 11.12e and f). The fluorescence switches associated with PN oxidation–GSH reduction cycles in mice indicated the superiority of near-infrared fluorescent probes in deep tissue imaging.

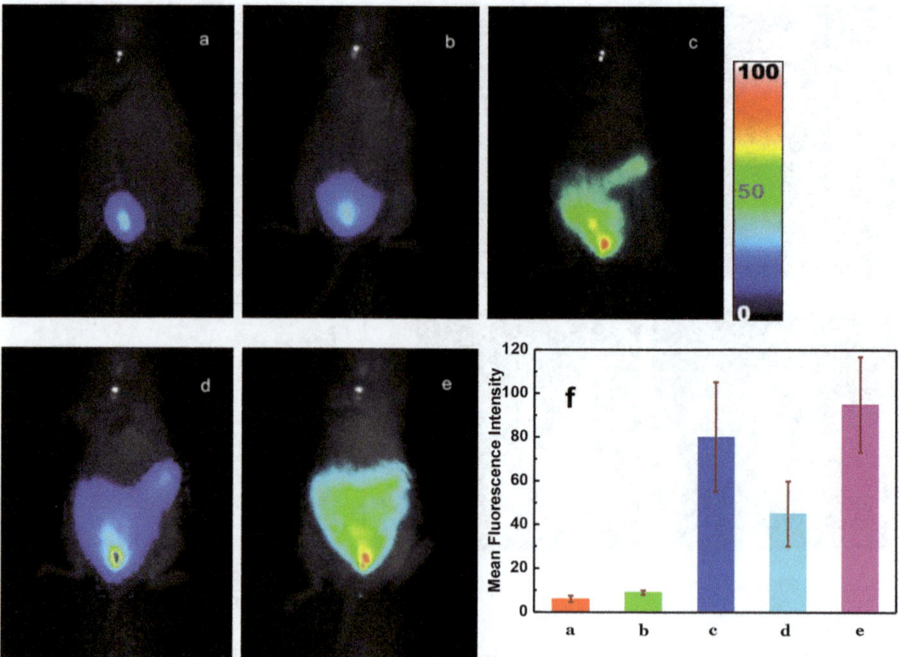

Figure 11.12 Imaging of redox cycles between PN and GSH in the peritoneal cavities of BALB/c mice. (a) Cy-NTe (1 μM, 50 μL in 1:9 DMSO:saline v/v) and L-cysteine (1 mM, 100 μL in saline) were injected in the intraperitoneal cavity. (b) Mice injected intraperitoneally (IP) with Cy-NTe (1 μM, 50 μL in 1:9 DMSO:saline v/v). (c) Mice injected IP with LPS (10 μg mL^{-1}) and IFN-γ (200 ng mL^{-1}) for 4 h, then with PMA (100 nM) for 0.5 h, and finally loaded with 1 μM Cy-NTe for 30 min. (d) Mice treated as described in (c), then injected IP with AG (1 mM, 100 μL in 1:9 DMSO:saline v/v), GST (250 U mL^{-1}, 300 μL in saline) and L-cysteine (1 mM, 200 μL in saline). (e) Mice treated as described in (d), and then injected IP with SIN-1 (1 mM, 1 mL in saline) for 20 min. (f) Quantification of the total photon flux from each mouse in (a)–(e). The total number of photons from the entire peritoneal cavity of the mice in (a)–(e) was integrated. Images constructed from a 790 to 860 nm fluorescence collection window, λ_{ex} = 735 nm.

11.5 Conclusions and Perspectives

In conclusion, the authors have developed two reversible near-infrared fluorescent probes that respond to PN selectively over other reactive oxygen species such as $O_2{}^{\bullet-}$, H_2O_2, ClO^-, NO, $^{\bullet}OH$ and $^1O_2$. The probes were designed by mimicking GPx activity and functionalizing heptamethine dyes with selenium/tellurium moieties. These probes work in a ping-pong-like mechanism and modulate the dyes' fluorescence through redox changes in the selenium/tellurium center in response to PN-related redox events. The near-infrared fluorescence of the probes effectively escaped interference caused by autofluorescence of biological samples and could penetrate into deep tissues.

The authors' results confirmed that both Cy-PSe and Cy-NTe were viable fluorescent probes for imaging PN and PN/GSH homeostasis in living cells and animals with negligible cytotoxicity. Looking forward, it is important to develop reversible near-infrared fluorescent probes that could respond faster to both PN and GSH in order to achieve real-time imaging of PN/GSH redox homeostasis. These probes could facilitate the depiction of the kinetics of PN bursts and GSH reparation, and thus help us understand the physiological and pathological roles of this active species. These probes could also be of use in theranostics of PN-related diseases because they are subjected to high GPx-like activity and could scavenge and image PN simultaneously.

References

1. C. Szabo, H. Ischiropoulos and R. Radi, *Nat. Rev. Drug Discovery*, 2007, **6**, 662–680.
2. N. Ieda, H. Nakagawa, T. Peng, D. Yang, T. Suzuki and N. Miyata, *J. Am. Chem. Soc.*, 2012, **134**, 2563–2568.
3. R. J. Gryglewski, R. M. J. Palmer and S. Moncada, *Nature*, 1986, **320**, 454–456.
4. P. Pacher, J. S. Beckman and L. Liaudet, *Physiol. Rev.*, 2007, 315–424.
5. G. Ferrer-Sueta and R. Radi, *ACS Chem. Biol.*, 2009, **4**, 161–177.
6. R. Radi, A. Cassina, R. Hodara, C. Quijano and L. Castro, *Free Radical Biol. Med.*, 2002, **33**, 1451–1464.
7. N. B. Surmeli, N. K. Litterman, A.-F. Miller and J. T. Groves, *J. Am. Chem. Soc.*, 2010, **132**, 17174–17185.
8. N. A. Sieracki, B. N. Gantner, M. Mao, J. H. Horner, R. D. Ye, A. B. Malik, M. E. Newcomb and M. G. Bonini, *Free Radical Biol. Med.*, 2013, **61**, 40–50.
9. H. Kawasaki, K. Ikeda, A. Shigenaga, T. Baba, K. Takamori, H. Ogawa and F. Yamakura, *Free Radical Biol. Med.*, 2011, **50**, 419–427.
10. T. Sawa, M. H. Zaki, T. Okamoto, T. Akuta, Y. Tokutomi, S. Kim-Mitsuyama, H. Ihara, A. Kobayashi, M. Yamamoto, S. Fujii, H. Arimoto and T. Akaike, *Nat. Chem. Biol.*, 2007, **3**, 727–735.
11. H. R. Kermis, Y. Kostov, P. Harms and G. Rao, *Biotechnol. Prog.*, 2002, **18**, 1047–1053.
12. X. Chen, Y. Zhou, X. Peng and J. Yoon, *Chem. Soc. Rev.*, 2010, **39**, 2120–2135.
13. J. Chan, S. C. Dodani and C. J. Chang, *Nat. Chem.*, 2012, **4**, 973–984.
14. Y. Yang, Q. Zhao, W. Feng and F. Li, *Chem. Rev.*, 2012, **113**, 192–270.
15. D. Yang, H. L. Wang, Z. N. Sun, N. W. Chung and J. G. Shen, *J. Am. Chem. Soc.*, 2006, **128**, 6004–6005.
16. Z. N. Sun, H. L. Wang, F. Q. Liu, Y. Chen, P. K. Tam and D. Yang, *Org. Lett.*, 2009, **11**, 1887–1890.
17. T. Ueno, Y. Urano, H. Kojima and T. Nagano, *J. Am. Chem. Soc.*, 2006, **128**, 10640–10641.
18. F. Yu, P. Song, P. Li, B. Wang and K. Han, *Analyst*, 2012, **137**, 3740–3749.

19. B. Wang, F. Yu, P. Li, X. Sun and K. Han, *Dyes Pigm.*, 2013, **96**, 383–390.
20. R. Weissleder and V. Ntziachristos, *Nat. Med.*, 2003, **9**, 123–128.
21. C.-H. Tung, Y. Lin, W. K. Moon and R. Weissleder, *ChemBioChem*, 2002, **3**, 784–786.
22. K. Kiyose, H. Kojima and T. Nagano, *Chem.-Asian J.*, 2008, **3**, 506–515.
23. S. A. Hilderbrand and R. Weissleder, *Curr. Opin. Chem. Biol.*, 2010, **14**, 71–79.
24. J. O. Escobedo, O. Rusin, S. Lim and R. M. Strongin, *Curr. Opin. Chem. Biol.*, 2010, **14**, 64–70.
25. R. Weissleder, *Nat. Biotech.*, 2001, **19**, 316–317.
26. D. D. Nolting, J. C. Gore and W. Pham, *Curr. Org. Synth.*, 2011, **8**, 521–534.
27. F. Yu, P. Li, G. Li, G. Zhao, T. Chu and K. Han, *J. Am. Chem. Soc.*, 2011, **133**, 11030–11033.
28. F. Yu, P. Li, B. Wang and K. Han, *J. Am. Chem. Soc.*, 2013, **135**, 7674–7680.
29. L. Yuan, W. Lin, K. Zheng, L. He and W. Huang, *Chem. Soc. Rev.*, 2013, **42**, 622–661.
30. J. V. Frangioni, *Curr. Opin. Chem. Biol.*, 2003, **7**, 626–634.
31. M. Lipowska, G. Patonay and L. Strekowski, *Heterocycl. Commun.*, 1995, **1**, 427.
32. L. Strekowski, M. Lipowska and G. Patonay, *J. Org. Chem.*, 1992, **57**, 4578–4580.
33. X. Peng, F. Song, E. Lu, Y. Wang, W. Zhou, J. Fan and Y. Gao, *J. Am. Chem. Soc.*, 2005, **127**, 4170–4171.
34. A. J. Meyer, T. Brach, L. Marty, S. Kreye, N. Rouhier, J. P. Jacquot and R. Hell, *Plant J.*, 2007, **52**, 973–986.
35. A. J. Mukherjee, S. S. Zade, H. B. Singh and R. B. Sunoj, *Chem. Rev.*, 2010, **110**, 4357–4416.
36. G. Mugesh and H. B. Singh, *Chem. Soc. Rev.*, 2000, **29**, 347–357.
37. H. Xu, W. Cao and X. Zhang, *Acc. Chem. Res.*, 2013, **46**, 1647–1658.

Subject Index